Nonlinear Programming

Edited by J. B. Rosen
O. L. Mangasarian
K. Ritter

Proceedings of a Symposium
Conducted by the Mathematics Research Center,
The University of Wisconsin, Madison
May 4-6, 1970

Academic Press
New York · London   1970

Copyright © 1970, by Academic Press, Inc.
ALL RIGHTS RESERVED
NO PART OF THIS BOOK MAY BE REPRODUCED IN ANY FORM,
BY PHOTOSTAT, MICROFILM, RETRIEVAL SYSTEM, OR ANY
OTHER MEANS, WITHOUT WRITTEN PERMISSION FROM
THE PUBLISHERS.

ACADEMIC PRESS, INC.
111 Fifth Avenue, New York, New York 10003

United Kingdom Edition published by
ACADEMIC PRESS, INC. (LONDON) LTD.
Berkeley Square House, London W1X 6BA

LIBRARY OF CONGRESS CATALOG CARD NUMBER: 75-132012

PRINTED IN THE UNITED STATES OF AMERICA

NONLINEAR PROGRAMMING

Publication No. 25
of the Mathematics Research Center
The University of Wisconsin

Contents

FOREWORD ix
PREFACE xi

A Method of Centers by Upper-Bounding Functions with
Applications 1
   P. Huard
      Electricite de France, Paris, France

A New Algorithm for Unconstrained Optimization 31
   M. J. D. Powell
      A. E. R. E. Harwell, Didcot, Berkshire, England

A Class of Methods for Nonlinear Programming
II Computational Experience 67
   R. Fletcher
      A. E. R. E. Harwell, Didcot, Berkshire, England
   Shirley A. Lill
      University of Leeds, Leeds, England

Some Algorithms Based on the Principle of Feasible Directions . . . 93
   G. Zoutendijk
      Instituut der Rijksuniversiteit te Leiden, The Netherlands

Numerical Techniques in Mathematical Programming 123
   R. H. Bartels
      University of Texas, Austin, Texas
   G. H. Golub
      Stanford University, Stanford, California
   M. A. Saunders
      Stanford University, Stanford, California

CONTENTS

A Superlinearly Convergent Method for Unconstrained
Minimization 177
   K. Ritter
      University of Wisconsin, Madison, Wisconsin

A Second Order Method for the Linearly Constrained
Nonlinear Programming Problem 207
   Garth P. McCormick
      University of Wisconsin, Madison, Wisconsin

Convergent Step-Sizes for Gradient-Like Feasible Direction
Algorithms for Constrained Optimization 245
   James W. Daniel
      University of Wisconsin, Madison, Wisconsin

On the Implementation of Conceptual Algorithms 275
   E. Polak
      University of California, Berkeley, California

Some Convex Programs Whose Duals Are Linearly Constrained . . . 293
   R. Tyrrell Rockafellar
      University of Washington, Seattle, Washington

Sufficiency Conditions and a Duality Theory for Mathematical
Programming Problems in Arbitrary Linear Spaces 323
   Lucien W. Neustadt
      University of Southern California, Los Angeles, California

Recent Results on Complementarity Problems 349
   C. E. Lemke
      Rensselaer Polytechnic Institute, Troy, New York

Nonlinear Nondifferentiable Programming in Complex Space 385
   Bertram Mond
      LaTrobe University, Bundoora, Victoria, Australia

Duality Inequalities of Mathematics and Science 401
   R. J. Duffin
      Carnegie-Mellon University, Pittsburgh, Pennsylvania

CONTENTS

Programming Methods in Statistics and Probability Theory 425
   Olaf Krafft
      Institüt für Mathematische Statistik der
      Universität Münster, Roxeler Strasse, West Germany

Applications of Mathematical Programming to ℓ_p Approximation . . 447
   I. Barrodale
      University of Victoria, Victoria, British Columbia, Canada
   F. D. K. Roberts
      University of Victoria, Victoria, British Columbia, Canada

Theoretical and Computational Aspects of Nonlinear Regression . . . 465
   R. R. Meyer
      Shell Development Company, Emeryville, California

INDEX 487

Foreword

This volume contains the proceedings of a Symposium on Nonlinear Programming held in Madison, Wisconsin, on May 4-6, 1970, and sponsored by the Mathematics Research Center, University of Wisconsin.

The organizing committee and editors of these proceedings consisted (in addition to myself as chairman) of my colleagues here in Madison, Olvi Mangasarian and Klaus Ritter. The Symposium consisted of five sessions. Sessions were chaired by R. R. Hughes and D. L. Russell, also both at Madison, in addition to the three of us on the organizing committee.

The Symposium was attended by 213 registrants. Its success was due in large measure to the previous experience, hard work, and improvisation, when needed, of Mrs. Gladys Moran as Symposium secretary and Steve Robinson of the MRC staff. The prompt publication of these proceedings is due largely to the fast and accurate typing of the manuscripts by Mrs. Carol Chase.

In view of the unusual circumstances during, and subsequent to, this Symposium, I feel that some additional remarks are in order. These remarks, of course, represent my own views.

The difficulties in holding a scientific conference (or any other intellectual activity) on many university campuses during May, 1970, were made painfully clear to many who attended this Symposium. These difficulties strongly suggest that the era of sheltered pursuit of academic research interests has ended and that (justified or not) a scientist is now likely to be held accountable for any ultimate use to which his research is put.

Violence on the Madison campus reached its climax in the early morning of August 24, when the bombing of Sterling Hall was carried out with the apparent intention of destroying the Mathematics Research Center. The well-known results were tragic for the faculty and graduate students of the Physics Department who occupied the basement and first floor of Sterling Hall. The final typed manuscripts for these proceedings were in the secretarial office on

FOREWORD

the second floor just above the blast area. Fortunately, they were all recovered essentially intact, as were all other manuscripts and research reports at the Mathematics Research Center. In this instance, as in all too many others, it seems that resort to violence, while rarely achieving its intended objective, almost always leads to tragedy.

October, 1970

J. B. Rosen

Preface

It was the intention of the organizing and editorial committee for this Symposium on Nonlinear Programming to emphasize those algorithms and related theory which lead to efficient computational methods for solving nonlinear programming problems. Therefore one of the main purposes of this Symposium was to further strengthen the existing relationship between theory and computational aspects of this subject. I hope the reader will agree that the 17 papers in these Proceedings are sufficient evidence that we have been successful.

In view of this it is difficult to classify the papers here with regard to their theoretical or computational emphasis. However, for convenience we have attempted to present them according to three general groupings. The first nine papers are concerned primarily with computational algorithms.[1] The second four papers are devoted to theoretical aspects of nonlinear programming, while in the papers of Duffin, Krafft, Barrodale, and Meyer, certain applications of nonlinear programming are considered. The word applications is being used in a somewhat limited sense in connection with these four papers. They all represent applications to other basic areas (physics, statistics, approximation) which in turn may be used to solve more applied problems. An application of nonlinear programming (or for that matter any other mathematical or computational method) is usually interpreted in a broader sense to mean the use of the method to solve some specific scientific, economic, or even sociological problem. In principle, any such problem which can be formulated in terms of an objective function to be minimized or maximized, subject to various conditions or constraints, is one to which mathematical programming methods can be applied. Clearly the scope of such problems is very large and includes many of the technological problems facing our society today. In particular, many of the environmental problems can be formulated in terms of satisfying stated conditions at a minimum cost. The methods presented in these Proceedings can then be applied directly to solve these problems. An excellent survey of such applications in the broader sense, is given in the recent paper by Van Dyne, Frayer, and Bledsoe.[2]

PREFACE

Among the more active areas of research covered in these papers are algorithms for nonlinear constraint problems, investigation of convergence rates, and the use of nonlinear programming for approximation.[3] Computational results are included in several of the papers and played an important role in motivating many of the others. I believe that experimental computing remains an essential part of this field in developing new algorithms and comparing the performance of known methods on different types of problems. Areas in which significant work still remains to be done include a unified (and possibly simpler) theory of convergence rates for many of the existing algorithms, and the application of mathematical programming to generalized approximation, including boundary value problems.

J. B. Rosen

---

[1] The paper by A. M. Geoffrion titled "Generalized Benders' decomposition" which was presented at this Symposium will be published in the *Journal of Optimization Theory and Applications*.

[2] G. M. Van Dyne, W. E. Frayer, and L. J. Bledsoe, "Some optimization techniques and problems in the natural resource sciences," *Studies in Optimization 1*, Soc. Indust. Appl. Math., Philadelphia, 1970, pp. 95-124.

[3] An application of mathematical programming to a generalized approximation problem is illustrated by the jacket design. This shows an approximate solution to the Navier-Stokes equations on a square domain obtained by minimizing the maximum error in the differential equation over the domain. For details see J. B. Rosen, "Approximate solution to transient Navier-Stokes cavity convection problems," Computer Sciences Dept. Tech. Rep. No. 32, Univ. of Wisconsin, Nov., 1970.

A Method of Centers by Upper-Bounding Functions with Applications

P. HUARD

ABSTRACT

The convergence of the method of centers is obtained with approximate centres, if the corresponding errors tend to zero. A very general procedure is developed by using an upper-bound of the F-distance. Different applications of this procedure lead back to classical methods such as those of Zoutendijk, Frank-Wolfe, and Rosen.

Introduction

Many methods of solving nonlinear programming problems have been proposed. Although some of these methods are basically similar, it is not easy to classify them into a small number of families. Zangwill (13), Polak (7, 8), Chevassus (2), Topkis and Veinott (11), Roode (9), and the present author (5) have proposed very general algorithms each englobing a number of particular methods. The question is whether this synthesis work is of interest and, if so, why.

For the mathematician, this work is interesting in that it brings about unification in the theory; it becomes, sooner or later, something which he cannot dispense with. On the other hand, it is not very useful, from a practical standpoint, to establish simpler proofs of the convergence in a particular method: it is indeed easier to prove the convergence in a particular rather than general context. In this respect, it should be noted that while the proofs relating to the general algorithms are very often short, they rely on a number of necessary hypotheses, and when applying them to a particular method it usually requires a considerable effort to establish that these hypotheses are satisfied.

However, still from a practical standpoint, the theory of a general algorithm can be of great interest in discussing the conditions of convergence of a particular method, and in specifying what is necessary to this convergence and what is not. Thus, a method can be modified, with a view to accelerate its convergence, on a heuristic or practical basis (e.g. reduction of the accuracy of some computations), while making sure that the conditions for obtaining the optimal solution are not violated.

The subject of this article lies somewhere between both extremes: starting from the method of centers (1) a very general algorithm, a particular variant is derived.

In the method of centers, one has to determine at each step a feasible interior solution, or "center", by maximizing a somewhat arbitrary function, called "F-distance", which characterizes the distance from the boundary of the domain. The determination of such a center can be made approximately, provided that the error thus induced tends to zero during the iterative procedure. This possibility is used here, taking a particular F-distance and making the computations with simpler upper bounding functions.

The variant so obtained is still a fairly general algorithm, since by particularizing it still further we get at well-known methods such as those of Zoutendijk (14) (method of feasible directions), of Frank and Wolfe (4), and of Rosen (10) (gradient projection method).

In particular an "anti-zigzag" process slightly different from that proposed by Zoutendijk is obtained in a natural manner under our algorithm. It is interesting to note that between the method of feasible directions and the linearized method of centers (described in (6)) the difference lies only in the value of a scalar parameter.

For Rosen's method of projected gradient, a fairly simple procedure for ensuring the convergence of the method is found.

Notation

R   Set of real numbers

R^n n-dimensional euclidian space

$\wp(R^n)$ set of all subsets of R^n

N   set of positive integers

If $x \in R^n$, x_i i-th component of x

If $A \subset R^n$, Fr (A) boundary of A
$\overset{o}{A}$ interior of A

If $f: R^n \to R$ differentiable at x, $\nabla f(x)$ value of the gradient of f at x

1. The Method of Centers: A Summary with Modifications

1.1. F-distance, center, ε-center

Let $\mathcal{E} \subset \mathcal{P}(R^n)$ be a set whose elements are subsets of R^n, and let $d: R^n \times \mathcal{E} \to R$ be a real function.

1.1.1. **Definition:** d is called an F-distance on $R^n \times \mathcal{E}$ if it satisfies:
 (i) $d(x, E) = 0$, $\forall E \in \mathcal{E}$, $\forall x \in \underset{o}{Fr}(E)$
 (ii) $d(x, E) > 0$, $\forall E \in \mathcal{E}$, $\forall x \in \overset{o}{E}$
 (iii) $\forall E \in \mathcal{E}$, $\forall E' \in \mathcal{E}: E \subset E'$; $\forall x \in E$, there exists a scalar $\rho(x) > 0$ such that:

$$d(x, E) \leq \rho(x) \cdot d(x, E')$$

Remark: If we take $\rho(x) = c^{\underline{t}} = 1$ in property (iii) we find the definition already given in (1) . This weakening of the definition of an F-distance has been given by Tremolieres in (12).

1.1.2. **Definition.** An F-distance d is said to be <u>regular</u> if it satisfies in addition:
 (iv) $\forall$ sequences $\{E_k \in \mathcal{E} \mid k \in N\}$ and $\{\overset{k}{x} \in R^n \mid k \in N\}$ such that:

$$E_k \supset E_{k+1} \supset E \in \mathcal{E}, \quad E \neq \emptyset$$

$$\overset{k}{x} \in E_k, \quad \overset{k}{x} \notin \overset{o}{E}_{k+1}$$

we have $d(\overset{k}{x}, E_k) \to 0$ when $k \to +\infty$.

Remark: This property of "regularity" replaces the property of "compatibility" given in (1), and plays a similar role.

1.1.3. Definition: Given a F-distance d, defined on $R^n \times \mathcal{E}$, a set $E \in \mathcal{E}$, and a number $\mathcal{E}$ such that $0 \leq \mathcal{E} < \sup\{d(x, E) \mid x \in E\}$, we call $\mathcal{E}$-center of E (wtih respect to d) every point $c \in E$ such that:

$$d(c, E) \geq \sup\{d(x, E) \mid x \in E\} - \mathcal{E}$$

If $\mathcal{E} = 0$, such a point is called a center of E.

Remark: A set E may have a $\mathcal{E}$-center only if $\overset{o}{E} \neq \emptyset$, because a $\mathcal{E}$-center is always an interior point.

1.2. Two examples of regular F-distances

Proposition: Let $g_i: R^n \to R$, $i \in L = \{1, 2, \ldots, m\}$ be continuous real functions such that:
(1) $Fr\{x \mid g_i(x) \geq \lambda\} = \{x \mid g_i(x) = \lambda\}$, $\forall i \in L, \forall \lambda \in R$.
Let $E(b) = \{x \mid g_i(x) \geq b_i, i \in L\}$, with $b \in R^m$

$$\mathcal{E} = \{E(b) \mid b \in K \subset R^m\}$$

where K is such that only one value of b corresponds to a set E(b).

Under these conditions, each of the two following functions:
(2) $d(x, E(b)) = \min\{g_i(x) - b_i \mid i \in L\}$
(3) $d(x, E(b)) = \prod_{i \in L} (g_i(x) - b_i)$

is a regular F-distance, defined on $R^n \times \mathcal{E}$.

Proof: The following proof is valid for both the function defined by (2) and that defined by (3).

(i) From (1), $x \in Fr(E(b)) \implies \exists i \in L : g_i(x) - b_i = 0 \implies d(x, E(b)) = 0$

(ii) From (2), $x \in \overset{o}{E}(b) \implies g_i(x) - b_i > 0$, $\forall i \in L \implies d(x, E(b)) > 0$

(iii) $E(b) \subset E(b') \iff b \geq b' \implies g_i(x) - b_i \leq g_i(x) - b'_i$, $\forall i \in L$, $\forall x \implies d(x, E(b)) \leq d(x, E(b'))$, $\forall x : g_i(x) - b_i \geq 0$, $\forall i \in L$

(iv) Consider an infinite sequence $\{\overset{k}{b} \in K \mid k \in N\}$, the sequence of corresponding $E_k = E(\overset{k}{b})$, and an infinite sequence $\{\overset{k}{x} \in R^n \mid k \in N\}$, such that, $\forall k \in N$:

$$\overset{k}{b} \leq \overset{k+1}{b} \leq \overset{*}{b}, \ \overset{*}{b} \in K, \ \text{constant}$$

$$\overset{k}{x} \in E_k, \ \overset{k}{x} \notin \overset{o}{E}_{k+1}$$

Then, $\forall k \in N$:

$$\exists i_k \in L : g_{i_k}(\overset{k}{x}) - \overset{*}{b}_{i_k} \leq g_{i_k}(\overset{k}{x}) - \overset{k+1}{b}_{i_k} < 0 \leq g_{i_k}(\overset{k}{x}) - \overset{k}{b}_{i_k}$$

Since L is finite, this relation is true for at least one subscript $i \in L$, independent of k, for every k of a subsequence $S_{i_k} \subseteq L$.

Because b_i, which has an upper-bound $\overset{*}{b}_i < +\infty$, tends to a limit $\overline{b}_i \leq \overset{*}{b}_i$, we have:

$$\lim(g_i(\overset{k}{x}) - \overset{k}{b}_i) \to 0 \text{ when } k \to +\infty, \ k \in S_i$$

and hence $d(\overset{k}{x}, E_k) \to 0$ when $k \to +\infty$, $k \in N$.

1.3. Finite algorithm for finding a feasible point

Suppose $B \subset R^n$, $\mathcal{E}$ a family of sets of R^n, d a regular F-distance on $R^n \times \mathcal{E}$, and $E_* \in \mathcal{E}$. We propose to find a point $x \in E_* \cap B$, using the two sequences $\{\overset{k}{x} \mid k \in N\}$ and $\{E_k \in \mathcal{E} \mid k \in N\}$, defined by the following algorithm:

Algorithm: Choose a decreasing sequence of numbers $\mathcal{E}_k \geq 0$, "not too large", tending to zero when $k \to +\infty$, $k \in N$.

Step k: We have $E_k \in \mathcal{E}$ and $\overset{k}{x}$ such that:

$$E_k \supset E_*, \quad \overset{\circ}{E}_k \cap B \neq \emptyset$$

$$\overset{k}{x} \in E_k \cap B: \ d(\overset{k}{x}, E_k) \geq \sup\{d(x, E_k) \mid x \in E_k \cap B\} - \mathcal{E}_k$$

If $\overset{k}{x} \notin E_* \cap B$, choose $E_{k+1} \in \mathcal{E}$ such that:

$$E_k \supset E_{k+1} \supset E_*$$

$$\overset{k}{x} \notin \overset{\circ}{E}_{k+1} \cap B$$

If $\overset{k}{x} \in E_* \cap B$, choose $E_{k+1} = E_k$, $\overset{k+1}{x} = \overset{k}{x}$.

Proposition: Under the above conditions (1.3), if $\overset{\circ}{E}_* \cap B \neq \emptyset$, there exists a finite $k^* \in N$ such that:

$$k \geq k^* \implies \overset{k}{x} \in \overset{\circ}{E}_* \cap B$$

Remark: The $\mathcal{E}_k$ should not be chosen too large to ensure that we actually have

$$\overset{k}{x} \in \overset{\circ}{E}_k \cap B.$$

Proof: Suppose we have the contrary, that is:

$$\overset{k}{x} \notin \overset{o}{E_*} \cap B, \quad \forall k \in N$$

We show this is not possible.
Since d is a regular F-distance:

$$d(\overset{k}{x}, E_k) \to 0 \quad \text{when} \quad k \to +\infty.$$

Suppose $\overset{o}{x} \in \overset{o}{E_*} \cap B$ - We have, $\forall k \in N$:

$$0 < d(\overset{o}{x}, E_*) \quad \text{since} \quad \overset{o}{x} \in \overset{o}{E_*}$$

$$\leq \rho_0 \cdot d(\overset{o}{x}, E_k) \quad \text{with } \rho_0 > 0, \text{ since } E_* \subset E_k$$

$$\leq \rho_0 \left(d(\overset{k}{x}, E_k) + \epsilon_k \right) \text{ since } \overset{k}{x} \text{ maximizes d to within } \epsilon_k \text{ on } E_k \cap B$$

Finally:

$$0 < d(\overset{o}{x}, E_*) \leq \rho_0 \left(d(\overset{k}{x}, E_k) + \epsilon_k \right)$$

$$d(\overset{k}{x}, E_k) \to 0, \quad \epsilon_k \to 0,$$

a contradiction.

2. Method of Centers (General algorithm)

2.1. <u>Problem set. Hypotheses</u>. Suppose the following programming problem to be solved:

Maximize f(x) subject to
$x \in A \cap B$

(P)

where $A \subset R^n$, $\overset{o}{A} \neq \emptyset$, $B \subset R^n$, $A \cap B$ closed.

$f : R^n \to R$ a continuous function, with an upperbound on $A \cap B$ such that:

$$Fr\{x \mid f(x) \geq \lambda\} = \{x \mid f(x) = \lambda\}, \quad \forall \lambda < f(\hat{x})$$

It is supposed that f attains its maximum value on $A \cap B$ at a point $\hat{x}$.

The following hypohtesis is made for A and B:
(H) $\overset{o}{A} \cap B \cap 0 = \emptyset \Longrightarrow A \cap B \cap \overset{o}{0} = \emptyset$, $\forall\, 0 \subset R^n$ open (satisfied for instance if $\overset{o}{A} \cap B \neq \emptyset$ and $A \cap B$ convex).

We consider a family $\mathcal{G}$ of sets $A_j \in R^n$, $j \in J$, such that:

$$A_j \supset A, \quad \forall j \in J$$

$$A \in \mathcal{G}$$

We set $F(\lambda) = \{x \mid f(x) \geq \lambda\}, \quad \lambda \in R$

$$\mathcal{E} = \{E(\lambda, j) \mid \lambda < f(\hat{x}), j \in J\}$$

with $E(\lambda, j) = A_j \cap F(\lambda)$

We choose a regular F-distance d, defined on $R^n \times \mathcal{E}$, and an infinite decreasing sequence of numbers $\varepsilon_k \geq 0$, not too large, tending to zero when $k \to +\infty$, $k \in N$.

2.2. <u>Algorithm</u>.

Start with $\lambda_0 < f(\hat{x})$

<u>Step k</u>. We have $A_k \in \mathcal{G}$ such that $A_k \supset A$

$\overset{k}{x}$ and λ_k such that $\lambda_k \leq f(\overset{k}{x})$, $\lambda_k \leq f(\hat{x})$

Set $E_k = A_k \cap F(\lambda_k)$

Determine $x' \in E_k^{k+1} \cap B$ such that

$$d(\overset{k+1}{x'}, E_k) \geq \sup\{d(x, E_k) \mid x \in E_k \cap B\} - \epsilon_k$$

If $\overset{k+1}{x'} \notin A$, take
$$\left|\begin{array}{l} A_{k+1} \subset A_k : \overset{k+1}{x'} \notin \overset{\circ}{A}_{k+1} \\ \lambda_{k+1} = \lambda_k \\ \overset{k+1}{x} = \overset{k+1}{x'} \end{array}\right.$$

If $\overset{k+1}{x'} \in A$, take
$$\left|\begin{array}{l} A_{k+1} \subset A_k \quad \text{(for example, } A_{k+1} = A) \\ \lambda'_{k+1} = f(\overset{k+1}{x'}) \\ \overset{k+1}{x} \in A \cap F(\lambda'_{k+1}) \cap B \quad \text{(for ex.,} \\ \qquad\qquad\qquad\qquad\qquad \overset{k+1}{x} = \overset{k+1}{x'} \text{)} \\ \lambda_{k+1} = f(\overset{k+1}{x}) \end{array}\right.$$

And so on.

Proposition: If at some step we have $\overset{\circ}{E}_k \cap B = \emptyset$, then $\overset{k}{x}$ is an optimal solution of problem (P).

If not, the infinite sequence $\{\overset{k}{x} \mid k \in N\}$ given by the algorithm is such that:

$$\lim f(\overset{k}{x}) = f(\hat{x}), \quad k \to +\infty, \quad k \in N.$$

Moreover, in the second case, there exists an infinite subsequence of points $\overset{k}{x}$, $k \in S \subset N$, for which every accumulation point is an optimal solution for problem (P).

BOUNDED METHOD OF CENTERS

Proof: If, at some step k, $\overset{\circ}{E_k} \cap B = \emptyset$, we have:

$$\overset{\circ}{A_k} \cap \overset{\circ}{F(\lambda_k)} \cap B = \emptyset \implies \overset{\circ}{A} \cap \overset{\circ}{F(\lambda_k)} \cap B = \emptyset \text{ since } A_k \supset A$$

$$\implies A \cap \overset{\circ}{F(\lambda_k)} \cap B = \emptyset \text{ by hypothesis (H)}$$

$$\implies \not\exists\, x \in A \cap B : f(x) > f(\overset{k}{x})$$

and then, $\overset{k}{x}$ is an optimal solution.

Otherwise, an infinite sequence $\{(\overset{k}{x}, \lambda_k) \mid k \in N\}$ is obtained, such that:

$$\lambda_k \le f(\hat{x}) \qquad\qquad \forall k \in N$$

$$\lambda_k \le \lambda_{k+1}$$

Then $\lambda_k \to \lambda^* \le f(\hat{x})$ when $k \to +\infty$, $k \in N$
Setting $F_* = F(\lambda^*)$ and $E_* = A \cap F_*$, we have:

$$\overset{\circ}{E_*} \cap B = \emptyset$$

If not, since we have $\forall k$:

$$E_k \supset E_{k+1} \supset E_*, \quad \overset{k}{x} \in E_k \cap B, \quad \overset{k}{x} \notin \overset{\circ}{E_{k+1}} \cap B \text{ and from}$$

proposition 1.3, a solution $\overset{k}{x} \in \overset{\circ}{E_*} \cap B$ would be obtained after a finite number of steps, such that

$$f(\overset{k}{x}) > \lambda^*$$

which is impossible. Therefore:

$$\overset{\circ}{E_*} \cap B = \emptyset \implies A \cap \overset{\circ}{F_*} \cap B = \emptyset \text{ by (H)}$$

$$\implies \lambda^* = f(\hat{x})$$

$$\implies \lim f(\overset{k}{x}) = f(\hat{x}) \text{ since } \lambda_k \le f(\overset{k}{x}), \forall k \in N .$$

Let $S \subset N$ be the set of indices corresponding to the feasible solutions $\overset{k}{x} \in A$, given by the algorithm. Since it is supposed here that λ_k is never equal to $f(\hat{x})$, the following results are obtained, by considering the two possible cases, S finite and S infinite:

1) <u>S finite</u>: The last obtained feasible solution $\overset{k}{x}$, $k \in S$, is an optimal one, because after this step k, λ_k cannot increase any further.

2) <u>S infinite</u>: Every accumulation point of the sequence $\{\overset{k}{x} \mid k \in S\}$ is an optimal solution.

3. Method of Center by Upper-Bounding Functions

3.1. <u>Problem set. Hypotheses</u>

The very general method we describe here falls within the method of centers. The problem considered is the same, that is:

$$\boxed{\text{Maximize } f(x) \text{ subject to } x \in A \cap B} \quad (P)$$

The method described below is specified by (in addition to a more restrictive definition for F-distance) the calculation of ϵ-centers, a computation which uses bounding functions for the F-distance.

Suppose that:

- $A \subset R^n$ closed, $B \subset R^n$ convex compact, satisfying:

  $(H_1) \overset{\circ}{A} \neq \emptyset$; $\overset{\circ}{A} \cap B \cap 0 = \emptyset \Longrightarrow A \cap B \cap 0 = \emptyset$, $\forall 0 \subset R^n$ open

- $f : R^n \to R$ a continuous function, attaining its maximum value on $A \cap B$ at $\hat{x}$, and satisfying:

$$(H_2) F_r \{x \mid f(x) \geq \lambda\} = \{x \mid f(x) = \lambda\}, \forall \lambda \in R$$

12

- $E(\lambda) = A \cap \{x \mid f(x) \geq \lambda\}; \quad \mathcal{E} = \{E(\lambda) \mid \lambda < f(\hat{x})$
- $d : R^n \times \mathcal{E} \to R$ a continuous, regular F-distance on $R^n \times \mathcal{E}$ satisfying:

$$(H_3) \quad d(x, E) < 0, \forall x \notin E, \forall E \in \mathcal{E}.$$

For simplification, we shall write, in the following, $d(x, \lambda)$, instead of $d(x, E(\lambda))$, that is we shall consider a function $d : R^n \times R \to R$ instead of the function $d : R^n \times \mathcal{E} \to R$.

- $d' : R^n \times R^n \to R$ a continuous function, satisfying

(H_4) (i) $d'(x, y, \lambda) \geq d(x, \lambda)$

$$\forall x \in A \cap B, \forall y \in B, \forall \lambda \in R$$

(ii) $d'(x, x, f(x)) = 0$

(H_5) $\forall a \in B, \forall b \in B$ fixed:

$$d(x, f(a)) \leq 0, \forall x \in [a, b] \implies d'(x, a, f(a)) \leq 0,$$

$$\forall x \in [a, b]$$

Remark. The continuity of d' will be necessary but at some point (x^*, y^*, λ^*), defined by the algorithm.

3.2. Algorithm. Choose a constant $\alpha \geq 1$.

Step k. We have $\overset{k}{x} \in A \cap B$, $\lambda_k = f(\overset{k}{x})$, $E_k = E(\lambda_k)$

- Choose $\overset{k}{z} \in B$ such that:
- $\alpha \cdot d'(\overset{k}{z}, \overset{k}{x}, \lambda_k) \geq d'(x, \overset{k}{x}, \lambda_k), \forall x \in A \cap B$

(for instance, maximizing $d'(x, \overset{k}{x}, \lambda_k)$ on $A \cap B$ or on B)

- Determine $\overset{k+1}{x} \in [\overset{k}{x}, \overset{k}{z}]$ such that:

$$d(\overset{k+1}{x}, \lambda_k) \geq d(x, \lambda_k), \forall x \in [\overset{k}{x}, \overset{k}{z}]$$

. Choose x^{k+1} in $E'_{k+1} \cap B$, setting

$$E'_{k+1} = E(\lambda'_{k+1}) \quad \text{and} \quad \lambda'_{k+1} = f(x^{k+1})$$

(for instance, maximizing $f(x)$ on $[x^{k+1}, z^k] \cap E'_{k+1}$
And so on.

<u>Remark 1.</u> B convex $\Longrightarrow [x^k, z^k] \subset B$
<u>Remark 2.</u> $E'_{k+1} \cap B \neq \emptyset$, because $d(x^{k+1}, \lambda_k) \geq d(x^k, \lambda_k) = 0$, and then, by (H_3):
$x^{k+1} \in E_k \cap B \subset A \cap B$, that is $f(x^{k+1}) \leq f(\hat{x})$.

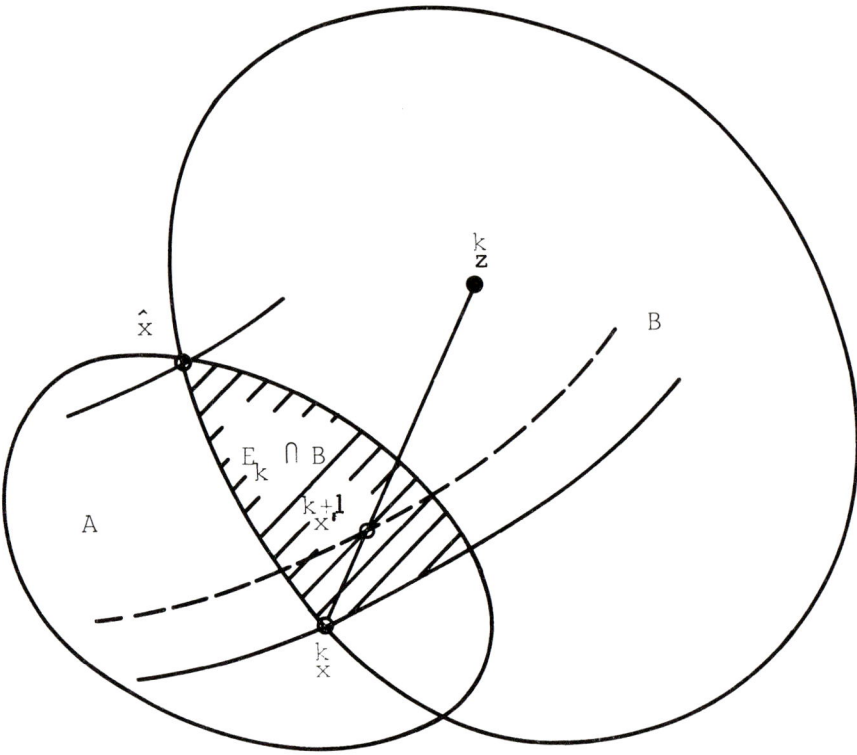

Figure 1

BOUNDED METHOD OF CENTERS

3.3. **Proposition.** The algorithm described in 3.2 gives a sequence of feasible solutions $\overset{k}{\hat{x}}$ and $\overset{k}{\hat{x}}'$. We have:

$$d(^{k+1}x', \lambda_k) \geq \sup\{d(x, \lambda_k) \mid x \in E_k \cap B\} - \epsilon_k \text{ with}$$

$$\epsilon_k \to 0$$

when $k \to +\infty$. Then, this algorithm falls within the method of centers, described in 2.2., and the results of the proposition 2.2 are valid herein.

Proof. We have to show that, in the infinite case, the approximation ϵ_k defined by:

$$\epsilon_k = \sup\{d(x, \lambda_k) \mid x \in E_k \cap B\} - d(^{k+1}x', \lambda_k)$$

tends to zero when $k \to +\infty$. The proof may be broken down into three steps. First, $d(^{k+1}x, \lambda_k) \to 0$, when $k \to +\infty$. Then there exists a subsequence for which $d'(\overset{k}{z}, \overset{k}{x}, \lambda_k) \to 0$ when $k \to +\infty$. Finally, $\epsilon_k \to 0$.

1. $\quad ^{k+1}x' \in F_k \implies f(\overset{k}{\hat{x}}) \leq f(^{k+1}x')$

    $^{k+1}x \in E'_{k+1} \implies f(^{k+1}x') \leq f(^{k+1}x) \Big\}$

    $^{k+1}x \in A \cap B \implies f(^{k+1}x) \leq f(\hat{x})$

$$\implies \lambda_k \leq f(\overset{k}{\hat{x}}) \leq f(^{k+1}x') \leq f(^{k+1}x) = \lambda_{k+1} \leq f(\hat{x}) \ldots$$

15

$$\ldots \Longrightarrow \boxed{\begin{aligned} f(^{k+1}x) \to \lambda^* &\leq f(\hat{x}) \qquad &(1)\\ f(^{k+1}x') \to \lambda^* \qquad &(1')\\ \lambda_k \to \lambda^* \qquad &(2)\\ f(^{k+1}x) - \lambda_k \to 0 \qquad &(3)\\ f(^{k+1}x') - \lambda_k \to 0 \qquad &(3') \end{aligned}}$$

when $k \to +\infty$, $k \in N$

On the other hand, $\forall k \in N$:

$$\left.\begin{aligned} E_k \supset E'_{k+1} \supset E_{k+1} \supset E(\lambda^*)\\ ^{k+1}x \in Fr(E_{k+1})\\ d \text{ is a regular F-distance} \end{aligned}\right\} \Longrightarrow \boxed{\begin{aligned} d(^{k+1}x, \lambda_k) \to 0,\\ \text{when } k \to +\infty, K \in N \end{aligned}} \quad (4)$$

Same results with $^{k+1}x'$:

$$\boxed{d(^{k+1}x', \lambda_k) \to 0, \text{ when } k \to +\infty, k \in N} \qquad (4')$$

2. The point $(^k\hat{x}, ^k\hat{z})$ belongs to the compact set $A \cap B \times B$. Hence, there exists $S \subset N$ such that:

$$\left.\begin{aligned} ^kx \to x^* \in A \cap B \qquad &(5)\\ ^kz \to z^* \in B \qquad &(6) \end{aligned}\right\} \text{ when } k \to +\infty, k \in S \subset N$$

On the other hand, $\forall \theta \in [0, 1]$ fixed, $\forall k \in N$, by the definition of x^{k+1}:

$$d(x^k + \theta(z^k - x^k), \lambda_k) \leq d(x^{k+1}, \lambda_k)$$

which gives, taking the limit $k \to +\infty$, $k \in S$, with fixed θ, d being continuous, and from (1'):

$$d(\overset{*}{x} + \theta(\overset{*}{z} - \overset{*}{x}), \overset{*}{\lambda}) \leq 0, \quad \forall \theta \in [0, 1] \tag{7}$$

Relation (7) and hypothesis (H5) give:

$$d'(\overset{*}{x} + \theta(\overset{*}{z} - \overset{*}{x}), \overset{*}{x}, \overset{*}{\lambda}) \leq 0, \quad \forall \theta \in [0, 1] \tag{8}$$

Setting $\theta = 1$, we have, since d' is continuous:

$$\lim d'(z^k, x^k, \lambda_k) \leq 0, \text{ when } k \to +\infty, k \in S \tag{9}$$

3. Let $\overset{k}{c}$ be a point of E_k, which maximizes $d(x, \lambda_k)$ on $E_k \cap B$. Such a point exists, because d is continuous, B is compact, $E_k \cap B \neq \emptyset$, and by (H3) $d(x, \lambda_k) < 0, \forall x \notin E_k$.

We have:

$$\epsilon_k = d(\overset{k}{c}, \lambda_k) - d(x^{k+1}, \lambda_k) \geq 0$$

We know from relation (1') that $d(x^{k+1}, \lambda_k) \to 0$, when $k \to +\infty$, $k \in N$. On the other hand, since $\overset{k}{c} \in A \cap B$:

$$0 \leq d(\overset{k}{c}, \lambda_k) \leq d'(\overset{k}{c}, x^k, \lambda_k) \quad \text{from (H4)}$$

$$\leq \alpha \cdot d'(z^k, x^k, \lambda_k) \quad \text{by definition of } z^k$$

Taking the limit, $k \to +\infty$, $k \in S$, because $\alpha \geq 1$ and from (9):

$$\lim_k d(\overset{k}{c}, \lambda_k) = 0$$

and

$$\boxed{\lim \epsilon_k = 0 \text{ when } k \to +\infty, \ k \in S} \qquad (10)$$

which completes the proof.

Remark. This proof remains valid if d' is not continuous everywhere, but only at the point $(\overset{*}{z}, \overset{*}{x}, \lambda^*)$. This weakening will be used further in the applications.

4. Applications of the Method of Centers by Upper-Bounding Functions

Let us consider the following programming problem:

$$\boxed{\text{Maximize} \quad f(x) \quad \text{subject to} \\ g_i(x) \geq 0, \ i \in L = \{1, 2, \ldots, m\}}$$

where $f: R^n \to R$ and $g_i: R^n \to R$, $i \in L$, are concave, continuously differentiable functions, upper-bounded on $\{x \mid g_i(x) \geq 0, \ i \in L\}$ - Moreover, f is supposed to attain its maximum value on this set, at a point $\hat{x}$.

In the following, we shall apply to this problem this particular method of centers in different ways. We shall find again well-known methods such as Zoutendijk's method of feasible directions, Frank and Wolfe's methods, and Rosen's gradient projection method. We shall distinguish two types of applications, depending on the roles played by the sets A and B used in the method of centers.

4.1. Applications type (A)

Set $A = \{x \mid g_i(x) \geq 0, \ i \in L\}$ (A is closed, $\overset{\circ}{A} \neq \emptyset$)

$B \subset R^n$, convex, compact, large enough to contain the optimal solution $\hat{x}$.

BOUNDED METHOD OF CENTERS

$$d(x, \lambda) = \min\{f(x) - \lambda, g_i(x) \mid i \in L\}, \forall \lambda \in R$$

$$E(\lambda) = \{x \mid f(x) - \lambda \geq 0, g_i(x) \geq 0, i \in L\}, \forall \lambda \in R.$$

The function $d: R^n \times R \to R$ is a continuous and concave one. From proposition 1.2, it is a regular F-distance defined for

$$\mathcal{E} = \{E(\lambda) \mid \lambda \in K \subset R\}$$

where

$$K = [\inf\{f(x) \mid x \in A\}, f(\hat{x})[$$

(we may verify that, choosing K in such a manner, the hypotheses of proposition 1.2 are satisfied).

We shall suppose here that $\overset{\circ}{A} \cap B \neq \emptyset$. Then, since A and B are convex sets, they satisfy hypothesis (H) of the method of Centers, defined in 2.1, and under these conditions, hypotheses (H1) to (H3) of the upper-bounded method of Centers, defined in 3.1, are satisfied.

Let us set

$$f'(x, y) = f(y) + \nabla f(y) \cdot (x - y)$$

$$g_i'(x, y) = g_i(y) + \nabla g_i(y) \cdot (x - y), \quad i \in L$$

and take $d'(x, y, \lambda) = \min\{f'(x, y) - \lambda, g_i'(x, y) \mid i \in J_\epsilon(y)\}$
with $J_\epsilon(y) = \{i \in L \mid g_i(y) < \epsilon\}$ ($\epsilon > 0$ being given).
Under these conditions (H4) and (H5) are satisfied, because:

$$\left.\begin{array}{l} f'(x, y) \geq f(x) \\[1em] g_i'(x, y) \geq g_i(x) \end{array}\right\} \forall x, \forall y \text{ since } f \text{ and } g_i \text{ are concave}$$

$$d'(x, y, \lambda) = \min\{f'(x, y) - \lambda, g'_i(x, y) \mid i \in J_\epsilon(y)\}$$

$$\geq \min\{f'(x, y) - \lambda, g'_i(x, y) \mid i \in L\} \text{ since } J_\epsilon(y) \subset L$$

$$\geq \min\{f(x) - \lambda, g_i(x) \mid i \in L\}$$

$$= d(x, \lambda)$$

which satisfies (H4)$_{(i)}$. On the other hand:

$$d'(x, x, f(x)) = \min\{0, g_i(x) \mid i \in J_\epsilon(x)\}$$

$$= 0, \quad \forall x \in A$$

which satisfies (H4)$_{(ii)}$.
For hypothesis (H5), let us consider $a \in B$, $b \in B$ and $\lambda \in R$ such that:

$$d(x, f(a)) \leq 0, \quad \forall x \in [a, b].$$

It is easy to verify that this relation implies:

$$\nabla f(a) \cdot (b-a) \leq 0 \text{ and/or } \exists i \in J_\epsilon(a) : \nabla g_i(a) \cdot (b-a) \leq 0$$

and hence, that $d'(x, a, f(a))$ is decreasing at point a on $[a, b]$. Since $d'(a, a, f(a)) = d(a, f(a)) = 0$, we have that:

$$d'(x, a, f(a)) \leq 0, \quad \forall x \in [a, b].$$

We still have to verify that $d'(x, y, \lambda)$ is actually a continuous function. This is evident if $J_\epsilon(y)$ is a constant, because the functions $\nabla f(y)$ and $\nabla g_i(y)$ are continuous ones by hypothesis.

The same is not true if $J_\epsilon(y)$ is not a constant. But, from the remark in 3.3 (end of the proof relative to the convergence of the algorithm), only continuity of d' at the point $(\overset{*}{z}, \overset{*}{x}, \lambda^*)$ is needed. So, it is clear that we have

$$J_\epsilon(y) = C^{te} = J_\epsilon(\overset{*}{x})$$

for every y sufficiently close to point $\overset{*}{x}$. It is necessary to take $\epsilon > 0$ and not $\epsilon = 0$.

Zoutendijk's method of feasible directions [14]

If we take for B a hyper-rectangle, very large with respect to the size of $A \cap \{x \mid f(x) \geq f(\overset{o}{x})\}$, this set being supposed compact, and if we choose a small ϵ, it is possible to determine $\overset{k}{z}$ by solving the following <u>linear programming</u> problem with variables $x \in R^n$ and $\mu \in R$:

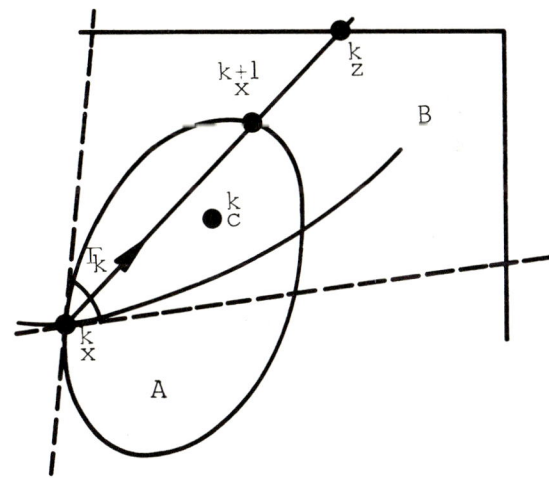

Figure 2

Maximize μ subject to

$$g'_i(x, \overset{k}{x}) \geq \mu, \quad i \in J_\epsilon(\overset{k}{x})$$

$$f'(x, \overset{k}{x}) - \lambda_k \geq \mu$$

$$x \in B$$

where $\lambda_k = f(\overset{k}{x})$ is fixed.
In fact, if we have for some point $\overset{k}{x}$:

$$J_\epsilon(\overset{k}{x}) = \{i \in L \mid g_i(\overset{k}{x}) = 0\}$$

and if B is chosen sufficiently large, it is not necessary to fully optimize linear programming problems. The domain of the linear program being the intersection of B and a polyhedral cone Γ_k with vertex $\overset{k}{x}$, we may take $\overset{k}{z}$ sufficiently far on the half-line interior to Γ_k, with initial point $\overset{k}{x}$, and direction that of steepest ascent at $\overset{k}{x}$ for $d'(x, \overset{k}{x}, \lambda_k)$. The faces of this cone are supporting planes for the binding constraints at point $\overset{k}{x}$ and for the supplementary constraint $f(x) \geq f(\overset{k}{x})$. we have then:

$$d'(\overset{k}{z}, \overset{k}{x}, \lambda_k) \geq d'(\overset{k}{c}, \overset{k}{x}, \lambda_k)$$

because $\|\overset{k}{z} - \overset{k}{x}\| \geq \|\overset{k}{c} - \overset{k}{x}\|$ and since the slope of $(\overset{k}{z} - \overset{k}{x})$ is the steepest. Hence, from a practical point of view, it is unnecessary to define the rectangle B, since the Simplex method determines a ray of unbounded feasible solutions, whose direction defines that of $(\overset{k}{z} - \overset{k}{x})$.

If $J_\epsilon(\overset{k}{x}) \neq \{i \in L \mid g_i(\overset{k}{x}) = 0\}$, the domain of the linear programming problem is not the intersection of B with a cone, and some preliminary steps (changes of basis of the Simplex method will be done before obtaining a ray whose direction corresponds to that of $\overset{k}{z}$.

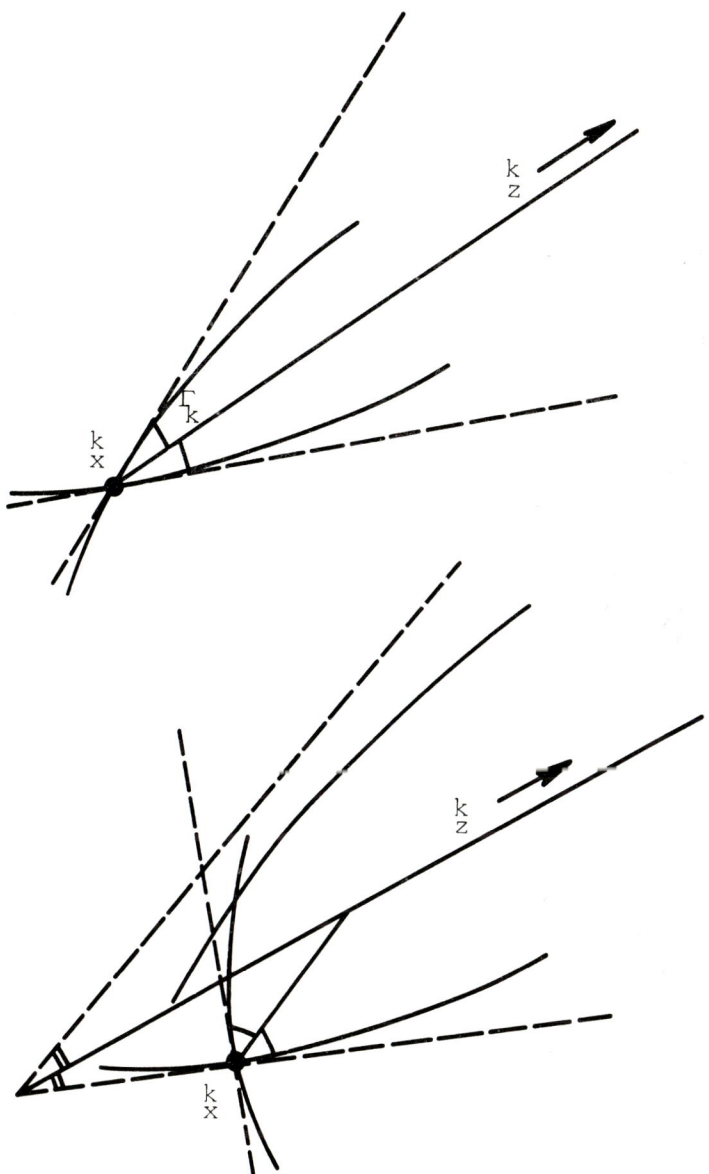

Figure 3

Zoutendijk proceeds here in a slightly different way: the linearized constraints with subscript i, such that:

$$i \in J_\epsilon(\overset{k}{x}), \quad g_i(\overset{k}{x}) \neq 0$$

are translated up to $\overset{k}{x}$. Then, under these conditions, the polyhedron of this new linear programming problem is always a cone (B being eliminated). But this cone may possibly reduce to a point $\overset{k}{x}$, and then we must diminish ϵ to obtain a smaller set $J_\epsilon(\overset{k}{x})$ and a cone of higher dimension. This procedure does not offer any advantage, at least in theory, over the general procedure described previously: the resolution of the linear programming problem is similar in both cases, and the preceding procedure has the advantage of not requiring ϵ to be diminished.

Once $\overset{k}{z}$ is determined, we may practically avoid the step corresponding to the search of $\overset{k+1}{x'}$, maximizing $d(x, \lambda_k)$ on $[\overset{k}{x}, \overset{k}{z}]$, and searching directly for $\overset{k+1}{x}$ on $A \cap [\overset{k}{x}, \overset{k}{z}]$, by solving:

$$\overset{k+1}{x} : f(\overset{k+1}{x}) = \max\{f(x) \mid x \in A \cap [\overset{k}{x}, \overset{k}{z}]\}$$

That is what Zoutendijk proposes: the computation is easy because it is just a maximization with one variable. But we also have the possibility, as indicated in 3.2, to choose $\overset{k+1}{x}$ arbitrarily in $E(\lambda'_{k+1})$, with $\lambda'_{k+1} = f(\overset{k+1}{x})$. The point $\overset{k+1}{x'}$ is determined, in these conditions, by maximization in one variable, with d instead of f:

$$\overset{k+1}{x'} : d(\overset{k+1}{x'}, \lambda_k) = \max\{d(x, \lambda_k) \mid x \in [\overset{k}{x}, \overset{k}{z}]\}$$

Linearized method of centers [6]

If $\mathcal{E}$ is chosen very large, i.e. if we set

$$J_\epsilon(x) = L, \quad \forall x$$

we obtain the linearized method of centers described in the article quoted above. The compact set B may be, in practice, eliminated, or its definition may allow for all constraints $g_i(x) \geq 0$ which are affine. What matters for the computations is that the condition $x \in B$ be represented by linear constraints.

All intermediates between the method of feasible directions and the linearized method of centers, represented by the different values chosen for ϵ, are possible. Of these two extreme methods the first chooses its direction of displacement using local data, whereas the second is more global, requiring more extensive linear programming problem solving.

4.2. <u>Applications type</u> (B)

Let us set $A = R^n$

$$B = \{x \mid g_i(x) \geq 0, \ i \in L\}$$

(B is here assumed to be compact and to have a nonempty interior. It is convex, since the g_i are concave)

$$d(x, \lambda) = f(x) \quad \lambda$$

$$E(\lambda) = \{x \mid f(x) \geq \lambda\}, \quad \lambda < \bar{\lambda} = \sup f(x)$$

$d : R^n \times R \to R$ is a continuous, concave function. It is a regular F-distance defined for $\mathcal{E} = \{E(\lambda) \mid \lambda < \bar{\lambda}\}$. The hypotheses (H_1) through (H_3) are satisfied.

Giving to $f'(x, y)$ the same definition as previously in 4.1, we take:

$$d'(x, y, \lambda) = f'(x, y) - \lambda$$

The hypotheses (H_4), (i) and (ii), and (H_5) are therefore satisfied.

Frank and Wolfe's method [4]

In this manner an algorithm is obtained which generalizes Frank and Wolfe's method which was established for the case where B is a linear polyhedron and f a quadratic function.

At each step, we have to maximize an affine function d' on B. The points $\overset{k+1}{x'}$ and $\overset{k+1}{x}$ are always coincident if we choose $\overset{k+1}{x}$ only on $[\overset{k}{x}, \overset{k}{z}]$. From a practical computational standpoint, this method is practicable only if B is a linear polyhedron:

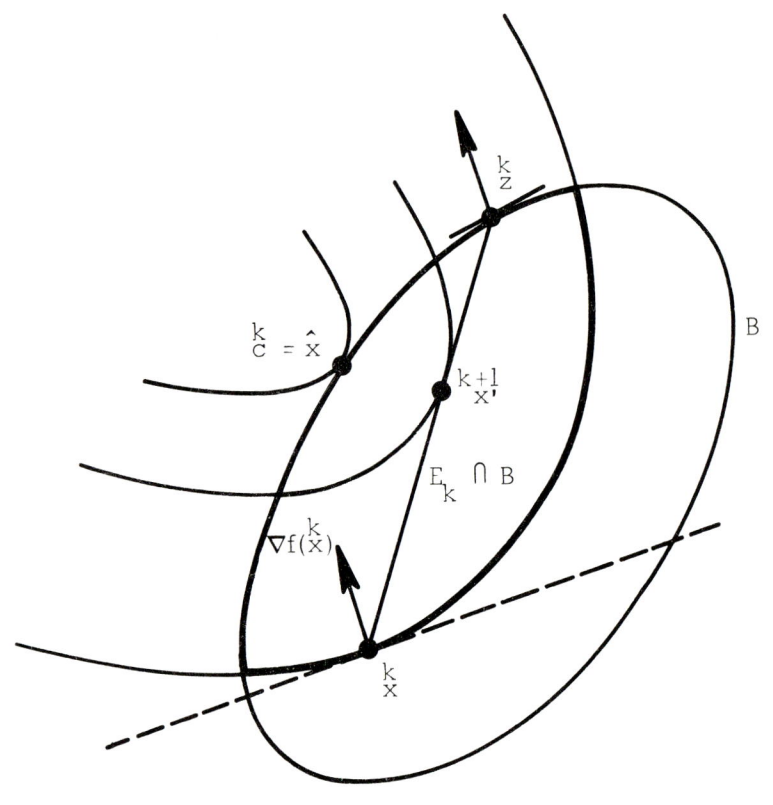

Figure 4

we are then led to solve linear programming problems. Note that the point $\overset{k}{c}$ maximizing $d(x, \lambda_k)$ on B is here a constant, optimal solution of the problem set.

If $B = R^n$, the problem set becomes that of the maximization of f in R^n and the algorithm becomes identical with the gradient method so-called "method of the steepest ascent".

Rosen's gradient projection method [10]

Given that B is a linear polyhedron, the computation of $\overset{k}{z}$ may be simplified considerably by solving only partially the corresponding linear programming problem. Like Rosen, we may confine ourselves to the first step of the path of steepest ascent on this polyhedron, with respect to d'. Practically, this comes to projecting the gradient of $d(x, \lambda_k)$, computed at point $\overset{k}{x}$ (i.e. $\nabla f(\overset{k}{x})$) on the cone tangent to $E_k \cap B$ at $\overset{k}{x}$. Let $\overset{k}{y}$ be this projection. The point $\overset{k}{z}$ is then taken at the end of the intersection of B with the half line having $\overset{k}{x}$ as its origin and $\overset{k}{y}$ as its direction. This

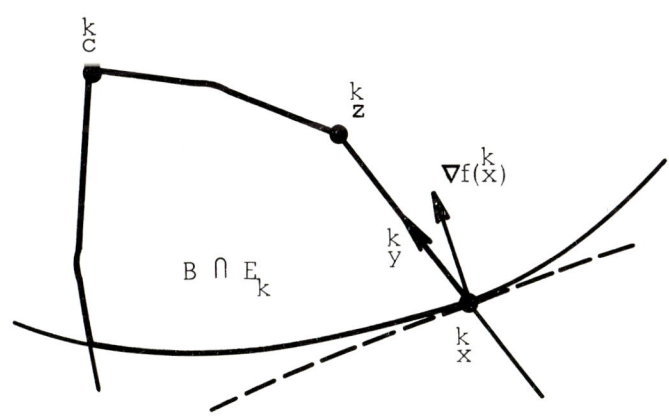

Figure 5

approximation $\overset{k}{z}$ is acceptable in defining the general algorithm if we have $|\overset{k}{x} - \overset{k}{z}| > \epsilon$, ϵ being a constant > 0 independent of k. As a matter of fact, we should have:

$$\alpha \cdot d'(\overset{k}{z}, \overset{k}{x}, \lambda_k) \geq \max\{d'(x, \overset{k}{x}, \lambda_k) \mid x \in B\} =$$

$$= d'(\overset{k}{c}, \overset{k}{x}, \lambda_k)$$

where $\alpha \geq 1$, is a constant independent of k. Noting that $\overset{k}{y}$ is the direction of steepest ascent on B, at point $\overset{k}{x}$, for $d'(x, \overset{k}{x}, \lambda_k)$, we may write:

$$d'(\overset{k}{z}, \overset{k}{x}, \lambda_k) \geq \frac{|\overset{k}{z} - \overset{k}{x}|}{|\overset{k}{c} - \overset{k}{x}|} \cdot d'(\overset{k}{c}, \overset{k}{x}, \lambda_k) .$$

Since B is bounded, $|\overset{k}{c} - \overset{k}{x}| \leq \beta$, a constant. It follows that if we take $\alpha = \beta/\epsilon \geq 1$, we actually obtain the condition sought.

If the point $\overset{k}{z}$ thus found is too close to $\overset{k}{x}$, i.e. if $|\overset{k}{z} - \overset{k}{x}| < \epsilon$, it is sufficient to continue along the path of steepest ascent without changing the gradient $\nabla f(\overset{k}{x})$ until we have $|\overset{k}{z} - \overset{k}{x}| \geq \epsilon$, or $d'(\overset{k}{z}, \overset{k}{x}, \lambda_k) = \max\{d'(x, \overset{k}{x}, \lambda_k) | x \in B\}$. We know as a matter of fact that such a path, which is piecewise linear, converges in a finite number of steps to an optimal solution for the linear programming problem, and that the average slope of the current solutions thus defined is decreasing. Therefore, we shall always have:

average slope of $[\overset{k}{x}, \overset{k}{z}] \geq$ average slope of $[\overset{k}{x}, \overset{k}{c}]$.

Acknowledgement. Je tiens à remercier vivement M. L. Balinski, qui m'a beaucoup aidé dans la traduction de ce texte, à l'exception de cette dernière phrase.

REFERENCES

1. Bui Trong Lieu et Huard (P), "La methode des centres dans un espace topologique" Numerische Math. (8) 1966, p. 56-67.

2. Chevassus (O. A.), "Condition suffisante de convergence pour les methodes iteratives de minimisation." Note E. D. F. N°HR 7713-8 Juin, 1967.

3. Fiacco (A. V.), Sequential unconstrained minimization methods for nonlinear programming", Thesis, Northwestern University, Evanston, Illinois, June, 1967.

4. Frank (M.) et Wolfe(Ph.), "An algorithm for quadratic programming", Nav. Res. Logist. Quarterly $\underline{3}$ ( ), 1956, p. 95-120.

5. Huard (P), "Resolution of Mathematical programming with nonlinear constraints by the method of Centres", in Nonlinear Programming (Ed. Abadie), p. 206-219, North Holland Publishing Co., Amsterdam, 1967.

6. Huard (P), "Programmation mathematique convexo", R.I.R.O. N°. 7, 1968, p. 43-59.

7. Polak (E), "Computational methods in discrete optimal control and nonlinear programming: a unified approach" University of California, Berkeley, Memo N°. ERL-M261, February 24, 1969, to be published in 1970.

8. Polak (E), "On the convergence of optimization algorithms", R.I.R.O. N° 16, 1969, p. 17-34.

9. Roode (J. D.), "Generalized Lagrangian functions in mathematical programming", Thesis, University of Leiden, October, 1968.

10. Rosen (J. B.), "The gradient projection method for nonlinear programming", Part I: Linear Constraints", SIAM Journal 8 (1), 1960, p. 181-217.

11. Topkis (M.) et Veinott (A), "On the convergence of some feasible directions algorithms for nonlinear programming", SIAM Control 5 (2), 1967, p. 268-279.

12. Tremolieres (R), Methode des Centres a troncatures variables", Bulletin de la Direction des Etudes et Recherches - E.D.F. - Serie C, N° 2, 1968, p. 57-64.

13. Zangwill (W. I.), "Convergence Conditions for nonlinear programming algorithms", Center for Research in Management Sciences, University of California, Berkeley, 1966. Working Paper N° 197.

14. Zoutendijk (G), "Methods of feasible directions", Elsevier Publ. Co., Amsterdam, 1960.

A New Algorithm for Unconstrained Optimization

M. J. D. POWELL

ABSTRACT

A new algorithm is described for calculating the least value of a given differentiable function of several variables. The user must program the evaluation of the function and its first derivatives. Some convergence theorems are given that impose very mild conditions on the objective function. These theorems, together with some numerical results, indicate that the new method may be preferable to current algorithms for solving many unconstrained minimization problems.

1. Introduction

We are concerned with the problem of calculating the least value of a given function $F(x_1, x_2, \ldots, x_n)$, in the case that just $F(\underline{x})$ and its first derivatives

$$g_i(\underline{x}) = \frac{\partial}{\partial x_i} F(\underline{x}), \quad i = 1, 2, \ldots, n, \qquad (1)$$

are available for any $\underline{x}$, where the notation $\underline{x}$ denotes the vector of variables $(x_1, x_2, \ldots, x_n)$. We expect $F(\underline{x})$ to have higher derivatives, but suppose that it is preferable to avoid the labor of calculating them.

The more successful current algorithms for this problem, for example Davidon (1959), Fletcher and Powell (1963), Davidon (1958), Pearson (1969), Murtagh and Sargent (1969), Powell (1969a) and Fletcher (1969), use calculated first derivatives to estimate a second derivative matrix of the objective function, for this leads to a quadratic approximation to $F(\underline{x})$, which is the simplest type of approximation that can have a minimum. These algorithms do not calculate first derivatives just for the purpose of estimating second derivatives, but instead they take advantage of the changes in gradient that happen to occur during the search for the least value of $F(\underline{x})$.

Two common formulae that are used to revise the approximation to the second derivative matrix are

$$G^* = G + \frac{(\underline{\gamma}-G\underline{\delta})(\underline{\gamma}-G\underline{\delta})^T}{(\underline{\gamma}-G\underline{\delta},\ \underline{\delta})} \qquad (2)$$

and

$$H^* = H + \frac{\delta \delta^T}{(\delta, \underline{y})} - \frac{H\underline{y}\,\underline{y}^T H}{(H\underline{y}, \underline{y})}, \qquad (3)$$

where $\underline{y}$ is the calculated change in gradient

$$\underline{y} = \underline{g}(\underline{x} + \underline{\delta}) - \underline{g}(\underline{x}), \qquad (4)$$

due to a change $\underline{\delta}$ in the vector $\underline{x}$. In the algorithms that use equation (2), the matrix G^* replaces G as an estimate of the second derivative matrix of the objective function, and in the algorithms that use equation (3), H^* is taken to be a better approximation than H to the _inverse_ of the second derivative matrix.

Note that in both these formulae there are divisions by scalar products, and this is worrying because of the possibility that two vectors in a scalar product may be nearly orthogonal. Therefore in this paper we offer a new formula for revising second derivative approximations, that is attractive because it does not involve divisions by scalar products of different vectors. The new formula is derived in Section 2.

An algorithm that uses the new formula is outlined in Section 3. The details of this algorithm, including a Fortran listing, will be published separately (Powell, 1970).

In fact the new algorithm is the result of an attempt to prove convergence of a successful current algorithm. Already we have proved (Powell, 1969b) that Davidon's (1959) variable metric method converges super-linearly, in the case that the objective function $F(\underline{x})$ satisfies a strict convexity condition, but this is inadequate, because frequently in real problems the objective function is not convex. In Section 4 we prove some convergence theorems for the new algorithm, that are valid for functions $F(\underline{x})$ that are bounded below, and that have continuous and bounded second derivatives. One theorem states that the algorithm will calculate a point $\underline{x}$ such that $\|\underline{g}(\underline{x})\| \le \varepsilon$, where ε is any prescribed positive number. Another states that if $\underline{g}(\underline{x})$ is small because $\underline{x}$ is close to a local minimum of $F(\underline{x})$, and if at the local minimum the second derivative matrix of $F(\underline{x})$ is positive definite, then the final rate of convergence of the algorithm is super-linear.

Recently I heard that McCormick (1969) has proved that the convergence properties of the "reset variable metric method" (McCormick and Pearson, 1969) are like those of the new algorithm. The reset variable metric method is an extension of Davidon's (1959) method, the extension being that the second derivative approximation is set to a prescribed, constant, positive definite matrix after every (n+1) iterations. The extension is a little unsatisfactory, for it causes accumulated information about second derivatives to be destroyed, but often it gives a substantial reduction in computing time.

Section 5 of this paper relates the new algorithm to some other published methods for minimization.

2. The Formula for Revising the Second Derivative Approximation

We are given a second derivative approximation G, and vectors $\underline{\delta}$ and $\underline{y}$ that are related by equation (4). We wish to use this information to calculate another second derivative approximation G^*, which is more accurate than G.

We note that if $F(\underline{x})$ is exactly a quadratic function, having the second derivative matrix $\overline{G}$, then the equation

$$\underline{y} = \overline{G}\underline{\delta} \qquad (5)$$

holds. Therefore we calculate G^* to satisfy the equation

$$G^* \underline{\delta} = \underline{y}. \qquad (6)$$

Expression (2) is one of many formulae that cause G^* to satisfy this equation, but we abandon equation (2) because its denominator can equal zero.

Broyden's (1965) paper on systems of nonlinear equations suggests the formula

$$G^{(1)} = G + \frac{(\underline{y} - G\underline{\delta})\underline{\delta}^T}{\|\underline{\delta}\|^2}, \qquad (7)$$

and this definition is attractive because its denominator is the square of the Euclidean length of $\underline{\delta}$. However the matrix $G^{(1)}$ is not symmetric. Therefore we consider letting G^* equal the matrix

$$\overline{G}^{(1)} = \frac{1}{2}[G^{(1)} + G^{(1)T}], \qquad (8)$$

where the superscript "T" denotes the transpose.

However the matrix (8) is also unsatisfactory, because it is not consistent with equation (6). Therefore, in order to obtain from Broyden's formula a symmetric matrix that satisfies equation (6), we let G^* be the limit of the infinite sequence $\overline{G}^{(1)}, \overline{G}^{(2)}, \ldots$, where, for $k = 1, 2, \ldots,$

$$G^{(k+1)} = \overline{G}^{(k)} + \frac{(\underline{y} - \overline{G}^{(k)}\underline{\delta})\underline{\delta}^T}{\|\underline{\delta}\|^2} \qquad (9)$$

and

$$\overline{G}^{(k+1)} = \frac{1}{2}[G^{(k+1)} + G^{(k+1)T}]. \qquad (10)$$

Straightforward algebra gives that the limit of the sequence is the matrix

$$G^* = G + \frac{(\gamma - G\underline{\delta})\underline{\delta}^T + \underline{\delta}(\gamma - G\underline{\delta})^T}{\|\underline{\delta}\|^2}$$

$$- \frac{\underline{\delta}\,\underline{\delta}^T(\gamma - G\underline{\delta},\,\underline{\delta})}{\|\underline{\delta}\|^4}, \quad (11)$$

and we note that G^* is symmetric and satisfies equation (6).

Because of the properties of Broyden's (1965) formula, we expect the definition (11) to provide some nice reduction of errors, in the ideal case when $F(\underline{x})$ is a quadratic function. This does happen, for, by using equation (5) to express γ in terms of $\underline{\delta}$, one can deduce from the definition (11) the relation

$$G^* - \overline{G} = (I - \frac{\underline{\delta}\,\underline{\delta}^T}{\|\underline{\delta}\|^2})(G - \overline{G})(I - \frac{\underline{\delta}\,\underline{\delta}^T}{\|\underline{\delta}\|^2}). \quad (12)$$

This equation shows that, when $F(\underline{x})$ is quadratic, the error in G^* is equal to the error in G multiplied by symmetric projection matrices. Therefore G^* is usually more accurate than G, so formula (11) is used in the new algorithm.

Moreover equation (12) shows another important property, which is that if equation (11) is applied iteratively n times, and if the n corresponding vectors $\underline{\delta}$ are mutually orthogonal, then, in the quadratic case, the final second derivative approximation is exact. The purpose of this remark is not that we intend to use sets of mutually orthogonal directions in the algorithm. It is to show that formula (11) can provide accurate second derivative approximations, even when the true second derivative matrix is not positive definite.

Some properties of equation (11), in the case that $F(\underline{x})$ is not a quadratic function, are given in Section 4.

3. An Outline of the New Algorithm

The algorithm is iterative, and calculates a sequence of points $\underline{x}^{(1)}$, $\underline{x}^{(2)}$, ..., which is intended to converge to the position of the minimum of $F(\underline{x})$. The kth iteration calculates $\underline{x}^{(k+1)}$ from $\underline{x}^{(k)}$. The initial vector $\underline{x}^{(1)}$ has to be set by the user of the algorithm to a guess of the position of the required minimum.

At the beginning of the kth iteration we require the point $\underline{x}^{(k)}$, the gradient vector $\underline{g}(\underline{x}^{(k)}) = \underline{g}^{(k)}$ say, an approximation, $G^{(k)}$, to the second derivative matrix at $\underline{x}^{(k)}$, and a bound $\Delta^{(k)}$ on the length of the change $\|\underline{x}^{(k+1)} - \underline{x}^{(k)}\|$. Except for some "special iterations", which will be discussed later, the kth iteration uses the quadratic function of $\underline{\delta}$

$$\Phi^{(k)}(\underline{\delta}) \equiv F(\underline{x}^{(k)}) + (\underline{g}^{(k)}, \underline{\delta}) + \frac{1}{2}(\underline{\delta}, G^{(k)}\underline{\delta}), \qquad (13)$$

which is an approximation to $F(\underline{x}^{(k)} + \underline{\delta})$, to calculate a displacement $\underline{\delta}^{(k)}$ that satisfies the conditions

$$\|\underline{\delta}^{(k)}\| \le \Delta^{(k)} \qquad (14)$$

and

$$\Phi^{(k)}(\underline{\delta}^{(k)}) < F(\underline{x}^{(k)}), \qquad (15)$$

so we expect the value of $F(\underline{x}^{(k)} + \underline{\delta}^{(k)})$ to be less than $F(\underline{x}^{(k)})$. The objective function and its gradient are calculated at $\underline{x}^{(k)} + \underline{\delta}^{(k)}$, and we define

$$\underline{y}^{(k)} = \underline{g}(\underline{x}^{(k)} + \underline{\delta}^{(k)}) - \underline{g}^{(k)}, \qquad (16)$$

in order to substitute $\underline{y}^{(k)}$ and $\underline{\delta}^{(k)}$ in equation (11) to define $G^{(k+1)}$. $\underline{x}^{(k+1)}$ is defined by the equation

$$\underline{x}^{(k+1)} = \begin{cases} \underline{x}^{(k)}, & F(\underline{x}^{(k)} + \underline{\delta}^{(k)}) \geq F(\underline{x}^{(k)}) \\ \underline{x}^{(k)} + \underline{\delta}^{(k)}, & F(\underline{x}^{(k)} + \underline{\delta}^{(k)}) < F(\underline{x}^{(k)}) \end{cases} \qquad (17)$$

The kth iteration then tests the convergence criterion

$$\|\underline{g}(\underline{x}^{(k+1)})\| \leq \varepsilon, \qquad (18)$$

where ε is a positive constant set by the user of the algorithm. If this inequality holds, then the algorithm finishes. Otherwise $\Delta^{(k+1)}$ is set, so that the algorithm can begin the $(k+1)^{st}$ iteration.

To calculate $\underline{\delta}^{(k)}$ we could follow Marquardt (1963), and define $\underline{\delta}^{(k)}$ to minimize expression (13) subject to condition (14). But in this case the condition (14) may necessitate of order n^3 computer operations, which is not acceptable because most other algorithms for unconstrained minimization use only of order n^2 operations per iteration. Therefore we follow the idea suggested by Powell (1968), and we force $\underline{\delta}^{(k)}$ to have the form

$$\underline{\delta}^{(k)} = \alpha^{(k)} \underline{g}^{(k)} + \beta^{(k)} [G^{(k)}]^{-1} \underline{g}^{(k)}, \qquad (19)$$

where $\alpha^{(k)}$ and $\beta^{(k)}$ are parameters. Thus we restrict $\underline{\delta}^{(k)}$ to the two-dimensional space containing both the gradient of $F(\underline{x})$ at $\underline{x}^{(k)}$, and also the displacement to the stationary point of the quadratic function (13).

In fact if the inequality

$$(\underline{g}^{(k)}, G^{(k)} \underline{g}^{(k)}) \Delta^{(k)} \leq \|\underline{g}^{(k)}\|^3 \qquad (20)$$

holds, we set

$$\left.\begin{array}{l}\alpha^{(k)} = -\Delta^{(k)}/\|\underline{g}^{(k)}\| \\ \\ \beta^{(k)} = 0\end{array}\right\}, \qquad (21)$$

and if the inequality (20) is not satisfied, the displacement $\underline{\delta}^{(k)}$ is the point on the straight line through $-[G^{(k)}]^{-1}\underline{g}^{(k)}$ and $-\underline{g}^{(k)}\|\underline{g}^{(k)}\|^2/(\underline{g}^{(k)}, G^{(k)}\underline{g}^{(k)})$ that gives the least value of $\Phi^{(k)}(\underline{\delta}^{(k)})$, subject to condition (14). Two important features of this choice of $\underline{\delta}^{(k)}$ are (i) a bias towards the direction $-\underline{g}^{(k)}$ if $\Delta^{(k)}$ is small, and (ii) if $\Delta^{(k)}$ is sufficiently large and $G^{(k)}$ is positive definite, then we set $\underline{\delta}^{(k)}$ to the vector that actually minimizes $\Phi^{(k)}(\underline{\delta})$, which is the "Newton correction"

$$\underline{\delta}^{(k)} = -[G^{(k)}]^{-1}\underline{g}^{(k)}. \qquad (22)$$

To calculate $\underline{\delta}^{(k)}$ in this way in only of order n^2 computer operations, it is necessary to keep the inverse of $G^{(k)}$, $H^{(k)}$ say. The formula for updating $H^{(k)}$ that corresponds to equation (11) is the equation

$$H^* = H - \{\underline{\eta}\underline{\eta}^T(\underline{\delta}, H\underline{\delta}) - [\underline{\eta}\underline{\delta}^T H + H\underline{\delta}\underline{\eta}^T](\underline{\delta}, H\underline{y})$$
$$+ H\underline{\delta}\underline{\delta}^T H(\underline{\eta}, \underline{y})\}/\{(\underline{\eta}, \underline{y})(\underline{\delta}, H\underline{\delta}) - (\underline{\delta}, H\underline{y})^2\}, \qquad (23)$$

where $\underline{\eta} = H\underline{y} - \underline{\delta}$. But this formula fails if its denominator is zero, which happens if G^* is singular. To avoid singularity we replace expression (11) by the matrix

$$G^{(k+1)} = G^{(k)}$$
$$+ \theta^{(k)} \frac{(\underline{y}^{(k)} - G^{(k)}\underline{\delta}^{(k)})\underline{\delta}^{(k)T} + \underline{\delta}^{(k)}(\underline{y}^{(k)} - G^{(k)}\underline{\delta}^{(k)})^T}{\|\underline{\delta}^{(k)}\|^2}$$
$$- [\theta^{(k)}]^2 \frac{\underline{\delta}^{(k)}\underline{\delta}^{(k)T}(\underline{y}^{(k)} - G^{(k)}\underline{\delta}^{(k)}, \underline{\delta}^{(k)})}{\|\underline{\delta}^{(k)}\|^4}, \quad (24)$$

and we let the value of $\theta^{(k)}$ be different from one if formula (11) would yield a singular matrix. Equation (24) is used because, if $F(\underline{x})$ is a quadratic function, then in place of equation (12) we have the identity

$$(G^{(k+1)} - \overline{G}) =$$
$$= (I - \theta^{(k)} \frac{\underline{\delta}^{(k)}\underline{\delta}^{(k)T}}{\|\underline{\delta}^{(k)}\|^2})(G^{(k)} - \overline{G})(I - \theta^{(k)} \frac{\underline{\delta}^{(k)}\underline{\delta}^{(k)T}}{\|\underline{\delta}^{(k)}\|^2}). \quad (25)$$

The value that we assign to $\theta^{(k)}$ is the number closest to one, such that the condition

$$|\det G^{(k+1)}| \geq 0.1 |\det G^{(k)}| \quad (26)$$

holds. It can be shown that (Powell, 1970) the calculation of $\theta^{(k)}$ requires only the solution of a quadratic equation, and also it can be shown that the bound $|1-\theta^{(k)}| \leq \sqrt{2/11}$ is always obtained.

A convenient way of applying formula (24), and of setting $H^{(k+1)} = [G^{(k+1)}]^{-1}$, is to substitute $\underline{\delta} = \underline{\delta}^{(k)}$ and

$$\underline{y} = \theta\underline{y}^{(k)} + (1-\theta)G^{(k)}\underline{\delta}^{(k)} +$$
$$+ \theta(1-\theta)(\underline{y}^{(k)} - G^{(k)}\underline{\delta}^{(k)}, \underline{\delta}^{(k)})\underline{\delta}^{(k)}/\|\underline{\delta}^{(k)}\|^2 \quad (27)$$

in equations (11) and (23), for then G^* and H^* are equal to the required matrices $G^{(k+1)}$ and $H^{(k+1)}$.

To calculate the bound $\Delta^{(k+1)}$ on $\|\underline{\delta}^{(k+1)}\|$, we compare the reduction in $F(\underline{x})$ due to the displacement $\underline{\delta}^{(k)}$ with the predicted change $\{\Phi^{(k)}(\underline{\delta}^{(k)}) - F(\underline{x}^{(k)})\}$. If the inequality

$$F(\underline{x}^{(k)} + \underline{\delta}^{(k)}) - F(\underline{x}^{(k)}) \leq 0.1 \{\Phi^{(k)}(\underline{\delta}^{(k)}) - F(\underline{x}^{(k)})\} \quad (28)$$

holds, we count that the iteration is successful, and the value of $\Delta^{(k+1)}$ is either $\|\underline{\delta}^{(k)}\|$ or $2\|\underline{\delta}^{(k)}\|$. But if inequality (28) fails, then it seems that $\|\underline{\delta}^{(k)}\|$ is too large, so we set $\Delta^{(k+1)} = \frac{1}{2}\|\underline{\delta}^{(k)}\|$. Because more details on the calculation of $\Delta^{(k+1)}$ are not important to the convergence theorems of Section 4, we will defer them to the report that will include a Fortran listing of the algorithm (Powell, 1970).

The initial bound $\Delta^{(1)}$ has to be set by the user of the algorithm. The initial matrices $G^{(1)}$ and $H^{(1)}$ are set to λI and $\lambda^{-1} I$, where I is the unit matrix, and λ is the scalar $0.01 \|\underline{g}^{(1)}\|/\Delta^{(1)}$.

At the beginning of this section it is stated that some "special iterations" choose $\underline{\delta}^{(k)}$ in a way that is different from the one that has been described. The reason is that numerical experimentation with formula (11) has shown that it is worthwhile to ensure that the directions $\underline{\delta}^{(k)}$ ($k = 1, 2, \ldots$), used to calculate the sequence of approximations $G^{(k)}$ ($k = 1, 2, \ldots$), satisfy a "strict linear independence condition". Therefore there is a device in the Fortran subroutine that causes the condition

$$\left\| \prod_{j=k}^{k+K} (I - \theta^{(j)} \frac{\underline{\delta}^{(j)} \underline{\delta}^{(j)T}}{\|\underline{\delta}^{(j)}\|^2}) \right\|_2 \leq c \quad (29)$$

to hold for all k, where K is a fixed integer, and where c is a constant whose value is less than one. Equation (25) suggests that this condition is helpful, for an immediate corollary of expressions (25) and (29) is that, if $F(\underline{x})$ is a quadratic function, then $G^{(k)}$ converges to $\overline{G}$.

In fact every third iteration of the algorithm is a "special iteration". Then $\underline{\delta}^{(k)}$ is calculated by a method that is like the one described in Section 7 of the report by Powell (1968). This definition of $\underline{\delta}^{(k)}$ is intended to provide the inequality (29), and $\underline{\delta}^{(k)}$ may even be such that $\Phi^{(k)}(\underline{\delta}^{(k)})$ exceeds $F(\underline{x}^{(k)})$. However the bound (14) is maintained. Unlike an ordinary iteration, a special iteration always sets $\Delta^{(k+1)} = \Delta^{(k)}$, but $G^{(k+1)}$, $H^{(k+1)}$ and $\underline{x}^{(k+1)}$ are calculated in the usual way, which is through formulae (11), (17), (23) and (27).

All other details of the algorithm are deferred to Powell (1970), because we have already given enough information to convey the main features of the algorithm, and to prove the convergence theorems.

4. Theorems on the New Algorithm

In this section we require $F(\underline{x})$ to have continuous bounded second derivatives

$$\|\overline{G}(\underline{x})\| \leq \overline{M}, \qquad (30)$$

where $\overline{G}(\underline{x})$ is the true second derivative matrix of $F(\underline{x})$ at the point $\underline{x}$, and where the matrix norm is induced by the Euclidean vector norm. Also we require the second derivative matrices to satisfy the Lipschitz condition

$$\|\overline{G}(\underline{x}) - \overline{G}(\underline{y})\| \leq L \|\underline{x} - \underline{y}\|. \qquad (31)$$

Theorem 1. Formulae (4) and (11) imply that the matrix A, defined by the equation

UNCONSTRAINED OPTIMIZATION

$$G^* = (I - \frac{\underline{\delta}\,\underline{\delta}^T}{\|\underline{\delta}\|^2})G(I - \frac{\underline{\delta}\,\underline{\delta}^T}{\|\underline{\delta}\|^2}) + A, \qquad (32)$$

is bounded by the inequality

$$\|A\| \leq 3\overline{M}, \qquad (33)$$

where $\overline{M}$ is the bound on $\|\overline{G}(\underline{x})\|$ given by condition (30).
Proof of Theorem 1: Equations (11) and (32) imply that A is the matrix

$$A = \frac{\underline{y}\,\underline{\delta}^T}{\|\underline{\delta}\|^2} + \frac{\underline{\delta}\,\underline{y}^T}{\|\underline{\delta}\|^2} - \frac{\underline{\delta}\,\underline{\delta}^T}{\|\underline{\delta}\|^4}(\underline{y},\underline{\delta}), \qquad (34)$$

so we deduce the bound

$$\|A\| \leq 3\|\underline{y}\|/\|\underline{\delta}\| . \qquad (35)$$

Now equation (4) implies the identity

$$\underline{y} = \int_0^1 \overline{G}(\underline{x} + \theta\underline{\delta})\underline{\delta}\, d\theta, \qquad (36)$$

so an extension of the triangle inequality for norms and condition (30) give the bound

$$\|\underline{y}\| \leq \int_0^1 \|\overline{G}(\underline{x} + \theta\underline{\delta})\|\, \|\underline{\delta}\|\, d\theta$$
$$\leq \overline{M}\|\underline{\delta}\| . \qquad (37)$$

Therefore Theorem 1 is a consequence of inequality (35).

Moreover, if we define $A^{(k)}$ by the equation

$$G^{(k+1)} = (I - \theta^{(k)} \frac{\underline{\delta}^{(k)} \underline{\delta}^{(k)T}}{\|\underline{\delta}^{(k)}\|^2}) G^{(k)} (I - \theta^{(k)} \frac{\underline{\delta}^{(k)} \underline{\delta}^{(k)T}}{\|\underline{\delta}^{(k)}\|^2}) + A^{(k)},$$

(38)

where $G^{(k+1)}$ is the matrix (24), then inequality (37) implies the bound

$$\|A^{(k)}\| \leq \{2\theta^{(k)} + [\theta^{(k)}]^2\} \overline{M}$$
$$< 4.89 \, \overline{M},$$

(39)

because we have already noted that $|1 - \theta^{(k)}|$ will not exceed $\sqrt{2/11}$. This inequality enables us to prove that the matrices $G^{(k)}$ ($k = 1, 2, \ldots$), generated by the algorithm, are uniformly bounded.

Theorem 2. Expressions (29), (38) and (39) imply that the inequality

$$\|G^{(k+K+1)}\| \leq \max[\|G^{(k)}\|, \; 4.89 \, (K+1)\overline{M}/(1-c^2)]$$

(40)

is satisfied for all positive integers k, where $c < 1$ is the constant of equation (29), and K is the integer of this equation.

Proof of Theorem 2: By using equation (38) to express $G^{(k+K+1)}$ in terms of $A^{(k)}$, $A^{(k+1)}$, ..., $A^{(k+K)}$ and $G^{(k)}$, by using the properties of matrix norms, and by using the inequality

$$\|(I - \theta^{(j)} \frac{\underline{\delta}^{(j)} \underline{\delta}^{(j)T}}{\|\underline{\delta}^{(j)}\|^2})\| \leq 1, \quad j = k, k+1, \ldots, k+K,$$

(41)

(in fact the left-hand side of this expression is equal to one unless we are minimizing a function of only one variable), we deduce the bound

$$\|G^{(k+K+1)}\| \leq \|\prod_{j=k}^{k+K}(I - \theta^{(j)}\frac{\underline{\delta}^{(j)}\underline{\delta}^{(j)T}}{\|\underline{\delta}^{(j)}\|^2})\|^2 \|G^{(k)}\|$$

$$+ \sum_{j=k}^{k+K}\|A^{(j)}\| \, .$$
(42)

Therefore expressions (29) and (39) imply the inequality

$$\|G^{(k+K+1)}\| \leq c^2 \|G^{(k)}\| + 4.89(K+1)\overline{M} \, . \qquad (43)$$

Theorem 2 is an immediate consequence of this inequality.
  Theorem 2 and expressions (38) and (39) imply that there exists a constant M such that the bound

$$\|G^{(k)}\| \leq M \qquad (44)$$

holds for all positive integers k.
  We now prove that the convergence criterion (18) will be satisfied for some value of k.
  Theorem 3. If $F(\underline{x})$ is bounded below, and if inequality (30) holds, then the algorithm will calculate a point $\underline{x}^{(k+1)}$ at which inequality (18) is satisfied, where ε is any positive constant.
  Proof of Theorem 3: We begin the proof by showing that, for the ordinary iterations of the algorithm, if the bound $\Delta^{(k)}$ satisfies the condition

$$\Delta^{(k)} \leq \|\underline{g}^{(k)}\|/\max[M, \overline{M}] \, , \qquad (45)$$

where M and $\overline{M}$ are the bounds (30) and (44), then the inequality (28) is satisfied.

Conditions (44) and (45) imply that inequality (20) holds, in which case the displacement $\underline{\delta}^{(k)}$ is defined by equations (19) and (21). Therefore the value of $\Phi^{(k)}(\underline{\delta}^{(k)})$ is given by the equation

$$\Phi^{(k)}(\underline{\delta}^{(k)}) = F(\underline{x}^{(k)}) - \Delta^{(k)} \|\underline{g}^{(k)}\| + \frac{1}{2}[\Delta^{(k)}]^2 \frac{(\underline{g}^{(k)}, G^{(k)}\underline{g}^{(k)})}{\|\underline{g}^{(k)}\|^2}. \quad (46)$$

Moreover, by applying Taylor's theorem to the function $F(\underline{x})$, we deduce the identity

$$F(\underline{x}^{(k)} + \underline{\delta}^{(k)}) = F(\underline{x}^{(k)}) - \Delta^{(k)} \|\underline{g}^{(k)}\| + \frac{1}{2}[\Delta^{(k)}]^2 \frac{(\underline{g}^{(k)}, \overline{G}(\underline{x}^{(k)} + \alpha^{(k)}\underline{\delta}^{(k)})\underline{g}^{(k)})}{\|\underline{g}^{(k)}\|^2}, \quad (47)$$

where $0 \leq \alpha^{(k)} \leq 1$. Now expressions (44), (45) and (46) give the bound

$$\Phi^{(k)}(\underline{\delta}^{(k)}) - F(\underline{x}^{(k)}) \geq -\frac{3}{2} \Delta^{(k)} \|\underline{g}^{(k)}\|, \quad (48)$$

and expressions (30), (45) and (47) give the bound

$$F(\underline{x}^{(k)} + \underline{\delta}^{(k)}) - F(\underline{x}^{(k)}) \leq -\frac{1}{2}\Delta^{(k)} \|\underline{g}^{(k)}\|. \quad (49)$$

Therefore inequality (45) implies that condition (28) is satisfied.

This remark is important because the algorithm sets $\Delta^{(k+1)} < \|\underline{\delta}^{(k)}\|$ only if condition (28) fails. We note that the parameters (21) give $\|\underline{\delta}^{(k)}\| = \Delta^{(k)}$, so we deduce that, if condition (45) holds, then the step-bound $\Delta^{(k+1)}$ is not less than $\Delta^{(k)}$.

If condition (45) does not hold, then the ordinary iterations of the algorithm provide the inequality $\Delta^{(k+1)} \geq \frac{1}{2}\|\underline{\delta}^{(k)}\|$. The right-hand side of this expression is not equal to $\frac{1}{2}\Delta^{(k)}$ only if $\underline{\delta}^{(k)}$ satisfies equation (22), in which case, from the identity $G^{(k)}\underline{\delta}^{(k)} = -\underline{g}^{(k)}$ and from the inequality (44), we deduce the bound $\|\underline{\delta}^{(k)}\| \geq \|\underline{g}^{(k)}\|/M$. Moreover we recall that every special iteration of the algorithm sets $\Delta^{(k+1)} = \Delta^{(k)}$, so we deduce that, until the convergence criterion (18) is satisfied, every step bound $\Delta^{(k)}$ is bounded away from zero by the inequality

$$\Delta^{(k)} \geq \min[\Delta^{(1)}, \frac{1}{2}\varepsilon/\max\{M, \overline{M}\}]$$
$$= \eta,$$
(50)

say.

To complete the proof of Theorem 3, we suppose that it is false, and deduce a contradiction.

The method used by an ordinary iteration to define $\underline{\delta}^{(k)}$ is such that the predicted change in $F(\underline{x})$, $\{\Phi^{(k)}(\underline{\delta}^{(k)}) - F(\underline{x}^{(k)})\}$, is a non-increasing function of $\Delta^{(k)}$. Therefore every value of $\Phi^{(k)}(\underline{x}^{(k)} + \underline{\delta}^{(k)})$ is less than or equal to the value that would have been obtained if $\Delta^{(k)}$ were equal to η. Thus we deduce from expressions (44), (46) and (50) that the condition

$$\Phi^{(k)}(\underline{\delta}^{(k)}) - F(\underline{x}^{(k)}) \leq -\eta\|\underline{g}^{(k)}\| + \frac{1}{2}M\eta^2$$
$$\leq -\frac{3}{4}\varepsilon\eta$$
(51)

is satisfied by every ordinary iteration of the algorithm. It follows that condition (28) holds only if the reduction in

$F(\underline{x})$ obtained by the kth iteration is bounded by the inequality

$$F(\underline{x}^{(k)} + \underline{\delta}^{(k)}) - F(\underline{x}^{(k)}) \leq -0.075 \, \eta \, \varepsilon \, . \qquad (52)$$

Now if $F(\underline{x})$ is bounded below the condition (52) is satisfied only a finite number of times. Therefore only a finite number of ordinary iterations set $\Delta^{(k+1)} = \|\underline{\delta}^{(k)}\|$ or $\Delta^{(k+1)} = 2\|\underline{\delta}^{(k)}\|$, while the remainder set $\Delta^{(k+1)} = \frac{1}{2}\|\underline{\delta}^{(k)}\|$. It follows that if the number of iterations of the algorithm were infinite, then the sequence of step-bounds $\Delta^{(k)}$ (k = 1, 2, ...) would converge to zero. But this would contradict inequality (50), so we deduce that Theorem 3 is true.

Note that Theorem 3 states only that the algorithm will finish because, for some k, $\|\underline{g}^{(k)}\| < \varepsilon$, where ε is a prescribed, positive tolerance. It does not claim that the sequence of points $\underline{x}^{(k)}$ (k = 1, 2, ...) would converge if ε were set to zero. Indeed if $\varepsilon = 0$, and the algorithm is applied to minimize the function of one variable $F(x) = e^{-x}$, then we will find that $x^{(k)} \to \infty$.

However when the algorithm is applied to real problems, it usually happens that $\|\underline{g}^{(k)}\|$ is small because $\underline{x}^{(k)}$ is close to a local minimum of $F(\underline{x})$. The condition $F(\underline{x}^{(k+1)}) \leq F(\underline{x}^{(k)})$ tends to prevent divergence from a local minimum, so our next theorem is easy to prove.

Theorem 4. If $F(\underline{x})$ satisfies the conditions of Theorem 3, if $\underline{x}^*$ is the position of a local minimum of $F(\underline{x})$, if $\underline{x}^*$ is contained in a closed convex region S such that $F(\underline{x})$ is strictly convex for $\underline{x} \varepsilon S$, and if there exists an integer σ such that the points $\underline{x}^{(k)}$, $k \geq \sigma$ (generated by the algorithm of this paper in the case $\varepsilon = 0$), all lie in S, then the sequence $\underline{x}^{(k)}$, k = 1, 2, ..., converges to $\underline{x}^*$.

Proof of Theorem 4: Define ρ to be the number

$$\rho = \inf \|\underline{x}^* - \underline{x}^{(k)}\| \, , \quad k \geq \sigma \, . \qquad (53)$$

UNCONSTRAINED OPTIMIZATION

If $\rho > 0$, then define

$$\overline{F} = \min F(\underline{x}), \quad \|\underline{x} - \underline{x}^*\| = \rho, \quad \underline{x} \in S. \tag{54}$$

In this case the strict convexity of $F(\underline{x})$ implies that $\overline{F} > F(\underline{x}^*)$, and it also implies the bound

$$F(\underline{x}^{(k)}) - F(\underline{x}^*) \geq \{\overline{F} - F(\underline{x}^*)\} \|\underline{x}^{(k)} - \underline{x}^*\|/\rho, \quad k \geq \sigma. \tag{55}$$

Another consequence of the convexity is the inequality

$$F(\underline{x}^*) \geq F(\underline{x}^{(k)}) + (\underline{x}^* - \underline{x}^{(k)}, \underline{g}^{(k)})$$

$$\geq F(\underline{x}^{(k)}) - \|\underline{x}^* - \underline{x}^{(k)}\| \|\underline{g}^{(k)}\|, \tag{56}$$

so from conditions (55) and (56) we deduce the bound

$$\|\underline{g}^{(k)}\| \geq \{\overline{F} - F(\underline{x}^*)\}/\rho, \quad k > \sigma. \tag{57}$$

But this statement contradicts Theorem 3, so we conclude that ρ, defined by expression (53), is equal to zero.
Therefore, because $\rho = 0$, because $F(\underline{x})$ is continuous at $\underline{x}^*$, because the sequence $F(\underline{x}^{(k)})$ ($k = 1, 2, \ldots$) decreases monotonically, and because the statement of the theorem implies that $F(\underline{x}^{(k)}) \geq F(\underline{x}^*)$, we deduce the limit

$$\lim_{k \to \infty} F(\underline{x}^{(k)}) = F(\underline{x}^*). \tag{58}$$

We now let η be a small positive number, and we define

$$\overline{F}(\eta) = \min F(\underline{x}), \quad \|\underline{x} - \underline{x}^*\| \geq \eta, \quad \underline{x} \in S. \qquad (59)$$

The value of $\overline{F}(\eta)$ must exceed $F(\underline{x}^*)$, so expression (58) implies that there exists an integer $\sigma(\eta) \geq \sigma$ such that $F(\underline{x}^{(k)}) < \overline{F}(\eta)$ for all $k \geq \sigma(\eta)$. It follows that $\|\underline{x}^{(k)} - \underline{x}^*\| < \eta$, $k \geq \sigma(\eta)$. Therefore, because the smallness of η is arbitrary the sequence $\underline{x}^{(k)}$ converges to $\underline{x}^*$. Theorem 4 is proved.

Theorem 4 suggests, correctly, that it is usual in practice for the points $\underline{x}^{(k)}$, generated by the algorithm, to converge to a point $\underline{x}^*$, which is the position of a local minimum of $F(\underline{x})$. We now prove that in this case the sequence of second derivative approximations $G^{(k)}$, $k = 1, 2, \ldots$, converges to the actual second derivative matrix at $\underline{x}^*$, namely $\overline{G}(\underline{x}^*)$.

Theorem 5. If the algorithm is applied with $\varepsilon = 0$, if the calculated sequence of points $\underline{x}^{(k)}$, $k = 1, 2, \ldots$, converges to $\underline{x}^*$, and if the derivatives of $F(\underline{x})$ satisfy conditions (30) and (31), then the matrices $G^{(k)}$ converge to $\overline{G}(\underline{x}^*)$.

Proof of Theorem 5: Because the sequence of points $\underline{x}^{(k)}$ converges, the differences $\{\underline{x}^{(k+1)} - \underline{x}^{(k)}\}$ must converge to zero. After an ordinary iteration, the value of $\Delta^{(k+1)}$ is not greater than $\max\{\frac{1}{2}\Delta^{(k)}, 2\|\underline{x}^{(k+1)} - \underline{x}^{(k)}\|\}$, and after a special iteration $\Delta^{(k+1)} = \Delta^{(k)}$. Therefore, because every special iteration is followed by an ordinary iteration, the sequence of bounds $\Delta^{(k)}$ ($k = 1, 2, \ldots$) also tends to zero. We let η be an arbitrarily small positive number, and we note that there exists an integer, $\sigma(\eta)$ say, such that the inequality

$$\|\underline{x}^{(k)} - \underline{x}^*\| + \Delta^{(k)} \leq \eta \qquad (60)$$

holds for all $k \geq \sigma(\eta)$.

UNCONSTRAINED OPTIMIZATION

We let $\overline{G}^*$ denote the matrix $\overline{G}(\underline{x}^*)$, and, following a method like the one that was used to deduce inequality (39), we define the matrix $B^{(k)}$ by the equation

$$(G^{(k+1)} - \overline{G}^*) = \qquad (61)$$
$$= (I - \theta^{(k)}\frac{\underline{\delta}^{(k)}\underline{\delta}^{(k)T}}{\|\underline{\delta}^{(k)}\|^2})(G^{(k)} - \overline{G}^*)(I - \theta^{(k)}\frac{\underline{\delta}^{(k)}\underline{\delta}^{(k)T}}{\|\underline{\delta}^{(k)}\|^2}) + B^{(k)},$$

where $G^{(k+1)}$ is the matrix (24). Therefore $B^{(k)}$ is the matrix

$$B^{(k)} = \theta^{(k)}\frac{(\underline{y}^{(k)} - \overline{G}^*\underline{\delta}^{(k)})\underline{\delta}^{(k)T} + \underline{\delta}^{(k)}(\underline{y}^{(k)} - \overline{G}^*\underline{\delta}^{(k)})^T}{\|\underline{\delta}^{(k)}\|^2}$$
$$- [\theta^{(k)}]^2 \frac{\underline{\delta}^{(k)}\underline{\delta}^{(k)T}(\underline{y}^{(k)} - \overline{G}^*\underline{\delta}^{(k)}, \underline{\delta}^{(k)})}{\|\underline{\delta}^{(k)}\|^4}, \qquad (62)$$

and its norm is bounded by the inequality

$$\|B^{(k)}\| \leq 4.89 \, \|\underline{y}^{(k)} - \overline{G}^*\underline{\delta}^{(k)}\| / \|\underline{\delta}^{(k)}\|. \qquad (63)$$

Now by using equation (36) to substitute for $\underline{y}^{(k)}$, we find the identity

$$\underline{y}^{(k)} - \overline{G}^*\underline{\delta}^{(k)} = \int_0^1 \{\overline{G}(\underline{x}^{(k)} + \theta\underline{\delta}^{(k)}) - \overline{G}^*\}\underline{\delta}^{(k)} d\theta. \qquad (64)$$

Moreover the Lipschitz condition (31), and expressions (14) and (60), give the bound

$$\|\overline{G}(\underline{x}^{(k)} + \theta\underline{\delta}^{(k)}) - \overline{G}^*\| \leq L\eta, \, k \geq \sigma(\eta). \qquad (65)$$

51

Therefore from inequality (63) we deduce that $B^{(k)}$ is bounded by the amount

$$\|B^{(k)}\| \leq 4.89 \, L\eta, \quad k \geq \sigma(\eta). \tag{66}$$

It follows that, by using the method of Theorem 2, we can deduce the inequality

$$\|G^{(k+K+1)} - \overline{G}^*\| \leq c^2 \|G^{(k)} - \overline{G}^*\| + 4.89(K+1)L\eta, \tag{67}$$

for $k \geq \sigma(\eta)$, in place of inequality (43). Further, if we apply inequality (67) p times in an iterative way, we find the bound

$$\|G^{(k+pK+p)} - \overline{G}^*\| \leq c^{2p} \|G^{(k)} - \overline{G}^*\|$$
$$+ 4.89(K+1)L\eta(1 + c^2 + \ldots + c^{2p-2}), \tag{68}$$

from which we deduce the inequality

$$\varlimsup_{k \to \infty} \|G^{(k)} - \overline{G}^*\| \leq 4.89(K+1)L\eta/(1-c^2). \tag{69}$$

Now η is any positive number, so this statement implies that $\|G^{(k)} - \overline{G}^*\|$ tends to zero. Theorem 5 is proved.

We now use this theorem to prove that usually the rate of convergence of the algorithm is super-linear.

<u>Theorem 6.</u> If the algorithm is applied with $\varepsilon = 0$, if the calculated sequence of points $\underline{x}^{(k)}$, $k = 1, 2, \ldots$, converges to $\underline{x}^*$, if the derivatives of $F(\underline{x})$ satisfy conditions (30) and (31), and if the second derivative matrix at $\underline{x}^*$, namely $\overline{G}^*$, is strictly positive definite, then the rate of convergence of the points $\underline{x}^{(k)}$ is super-linear.

UNCONSTRAINED OPTIMIZATION

Proof of Theorem 6: We let $d > 0$ be the least eigenvalue of the matrix $\overline{G}^*$. Because the left-hand side of expression (60) tends to zero, we can let J be an integer such that the inequality

$$\|\underline{x}^{(k)} - \underline{x}^*\| + \Delta^{(k)} \leq \frac{1}{2} d/L \qquad (70)$$

holds for all $k \geq J$. We consider only those iterations after the Jth iteration, so all the values of $\underline{x}$ that concern us are in the region

$$R = \{\underline{x} : \|\underline{x} - \underline{x}^*\| \leq \frac{1}{2} d/L\} . \qquad (71)$$

We have defined R in this way, because the Lipschitz condition (31) implies that if $\underline{x}$ is in R, then the least eigenvalue of the second derivative matrix $\overline{G}(\underline{x})$ is bounded away from zero, by the amount $\frac{1}{2}d$. Therefore for $\underline{x} \varepsilon R$ there are convenient relations between $\|\underline{x} - \underline{x}^*\|$, $F(\underline{x}) - F(\underline{x}^*)$ and $\|\underline{g}(\underline{x})\|$. Specifically Powell's (1969b) Lemma 2 and the bound (37) give the inequality

$$\frac{1}{2} d \|\underline{x} - \underline{x}^*\| \leq \|\underline{g}(\underline{x})\| \leq \overline{M} \|\underline{x} - \underline{x}^*\|, \quad \underline{x} \varepsilon R. \qquad (72)$$

Moreover the identity

$$F(\underline{x}) - F(\underline{x}^*) = \frac{1}{2}(\underline{x} - \underline{x}^*, \overline{G}\{\underline{x}^* + \alpha[\underline{x} - \underline{x}^*]\}[\underline{x} - \underline{x}^*]), \qquad (73)$$

$0 \leq \alpha \leq 1$, implies the inequality

$$\frac{1}{4} d \|\underline{x} - \underline{x}^*\|^2 \leq F(\underline{x}) - F(\underline{x}^*) \leq \frac{1}{2} \overline{M} \|\underline{x} - \underline{x}^*\|^2, \quad \underline{x} \varepsilon R . \qquad (74)$$

Of course the super-linear convergence of our algorithm is a consequence of the close relation between equation (22) and the classical "generalized Newton iteration", in which $\overline{G}(\underline{x}^{(k)})$ would replace $G^{(k)}$ in equation (22). But the proof of this theorem is complicated by the fact that our algorithm does not use equation (22) to define $\underline{\delta}^{(k)}$, if this would conflict with inequality (14). Indeed most of the proof is spent on showing that, for sufficiently large k, the bound (14) does not prevent equation (22) from being satisfied. First we show that, for sufficiently large k, the inequality (28) holds for all the ordinary iterations of the algorithm, which is relevant because inequality (28) governs the calculation of $\Delta^{(k+1)}$.

We let η be some number in the interval $0 < \eta \leq \frac{1}{2}d$, and we define $J(\eta)$ to be an integer such that the inequalities

$$\|\overline{G}^* - G^{(k)}\| \leq \eta \qquad (75)$$

and

$$\|\underline{x}^{(k)} - \underline{x}^*\| + \Delta^{(k)} \leq \eta/L \qquad (76)$$

hold, for all $k \geq J(\eta)$. This integer exists because of Theorem 5, and because the left-hand side of expression (60) tends to zero. We identify a value of η such that inequality (28) holds if $k \geq J(\eta)$.

The expression $\Phi^{(k)}(\underline{\delta}^{(k)}) - F(\underline{x}^{(k)})$, appearing in the right-hand side of inequality (28), has the value

$$\Phi^{(k)}(\underline{\delta}^{(k)}) - F(\underline{x}^{(k)}) = (\underline{g}^{(k)}, \underline{\delta}^{(k)}) + \frac{1}{2}(\underline{\delta}^{(k)}, G^{(k)}\underline{\delta}^{(k)}), \qquad (77)$$

and the definition of $\underline{\delta}^{(k)}$ is such that the inequality

UNCONSTRAINED OPTIMIZATION

$$(\underline{g}^{(k)} + G^{(k)}\underline{\delta}^{(k)}, \underline{\delta}^{(k)}) \leq 0 \qquad (78)$$

holds (Powell, 1970). Therefore, from expressions (75) and (77), we deduce the bound

$$\Phi^{(k)}(\underline{\delta}^{(k)}) - F(\underline{x}^{(k)}) \leq -\frac{1}{2}(\underline{\delta}^{(k)}, G^{(k)}\underline{\delta}^{(k)})$$

$$\leq -\frac{1}{2}(d-\eta)\|\underline{\delta}^{(k)}\|^2 . \qquad (79)$$

The left-hand side of inequality (28) is the quantity

$$F(\underline{x}^{(k)} + \underline{\delta}^{(k)}) - F(\underline{x}^{(k)}) = (\underline{g}^{(k)}, \underline{\delta}^{(k)}) +$$

$$+ \frac{1}{2}(\underline{\delta}^{(k)}, \overline{G}\{\underline{x}^{(k)} + \alpha^{(k)}\underline{\delta}^{(k)}\}\underline{\delta}^{(k)}) , \qquad (80)$$

where $0 \leq \alpha^{(k)} \leq 1$. By combining this equation with equation (77) we obtain the relation

$$F(\underline{x}^{(k)} + \underline{\delta}^{(k)}) - F(\underline{x}^{(k)}) = \Phi^{(k)}(\underline{\delta}^{(k)}) - F(\underline{x}^{(k)})$$

$$+ \frac{1}{2}(\underline{\delta}^{(k)}, [\overline{G}\{\underline{x}^{(k)} + \alpha^{(k)}\underline{\delta}^{(k)}\} - G^{(k)}]\underline{\delta}^{(k)}) . \qquad (81)$$

Now the inequalities (14), (31), (75) and (76) give the bound

$$\|\overline{G}\{\underline{x}^{(k)} + \alpha^{(k)}\underline{\delta}^{(k)}\} - G^{(k)}\| \leq$$

$$\leq \|\overline{G}\{\underline{x}^{(k)} + \alpha^{(k)}\underline{\delta}^{(k)}\} - \overline{G}^*\| + \|\overline{G}^* - G^{(k)}\| \qquad (82)$$

$$\leq 2\eta ,$$

so instead of equation (81) we can write the inequality

$$F(\underline{x}^{(k)} + \underline{\delta}^{(k)}) - F(\underline{x}^{(k)}) \leq \Phi^{(k)}(\underline{\delta}^{(k)}) - F(\underline{x}^{(k)}) + \quad (83)$$
$$+ \eta \|\underline{\delta}^{(k)}\|^2 .$$

Further, by eliminating $\|\underline{\delta}^{(k)}\|^2$ from expressions (79) and (83), we deduce the bound

$$(d-\eta)\{F(\underline{x}^{(k)} + \underline{\delta}^{(k)}) - F(\underline{x}^{(k)})\} \leq (d-3\eta)\{\Phi^{(k)}(\underline{\delta}^{(k)}) - F(\underline{x}^{(k)})\} . \quad (84)$$

Therefore inequality (28) holds for all $k \geq J(\eta)$, provided that $0 < \eta \leq 9d/29$.

This last remark implies that, for $k \geq J(9d/29)$, every ordinary iteration sets $\Delta^{(k+1)} \geq \|\underline{\delta}^{(k)}\|$. But the sequence $\Delta^{(1)}, \Delta^{(2)}, \ldots$ tends to zero. Therefore an infinite number of ordinary iterations make $\|\underline{\delta}^{(k)}\| < \Delta^{(k)}$, which is equivalent to the statement that an infinite number of ordinary iterations calculate $\underline{\delta}^{(k)}$ to satisfy equation (22).

For the next part of the proof we give η the value

$$\eta = \frac{1}{4} d[d/2\overline{M}]^{3/2}, \quad (85)$$

and again we consider the iterations with $k \geq J(\eta)$. The factor $[d/2\overline{M}]^{3/2}$ is present because expressions (17), (72) and (74) imply the bound

UNCONSTRAINED OPTIMIZATION

$$\|\underline{g}^{(j)}\| \leq \overline{M}\,\|\underline{x}^{(j)} - \underline{x}^*\|$$

$$\leq \overline{M}[\,4\{F(\underline{x}^{(j)}) - F(\underline{x}^*)\}/d\,]^{\frac{1}{2}}$$

$$\leq \overline{M}[\,4\{F(\underline{x}^{(k)}) - F(\underline{x}^*)\}/d\,]^{\frac{1}{2}} \qquad (86)$$

$$\leq \overline{M}[\,2\overline{M}/d\,]^{\frac{1}{2}}\,\|\underline{x}^{(k)} - \underline{x}^*\|$$

$$\leq [\,2\overline{M}/d\,]^{3/2}\,\|\underline{g}^{(k)}\|, \quad k \leq j.$$

Because an infinite number of iterations satisfy equation (22), we can let q be the least value of $k \geq J(\eta)$ such that equation (22) holds. We now show that the definition (85) is such that, after the qth iteration, <u>every</u> ordinary iteration of the algorithm calculates $\underline{\delta}^{(k)}$ from formula (22).

Equations (22) and (36) imply the identity

$$\underline{g}^{(q+1)} = \underline{g}^{(q)} + \int_0^1 \overline{G}(\underline{x}^{(q)} + \theta\underline{\delta}^{(q)})\underline{\delta}^{(q)}\,d\theta$$

$$= \int_0^1 \{G(\underline{x}^{(q)} + \theta\underline{\delta}^{(q)}) - G^{(q)}\}\underline{\delta}^{(q)}\,d\theta, \qquad (87)$$

so from inequality (82) we deduce the bound

$$\|\underline{g}^{(q+1)}\| \leq 2\eta\,\|\underline{\delta}^{(q)}\|. \qquad (88)$$

Therefore expressions (85) and (86) give the inequality

$$\|\underline{g}^{(j)}\| \leq [\,2\overline{M}/d\,]^{3/2}\,\|\underline{g}^{(q+1)}\|$$

$$\leq \tfrac{1}{2}d\,\|\underline{\delta}^{(q)}\|, \quad j \geq q+1. \qquad (89)$$

Now the algorithm sets $\Delta^{(q+1)} \geq \|\underline{\delta}^{(q)}\|$, and expressions (75) and (85) imply that the least eigenvalue of $G^{(j)}$, $j \geq q+1$, exceeds $\frac{1}{2}d$. Therefore from inequality (89) we deduce the bound

$$\|[G^{(j)}]^{-1}\| \, \|\underline{g}^{(j)}\| \leq \Delta^{(q+1)}, \quad j \geq q+1. \qquad (90)$$

It follows that the value of $\Delta^{(q+1)}$ allows the next ordinary iteration after the qth iteration to calculate $\underline{\delta}^{(k)}$ to satisfy equation (22). Further, by applying this argument in an inductive way, we conclude that, after the qth iteration, all ordinary iterations of the algorithm satisfy equation (22).

Note that in the above reasoning we needed to introduce inequality (86) only because some iterations of the algorithm are "special iterations".

For the next part of the proof we let η be any positive number, that is not greater than the quantity (85), and we consider the ordinary iterations of the algorithm for $k \geq \max[q, J(\eta)]$. By applying the method used to deduce inequality (88), we obtain the bound

$$\|\underline{g}^{(k+1)}\| \leq 2\eta \|\underline{\delta}^{(k)}\| . \qquad (91)$$

Therefore expression (72) implies the inequality

$$\frac{1}{2}d\|\underline{x}^{(k+1)} - \underline{x}^*\| \leq 2\eta\{\|\underline{x}^{(k)} - x^*\| + \|\underline{x}^{(k+1)} - \underline{x}^*\|\}, \qquad (92)$$

which gives the bound

$$\|\underline{x}^{(k+1)} - \underline{x}^*\| \leq \frac{2\eta}{\frac{1}{2}d - 2\eta} \|\underline{x}^{(k)} - \underline{x}^*\| . \qquad (93)$$

Thus we obtain the expression

$$\varlimsup_{k \to \infty} \frac{\|\underline{x}^{(k+1)} - \underline{x}^*\|}{\|\underline{x}^{(k)} - \underline{x}^*\|} \le \frac{2\eta}{\frac{1}{2}d - 2\eta} \quad . \quad (94)$$

But η can be chosen to be arbitrarily close to zero, so we deduce that the iterations that define $\underline{\delta}^{(k)}$ by equation (22) converge superlinearly.

To complete the proof it remains to show that the superlinear convergence is not damaged by the "special iterations" of the algorithm. Since every iteration applies equation (17), inequality (74) gives the bound

$$\frac{1}{4} d \|\underline{x}^{(k+1)} - \underline{x}^*\|^2 \le \frac{1}{2} \overline{M} \|\underline{x}^{(k)} - \underline{x}^*\|^2 , \quad (95)$$

which implies that a special iteration cannot increase the error of the estimate of $\underline{x}^*$ by more than the constant factor $\sqrt{2\overline{M}/d}$. Therefore, because every special iteration is followed by an ordinary iteration, the ratio $\|\underline{x}^{(k+p)} - \underline{x}^*\| / \|\underline{x}^{(k)} - \underline{x}^*\|$ tends to zero as k tends to infinity, for any fixed integer $p \ge 2$. Theorem 6 is proved.

These convergence theorems suggest that the algorithm will work well in practice, but they do not really reflect the power of the algorithm, because they depend on condition (29). We have noted, correctly, that the special iterations are needed to obtain this condition, but Theorems 3, 4, and 6 can be proved, even if no special iterations are present. Without the special iterations the proofs of the theorems are rather complicated, so, because numerical experimentation suggests that the special iterations are worthwhile, we have preferred to give the more direct proofs. Moreover Theorem 5 is valid only if the algorithm includes some special iterations.

5. Discussion

In the introduction to this paper we stated that the new algorithm is a result of an attempt to find a method for unconstrained optimization, with guaranteed fast convergence, that does not require the objective function $F(\underline{x})$ to be convex. It seems that the project has been fulfilled, but it is important to compare the new algorithm with the current methods. Because we have been careful to keep the number of computer operations required by the new algorithm to of order n^2 for each function and gradient evaluation, it is sensible to base the comparison on the number of times the new algorithm requires $F(\underline{x})$ and $\underline{g}(\underline{x})$ to be calculated. Therefore in this section we report the number of function evaluations that are needed to solve three usual test problems.

These problems are the minimization of Rosenbrock's (1960) function

$$F(x_1, x_2) = 100(x_2 - x_1^2)^2 + (1 - x_1)^2 , \qquad (96)$$

starting from $\underline{x}^{(1)} = (-1.2, 1.0)$, the minimization of Wood's (Colville, 1968) function

$$F(x_1, x_2, x_3, x_4) = 100(x_2 - x_1^2)^2 + (1 - x_1)^2 + 90(x_4 - x_3^2)^2 \qquad (97)$$
$$+ (1 - x_3)^2 + 10.1\{(x_2 - 1)^2 + (x_4 - 1)^2\} + 19.8(x_2 - 1)(x_4 - 1) ,$$

starting from $\underline{x}^{(1)} = (-3, -1, -3, -1)$, and the minimization of the trigonometric functions

$$F(\underline{x}) = \sum_{i=1}^{n} \{E_i - \sum_{j=1}^{n} A_{ij} \sin x_j + B_{ij} \cos x_j\}^2 , \qquad (98)$$

whose parameters are defined by Fletcher and Powell (1963). For this last function ten sets of parameters were chosen, two for each of five values of n, namely $n = 3$, $n = 5$, $n = 10$, $n = 20$ and $n = 30$. For each of the test functions we report the number of function evaluations that are needed to calculate a point $\underline{x}^{(k)}$, such that $F(\underline{x}^{(k)}) \leq 10^{-5}$.

The number of evaluations of $F(\underline{x})$ and $\underline{g}(\underline{x})$ is given in the second column of Table 1. The first column distinguishes the three test functions, R denoting Rosenbrock's function, W denoting Wood's function, and the values of n in the last five rows being relevant to the function (98). In the last three columns the number of evaluations made by V. M. (Davidon's (1959) variable metric algorithm), by R. O. (the rank one algorithm described by Powell (1969a)), and by F (Fletcher's (1969) algorithm) are quoted. The two figures quoted for the variable metric method applied to Wood's function are those stated by Colville (1968) and by Fletcher (1969). Discrepancies like this one are not uncommon, due to differences in coding, and due to differences in selecting program parameters. The numbers given for Fletcher's method in the cases $n = 3$ and $n = 5$ are estimated from the figures that he gives for $n = 2$, $n = 4$ and $n = 6$.

We note that, except for the last row of the table, the new method does appear to be as efficient as other algorithms, if efficiency is measured by a count of function evaluations.

The poor figure in the last row of the table is due to the fact that the condition number of the final second derivative matrix $\overline{G}(\underline{x}^*)$ is very large. To understand why this condition number is important, we let $\lambda_i^{(k)}$ and $\underline{v}_i^{(k)}$ $(i = 1, 2, \ldots, n)$ be the eigenvalues and vectors of the matrix $G^{(k)}$, and we write the identity

$$G^{(k)} = \sum_i \lambda_i^{(k)} \underline{v}_i^{(k)} \underline{v}_i^{(k)T} . \qquad (99)$$

Now the algorithm tries to make the error $\|G^{(k)} - \overline{G}(\underline{x}^*)\|$ small, although equation (22), which governs the final convergence of the algorithm, uses the matrix

$$H^{(k)} = \sum_i [\lambda_i^{(k)}]^{-1} \underline{v}_i^{(k)} \underline{v}_i^{(k)T} . \qquad (100)$$

Therefore it would be preferable for the algorithm to contain a device that automatically provides higher accuracy in the smaller values of λ_i, but there is no device to emphasize the accuracy of the small eigenvalues, except that there is a tendency for the directions $\underline{\delta}^{(k)}$, generated by the ordinary iterations of the algorithm, to be orthogonal to the eigenvectors with large eigenvalues. Therefore, when the condition number of $\overline{G}(\underline{x}^*)$ is large, there is a tendency for the algorithm to use more iterations to calculate a matrix $H^{(k)} = [G^{(k)}]^{-1}$ that is adequate for equation (22).

Note that the difficulty of the last paragraph can sometimes be mitigated by scaling the variables of the objective function so that their magnitudes are nearly equal. However it is reassuring that the algorithm treated Wood's function (97) well, for in this example the condition number of the final second derivative matrix is about 1400.

Another unsatisfactory feature of the new algorithm is that the convergence criterion (18) is seldom the one that computer users want. Instead they would prefer the algorithm to finish when either $F(\underline{x}^{(k)}) - F(\underline{x}^*)$ or $\|\underline{x}^{(k)} - \underline{x}^*\|$ is less than a pre-set amount, but it is awkward to meet these criteria, because they involve estimates of the behavior of $F(\underline{x})$. It would be straightforward to base these estimates on the quadratic function (13), but one would have to give up a nice feature of the present algorithm, which is that it is guaranteed to finish after a finite number of iterations, even for functions like $F(x_1) = x_1^4$, $F(x_1) = \exp(x_1)$ and $F(x_1, x_2) = (x_1 - x_2)^2$.

The given algorithm is intended to be suitable for incorporation in a library of subroutines that is available to computer users.

Acknowledgements

I am most grateful to R. Fletcher and J. K. Reid for studying early drafts of this paper very carefully. They made many valuable suggestions that led to improvements to the paper.

TABLE I

Numerical Results

	New	V.M.	R.O.	F
R	41	64	50	47
W	99	114/154	-	136
n = 3	13/13	-	12/13	10
n = 5	16/16	19/23	16/22	19
n = 10	31/41	29/36	30/44	18
n = 20	53/65	68-121	-	51
n = 30	93/167	86-118	-	75

REFERENCES

1. Broyden, C. G. (1965) "A class of methods for solving non-linear simultaneous equations", Math. Comp., Vol. 19, pp. 577-593.

2. Colville, A. R. (1968) "A comparative study of non-linear programming codes", I.B.M. Technical Report No. 320-2949.

3. Davidon, W. C. (1959) "Variable metric method for minimization", A.E.C. Research and Development Report, ANL-5990 (Rev.).

4. Davidon, W. C. (1968) "Variance algorithm for minimization", Computer Journal, Vol. 10, pp. 406-410.

5. Fletcher, R. (1969) "A new approach to variable metric algorithms", Report No. T.P. 383, A.E.R.E., Harwell.

6. Fletcher, R. and Powell, M.J.D. (1963) "A rapidly convergent descent method for minimization", Computer Journal, Vol. 6, pp. 163-168.

7. McCormick, G. P. (1969) "The rate of convergence of the reset Davidon variable metric method", namuscript.

8. McCormick, G. P. and Pearson, J. D. (1969) "Variable metric methods and unconstrained optimization", in "Optimization", edited by R. Fletcher, Academic Press.

9. Marquardt, D. W. (1963) "An algorithm for least squares estimation of non-linear parameters", J.S.I.A.M., Vol. 11, pp. 431-441.

10. Murtagh, B. A. and Sargent, R. W. H. (1969) "A constrained minimization method with quadratic convergence", in "Optimization", edited by R. Fletcher, Academic Press.

11. Pearson, J. D. (1969) "Variable metric methods of minimization", Computer Journal, Vol. 12, pp. 171-178.

12. Powell, M. J. D. (1968) "A Fortran subroutine for solving systems of non-linear algebraic equations", Report No. R-5947, A.E.R.E., Harwell.

13. Powell, M. J. D. (1969a) "Rank one methods for unconstrained optimization", Report No. T.P. 372, A.E.R.E., Harwell.

14. Powell, M. J. D. (1969b) "On the convergence of the variable metric algorithm", Report No. T.P. 382, A.E.R.E., Harwell.

15. Powell, M. J. D. (1970) "A Fortran subroutine for unconstrained minimization, requiring first derivatives of the objective function", to be published.

16. Rosenbrock, H. H. (1960) "An automatic method for finding the greatest or the least value of a function", Computer Journal, Vol. 3, pp. 175-184.

A Class of Methods for Nonlinear Programming
II Computational Experience

R. FLETCHER AND SHIRLEY A. LILL

ABSTRACT

This work arises from the theoretical development of a class of methods for nonlinear programming, presented at the NATO summer school 1969. Consideration is first given to the problem in which all the constraints are equalities, and it is found that firm recommendations can be made as to the best computational scheme. Various developments to the more general problem involving inequalities are considered and favorable comparisons are made with other well known methods.

Introduction

This paper aims to extend work presented in Part I of this series (Fletcher (1969a)) referred to as (I), in which a class of methods for solving nonlinear programming problems was described. In that paper it was shown that for any problem in which all the constraints were equalities, it is possible to construct a penalty function which has an unconstrained minimum at the solution of the problem. Thus the class of methods has a considerable advantage over conventional penalty function methods in that only one application of an unconstrained minimization package is required, as against the number required for conventional penalty functions. It also has the advantage that the penalty function is well-conditioned (as against the increasing ill-conditioning associated with the conventional penalty functions), so that the unconstrained optimization can be expected to converge rapidly. In particular if the object function in the problem is quadratic and the constraints are linear, then the penalty function is quadratic and its matrix of second derivatives is bounded. The penalty functions also have an interpretation as Lagrangian functions and an approximation is always available to the Lagrange multipliers. This feature enables the method to be extended to problems in which inequality constraints appear by using the signs of the approximate Lagrange multipliers to help determine the effective constraints.

The paper I concerned itself with describing the method and proving a number of theoretical properties. Only a brief sketch was given of practical possibilities and no results were given. This paper will attempt to redress the balance by giving only a brief sketch of the methods, and by concentrating on presenting results on practical problems which resolve most of the choices which have to be made when writing a computer program.

Assume that the problem is to

$$\text{minimize } F(\underline{x}) \qquad \underline{x} \in E^n \tag{1.1}$$

$$\text{subject to } c_i(\underline{x}) = \underline{0} \quad (i = 1, 2, \ldots, k \leq n)$$

with solution $\underline{x} = \underline{\xi}$. The first derivatives of F with respect to $\underline{x}$ will be denoted by $\underline{g}$ (gradient) and the hessian matrix of second derivatives by $\underline{G}$. The first derivatives of each constraint c_i with respect to $\underline{x}$ are denoted by $\underline{n}_i$ (normal vectors), and $\underline{N} = (\underline{n}_1, \underline{n}_2, \ldots, \underline{n}_k)$ is the $n \times k$ matrix whose columns are the normals. It is assumed that the normals are always independent so that $\underline{N}$ is of full rank (k). The matrix $\underline{N}^+$ is used to denote the generalized inverse of $\underline{N}$, that is $(\underline{N}^T\underline{N})^{-1}\underline{N}^T$ (T = transpose). An important concept is the projection matrix $\underline{P} = \underline{N}\underline{N}^+$ which projects vectors into the subspace spanned by the constraint normals, and the complementary projection matrix $\hat{\underline{P}}$ ($\hat{\underline{P}} = \underline{I} - \underline{P}$) which projects vectors so that their direction lies in the manifold formed by the intersection of the constraints.

The class of methods are based on the observation that the function $\psi(\underline{x})$ defined by

$$\psi(\underline{x}) = F - \underline{c}^T \underline{N}^+ \underline{g}$$

has a stationary point at $\underline{\xi}$, the solution to (1.1). The gradient of ψ can be written

$$\underline{\nabla} \psi(\underline{x}) = \underline{g} - \underline{N}\underline{N}^+\underline{g} - \underline{\nabla}(\underline{N}^+\underline{g})\underline{c}$$

where $\nabla(N^+g)$ is an $n \times k$ matrix, and the result follows by applying the Kuhn-Tucker conditions $\hat{P}g = 0$ and $c = 0$, which hold at ξ.

It is important to know whether the stationary point is likely to be a minimum, in which case it can be located by an optimization routine. To this end it is useful to examine the second derivatives of ψ in the case in which F is quadratic and c is linear, (the quadratic/linear problem). Then N, N^+ and G are all constant, so following (I) the result

$$\nabla^2 \psi = \hat{P}G\hat{P} - PGP$$

is obtained. This shows that for small variations about ξ which keep the constraints satisfied, the curvature of ψ is the same as that of F, which must be positive else a solution would not exist. Thus ψ has suitable curvature in one subspace. However for variations about ξ in the direction of the normals, the curvature is opposite to that of F. Thus ψ only has a minimum at ξ if the curvature in the subspace spanned by the normals is negative (i.e. PGP negative semi-definite and of rank k). A similar conclusion holds in the general problem, and emphasizes the fact that ψ is unlikely to provide a suitable penalty function although there are problems for which it would be suitable.

However this discussion about curvature indicates that if a term of the form $\frac{1}{2}c^T Q c$, were added to ψ, where Q is a sufficiently large positive definite matrix, then a suitable penalty function $\phi_1(x)$ might be obtained

$$\phi_1 = F - c^T N^+ g + \tfrac{1}{2} c^T Q c ,$$

the subscript differentiating ϕ_1 from other possible penalty functions which follow. Two more simple penalty functions which could be derived from this are

$$\phi_2 = F - \underset{\sim}{c}^T \underset{\sim}{N}^+ \underset{\sim}{g} + \tfrac{1}{2} q \underset{\sim}{c}^T \underset{\sim}{c}$$

and

$$\phi_3 = F - \underset{\sim}{c}^T \underset{\sim}{N}^+ \underset{\sim}{g} + \tfrac{1}{2} q \underset{\sim}{c}^T \underset{\sim}{N}^+ \underset{\sim}{N}^{+T} \underset{\sim}{c} .$$

In fact, the following result specific to more general problems than the quadratic/linear problem can be proved.

Theorem. If, at $\underset{\sim}{\xi}$, F and $\underset{\sim}{c}$ are twice differentiable functions of $\underset{\sim}{x}$, if $\underset{\sim}{N}(\underset{\sim}{\xi})$ is of full rank and if $\underset{\sim}{N}$ is a differentiable function of $\underset{\sim}{x}$ at $\underset{\sim}{\xi}$; if also

$$\nabla_{\underset{\sim}{x}} h = 0$$

$$\underset{\sim}{y}^T \nabla_{\underset{\sim}{x}}^2 h \, \underset{\sim}{y} > 0$$

at $\underset{\sim}{x} = \underset{\sim}{\xi}$, where $h(\underset{\sim}{x}, \underset{\sim}{\lambda}) = F - \underset{\sim}{c}^T \underset{\sim}{\lambda}$ and where $\underset{\sim}{y}$ is such that $\underset{\sim}{y} \neq 0$ and $\underset{\sim}{N}^T \underset{\sim}{y} = 0$;
then there exists a finite q_0 such that for all $q > q_0$,

$$\nabla \phi_3(\underset{\sim}{\xi}) = 0$$

$$\underset{\sim}{v}^T \nabla^2 \phi_3(\underset{\sim}{\xi}) \underset{\sim}{v} > 0,$$

where $\underset{\sim}{v} \neq 0$.

Proof: As in I.

The implications of this result are that in general it is likely that a value of q can be chosen so that $\phi_3(\underset{\sim}{x})$ has an unconstrained minimum at $\underset{\sim}{\xi}$. The conditions above are in reality not at all restrictive (those on h being the least restrictive sufficient conditions known for a solution of (1.1)) and so the function ϕ_3 is of quite general applicability. A similar result can be proved for ϕ_1 and ϕ_2.

Also of interest is the question of whether any local minima $\hat{\underset{\sim}{x}}$ of ϕ_3 correspond to local solutions to (1.1). For this to be true, a more restrictive condition on $\underset{\sim}{c}$ must be imposed. However the result can be proved either by assuming that $\underline{c}$ is linear or by assuming that $\underline{c}(\hat{\underline{x}}) = \underline{0}$.

It will be noticed that the parameter q is not the sort which occurs in conventional penalty function methods. In the latter the parameter is varied to the limiting case and a minimization carried out for each value of the parameter. In this new class of methods, any $q > q_0$ will suffice and one minimization of ϕ solves the problem. An upper bound on q_0 is a consequence of the proof of the above theorem. A particular case of this when the constraints are linear is that any

$$q > \|\underline{G}\| \qquad (1.2)$$

is satisfactory.

Finally it is pointed out in I that h above is related to ψ, such that ψ can be derived from h by replacing $\underline{\lambda}$ by the continuous function of $\underline{x}$ given by $\underline{N}^+\underline{g}$. In fact $\hat{\underline{\lambda}} = \underline{N}^+\underline{g}$ has to hold at $\hat{\underline{\xi}}$ by virtue of the Kuhn-Tucker conditions. Subsequently in this paper $\underline{\lambda}$ will refer to $\underline{N}^+\underline{g}$ and not as a vector of parameters independent of $\underline{x}$.

The best practical approach to minimizing the penalty functions is by no means obvious and the purpose of this paper is to explore in detail the possibilities which exist. The theorem stated above implies that it is possible to prove convergence in the general case by choosing a minimization method, such as conjugate gradients with restarts, for which convergence can be proved. This fact is brought out in I, but it has been found that there are good reasons for not adhering to this property without question. One point is that there are only a few methods for which ultimate convergence has been proved, and these are not the ones which have performed best in practice. Even Powell's (1969c) results on the variable metric algorithm assume that the linear search can be carried out exactly and do not refer to any algorithm which can be implemented in practice, valuable and impressive though the results are. However the main point for not retaining convergent methods is that superior performance in practice can be obtained by making certain approximations which make analysis of the general case too difficult.

Another property which is proved in I is that if a minimization method is used which terminates when the object function is quadratic, then this termination property is retained for the quadratic/linear equality problem. Many good minimization methods have been developed by paying attention to termination properties, and in fact they apply to all the methods to be used in this paper.

The information required to determine ϕ_1, ϕ_2, ϕ_3 is F, $\underset{\sim}{g}$, $\underset{\sim}{c}$ and $\underset{\sim}{N}$, that is all the functions and their first derivatives. It is assumed that this information is available^ it seems otherwise unlikely that the method will be practicable. If $\underset{\sim}{G}$ is readily available it is possible to use it conveniently, in which case some of the other penalty functions given in I may also be useful. However it is demonstrated in section 3 that little is lost by updating an approximation to $\underset{\sim}{G}$ at each iteration. It is also necessary to calculate the generalized inverse $\underset{\sim}{N}^+$ for each $\underset{\sim}{x}$. One difficulty here is that $\underset{\sim}{N}^+$ (as defined) is infinite if $\underset{\sim}{N}$ is not of full rank, although this situation cannot occur in the neighborhood of $\underset{\sim}{\xi}$ by assumption. In general it suggests that ϕ is infinite, and so is unlikely to occur when applying a minimization method which reduces ϕ at each iteration. It is important however that N is of full rank at the initial approximation. A more important difficulty is that the computation of $\underset{\sim}{N}^+$ might be excessively long as this is an $O(nk^2)$ process. However it is shown in I that simplifications occur with linear constraints which enable the order to be reduced to $O(nk)$ multiplied by the number of nonlinear constraints. Thus as long as the number of nonlinear constraints is not too large (or if the calculation of the functions and their derivatives is relatively expensive) calculation of the generalized inverse is acceptable.

A choice for q has to be made initially, and the simplest way to do this is to use the inequality (1.2). This involves using ϕ_3 rather than ϕ_2, a choice which is made throughout. However it ignores the nonlinearity of the constraints so cannot be used as it stands. Two approaches have been used; one is to set $q = \max(p\|G\|, r)$, the other to set $q = p\|G\| + r$. An initial value of $r = 1$ is taken, and p is set so as to give a safety margin above the threshold. Values of p = 2 and p = 10 have both been used: for p = 2 there is more likelihood that q will become unsatisfactory at a later stage if $\|\underset{\sim}{G}\|$ increases, whilst for p = 10 the penalty function is more ill-conditioned. In each case it is necessary to reset q either if $\|\underset{\sim}{G}\|$ increases too much or if negative curvature is detected (and possibly in some other circumstances). In the case of negative curvature it is desirable to increase r, say by a factor of 10. These approaches have all been used successfully, so that it is rarely necessary to reset q, in fact never more than once in the cases which we have considered.

The simplest way to minimize ϕ_3 is to use an unconstrained minimization method which does not require derivatives to be calculated. This approach is investigated in section 2 and it is shown that the efficiency of this process is about the same as that of conventional penalty function methods. However a great deal is known about the gradient of ϕ_3 and at the expense of making some approximations, it is possible to adapt variable metric methods to minimize ϕ_3. In section 3 the various possibilities of doing this are considered, and practical results are given which indicate the approximations which ought to be made. It is also shown that a considerable increase in efficiency is obtained over the simple method of section 2.

In section 4, consideration is given to the problem of extending the penalty function to deal with inequality constraints. An exchange type algorithm is discussed, which will terminate in a finite number of iterations if the object function is a positive definite quadratic, and if the constraint are linear. The results of practical calculations, although incomplete, show that considerable gains in efficiency can be obtained over methods which apply conventional penalty functions to inequality constrained problems.

2. A Basic Approach

In order to evaluate methods for solving equality constrained optimization problems, it is necessary to have various test problems of reasonable difficulty available. There is no well known collection of problems, and the only one which springs to mind is that of Powell (1969a) with 5 variables and 3 nonlinear constraints. Powell does not supply an initial point, so various of these have been suggested, viz: (1) the solution rounded to one decimal place $\underset{\sim}{x}^T = (-1.7, 1.6, 1.8, -.8, -.8)$, and (2) the solution rounded to integers $\underset{\sim}{x}^T = (-2, 2, 2, -1, -1)$. Another which is used by Lootsma is (1a) the solution rounded to two significant figures which differs from (1) only in that $x_4 = x_5 = -0.76$. In order to obtain more test problems, a number of well known inequality constraint problems have been modified so that they involve only equality constraints. This is done by examining the solution to the inequality constrained problem, isolating the active constraints, and treating these as equalities, neglecting the rest. Some good inequality problems are given by Colville (1968). In test problem No. 1 (TP1) none of the bounds are active and only constraints 3, 5, 6 and 9 of the general linear constraints are active. Thus a problem in 5 variables and 4 equality constraints is constructed. In test problem No. 2 (TP2), the 5 nonlinear constraints are all active, and also the constraints $x_i = 0$ for $i = 1, 2, 4, 7, 8, 10$, so that the resulting equality problem in 15 variables has 11 equalities. The equality problem (TP3) which results from the third of Colville's test problems has active constraints $x_1 = 78$, $x_2 = 33$, $x_4 = 45$ and constraints 2 and 5 of the quadratic constraints. Because this problem has 5 variables and 5 constraints it is somewhat special, and another related problem (TP3') has been constructed by adding the terms $(x_1 - 78)^2$ and $(x_4 - 45)^2$ to the definition of the function in TP3 and removing the equalities $x_1 = 78$ and $x_4 = 45$. The resulting problem has only 3 constraints and the minimum is at (46.51, 33, 29.65, 50.37, 40.62) to four significant figures. Finally the parcel problem of Rosenbrock (1960) was considered, giving the equality problem (PP1) with $F(\underset{\sim}{x}) = -x_1 x_2 x_3$ and one equality $x_1 + 2x_2 + 2x_3 = 72$. There are other versions (PP2, PP3) in which the constraint $x_1 = 20$ is active and then also the constraint $x_2 = 11$. These latter problems proved so simple to solve that they are not reported.

To assess the results of applying the new techniques, some comparative results from well known penalty function techniques are required. The results have to be obtained using the same information (functions and first derivatives) as the new method, which rules out SUMT for instance which requires second derivatives in addition. Lootsma (private communication) gives the result of applying Davidon's method to the SUMT penalty function for Powell's problem, and Fletcher and McCann (1969) give similar results on the other problems. The SUMT acceleration techniques are also used to improve the rate of convergence so the results can be taken as typical os what can be done with conventional penalty functions. The results are relevant to the original problems and not to the amended equality problems above, so only Powell's problem is directly comparable. However because of the nature of penalty functions, there is unlikely to be a vast difference in the number of times the function and gradient are calculated, whether an equality or inequality penalty function is used, (although this would not be true of the number of constraint evaluations).

The simplest way of using ϕ_3 is to estimate a suitable q from $\|G\|$ as has been described, having estimated G by differences, and then to minimize the penalty function ϕ_3 by using a method which does not require $\nabla \phi_3$ to be calculated. Typical of suitable minimization methods is that given by Powell (1964), which was in fact used, although more recent results (Fletcher (1969b)) indicate that Stewart's modification of Davidon's variable metric method is about twice as fast. The results on some of the test problems above are set out in Table 1 together with the comparative results of Lootsma and of Fletcher and McCann.

The results indicate that the basic approach is a workable process for obtaining solutions to equality problems, although one with no significant advantage over conventional penalty function methods. The slow convergence for TP2 is more likely to be due to deficiencies in the minimization method rather than to any difficulties in the function being minimized. It has previously been observed that some loss of efficiency is noticed with Powell's method as the number of variables is increased. However the main conclusion from these results is that the basic approach does not use the information provided at each point to the best advantage. The next section describes how this can be done, and shows that a significant improvement in rate of convergence can be obtained.

Table 1

The basic approach and comparison with conventional penalty functions

Problem	Basic approach with new method		Conventional penalty functions	
TP1	41*	104* 2.10^{-5*}	66	184 $\sim 10^{-8}$
TP2	>400	>1200 —	159	468 $\sim 10^{-8}$
TP3	not tried		61	194 $\sim 10^{-6}$
PP1	15	46 1.10^{-5}	33	105 $\sim 10^{-6}$
Powell (1a)	27	72 5.10^{-4}	25	— 1.10^{-5}

* Entries are "numbers of iterations", "number of evaluations of F and g together" and "absolute accuracy obtained in x_i".

3. Algorithms based on Variable Metric methods

In order to use a Variable Metric method to minimize the penalty function ϕ_3 it is necessary to be able to compute the gradient $\nabla \phi_3$. It is clear that this presents some difficulties because

$$\nabla \phi_3 = g - NN^+ g - GN^{+T} c - c^T [\nabla N^+] g + \qquad (3.1)$$
$$+ qN^{+T} c + qc^T [\nabla N^+] N^{+T} c .$$

In the computation of ϕ_3 itself, N^+ and terms involving N^+ will have been calculated, so the difficulties lie in the two new quantities G and $[\nabla N^+]$ which arise. It is necessary in estimating q that G be approximated by differences in g, and it would be possible to update this approximation after each iteration; thus the main difficulty lies with $[\nabla N^+]$. This is a three suffix quantity and it has been written in this form for convenience although the exact position of the suffices is somewhat obscured.

Although it is possible to compute $[\nabla N^+]$ (see (E) below), it would seem that more is likely to be gained by making an approximation to the gradient. If the last term in (3.1) is examined, then it will be seen that this is of order of magnitude $\|c\|^2$. Now as the solution is neared, $\|c\| \to 0$ and $\|\hat{P}g\| \to 0$, and the other terms in $\nabla \phi_3$ are of order $\|c\|$ or $\|\hat{P}g\|$. Thus the last term will be negligible near the solution and also at any points which nearly satisfy the constraints. As the usual situation is that q is over-estimated, so that it is most favourable for the algorithm to satisfy the constraints, it is clear that the term can be neglected in safety. It will be noticed that if ϕ_2 is used as a penalty function, then the term does not appear at all.

The remaining difficulty lies with the term $c^T[\nabla N^+]g$ which cannot be neglected on the above grounds, because it is only of order $\|c\|$. Various possibilities exist:

(A) To neglect it: this is equivalent to neglecting all curvature of the constraints. It would be expected that this approach would only be satisfactory for problems in which the curvature of function is the dominant factor in locating the minimum point which satisfies the constraints. In this case the gradient becomes

$$\nabla \phi_3 \approx g - NN^+ g - GN^{+T} c + qN^{+T} c \;.$$

(B) To collect it with the term $GN^{+T} c$ writing $\nabla \phi_3$ as

$$\nabla \phi_3 \approx g - NN^+ g - \Lambda c + qN^{+T} c$$

where $\Lambda = \nabla \lambda$ is the $n \times k$ matrix whose elements are the derivatives of $\lambda = N^+ g$, the continuous approximation to the Lagrange multipliers. Λ can then be updated at each iteration by the well known rank 1 formula

$$\Lambda \leftarrow \Lambda + \Delta x (\Delta \lambda - \Lambda^T \Delta x)^T / (\Delta x^T \Delta x) \;,$$

which preserves the relationship $\Lambda^T \Delta x = \Delta \lambda$, where Δx and $\Delta \lambda$ are the changes in x and λ on an iteration. Λ can be set initially as GN^{+T}, based on a quadratic/linear approximation (B1), or at more expense it can be calculated by differences in λ (B2).

(C) The term $c^T [\nabla N^+] g$ can be collected with $NN^+ g$ and written as $(\nabla u) g$ where ∇u is the $n \times n$ matrix whose elements are the derivatives of the vector $u = N^{+T} c$. The gradient then becomes

79

$$\nabla \phi_3 = g - \nabla u\, g - GN^{+T} c + q\, \nabla u\, N^{+T} c,$$

where it will be noticed that it is not necessary to neglect any terms at all. ∇u can be updated by rank 1 corrections as before.

(D) In a similar vein it is possible to consider the term $c^T[\nabla N^+]$ as an $n \times n$ matrix K, which can also be updated by rank 1 corrections. In this case no terms need again be neglected in $\nabla \phi_3$ as given by (3.1).

(E) On certain problems where second derivatives of c are readily available, it might be advantageous to consider computing the three suffix tensor $[\nabla N^+]$. In fact the ith element of the vector $c^T[\nabla N^+]g$ is specifically given by

$$(c^T[\nabla N^+]g)_i = c^T N^+ N^{+T} \frac{\partial N^T}{\partial x_i}(I - NN^+)g + $$

$$+ c^T N^+ \frac{\partial N}{\partial x_i} N^+ g \qquad (3.2)$$

with a similar expression for the other second order term in $[\nabla N^+]$ which was neglected. This is computable although cumbersome, and a more profitable approach might be to continue the philosophy of neglecting second order terms in $\|c\|$ and $\|\hat{P}g\|$. Now the middle term in (3.2) contains both c and $\hat{P}g$, so on neglecting this, and rearranging the last term, the approximation

$$c^T[\nabla N^+]g \approx [\sum_i \lambda_i \nabla^2 c_i] u$$

is obtained, where $\nabla^2 c_i$ is the matrix of second derivatives of the constraint function $c_i(\underset{\sim}{x})$. It would be possible to arrange to compute this without using $n^2 k$ storage locations which is advantageous. However it was decided that it was more important in the first instance to assess the approximations which only involved first derivatives and so the possibility is not considered further in this paper.

The effect of the various approximations A-D was tested on the problem

Minimize $F(\underset{\sim}{x}) = 4x_1^2 + 2x_2^2 + 2x_3^2 - 33x_1 + 16x_2 - 24x_3$

Subject to $c_1(\underset{\sim}{x}) = 2x_2^2 + 3x_1 - 7$

$c_2(\underset{\sim}{x}) = x_3^2 + 4x_1 - 11$

with solution at $\underset{\sim}{x} = (5.32677, -2.11900, 3.21046)$ to 6 significant figures.

This function was chosen so that terms like $\underset{\sim}{\Lambda}$ and $\nabla \underset{\sim}{u}$ could be calculated exactly at any point, and the adequacy of the approximation to them examined. The results obtained when starting from three different initial approximations are set out in Table 2. It is clear from the results that approximations C and D are most unsatisfactory. The reason for this is not too hard to see. The result of an error in $\nabla \underset{\sim}{u}$ or $\underset{\sim}{K}$, because they multiply $\underset{\sim}{g}$ which is generally non-zero at the solution, is that the error is passed on directly to $\nabla \phi_3$. The result of an error in $\underset{\sim}{\Lambda}$, because this multiplies $\underset{\sim}{c}$ which is zero at the solution, is that the resulting error in $\nabla \phi_3$ is reduced by an order of magnitude. That is to say the approximations to $\nabla \phi_3$ are much more sensitive to errors in $\nabla \underset{\sim}{u}$ or $\underset{\sim}{K}$ than in $\underset{\sim}{\Lambda}$, a feature which is reflected in the results. In fact because C and D are somewhat less efficient in housekeeping, it would have been necessary to show an overall improvement in rate of convergence to justify their use. As it is they will not be considered further.

Table 2

Different Approximations for $\nabla\phi_3$

Approximation	$x_0 = (4, -3, 4)$		$x_0 = (1, -4, 5)$		$x_0 = (1, -6, 7)$	
A	7*	22* 3.10^{-5}	7	21 5.10^{-5}	7	22 3.10^{-5}
B1	5	16 4.10^{-5}	7	22 1.10^{-5}	7	22 1.10^{-5}
C	failed		failed		16	61 5.10^{-6}
D	18	61 1.10^{-5}	16	72 2.10^{-5}	12	49 1.10^{-4}

*Entries are "number of iterations", "number of evaluations of ϕ_3 and $\nabla\phi_3$ together" and $\|\nabla\tilde\phi\|_2$ where $\nabla\tilde\phi$ is the final approximation to the gradient obtained by the program.

Table 3

Comparison of approximations A and B. (H updated by rank 2)

Problem	Approximation A	Approximation B1	Approximation B2	Comments
TP1	3* 10* 2.10^{-5}*	3 10 9.10^{-7}	3 10 9.10^{-7}	$p = 2$ approximate $\tilde{G}$
TP2	65 213 3.10^{-4} q reset once	63 208 2.10^{-4} q reset once	failed	$p = 10$ exact $\tilde{G}$
TP3	4 11 4.10^{-8}	5 14 3.10^{-5}	6 16 1.10^{-5}	$p = 2$ $\tilde{G}$ exact
TP3'	6 19 2.10^{-4}	7 22 2.10^{-4}	7 20 2.10^{-4}	because $F(x)$ is quadratic
PP1	5 16 8.10^{-5}	5 16 1.10^{-5}	5 16 1.10^{-5}	$p = 2$ approximate $\tilde{G}$
Powell 1	4 13 5.10^{-6}	4 13 6.10^{-6}	4 13 5.10^{-6}	$p = 10$ approximate $\tilde{G}$
Powell 2	9 35 2.10^{-6} q reset once	9 34 1.10^{-5} q reset once	8 29 6.10^{-6} q reset once	

* Entries as Table 2.

In an attempt to distinguish between approximations A and B, a wide selection of the test problems described in section 2 was considered. The results are set out in Table 3. Again it would seem that there is little to be gained by using approximation B rather than approximation A. This is not surprising for those problems in which the curvature of the constraints can be neglected. However TP2 is a problem which is linear in many of the variables, and in which the curvature of the constraints is essential in determining the solution. Yet no significant advantage for approximation B, which takes curvature into account, has been demonstrated for this problem. The reason may be either that the rate of convergence of $\underset{\sim}{\Lambda}$ near the solution is not sufficiently rapid, or that the approximation is too crude remote from the solution. In view of the simplicity of approximation A, we have concluded that this approximation should be used.

It might be thought that the use of an approximation to $\underset{\sim}{G}$, rather than the exact $\underset{\sim}{G}$, might be obscuring some of the conclusions. To check this, some cases in which the approximate and exact $\underset{\sim}{G}$ differed (i.e. non-quadratic functions) were run in both ways. The results are given in Table 4.

Table 4

The effect of approximating the Hessian matrix $\underset{\sim}{G}$

Problem	Exact $\underset{\sim}{G}$			Approximate $\underset{\sim}{G}$		
PP1	5*	15	5.10^{-6}	5	16	8.10^{-5}
TP1	3	10	5.10^{-6}	3	10	9.10^{-7}
TP2	63	208	2.10^{-4} q reset once	56	183	2.10^{-4} q reset once

*Entries as Table 2.

It seems safe to conclude that the effect of this approximation is less than that of other approximations which are made in solving the problem. Consequently use of the exact $\underset{\sim}{G}$ is only recommended if it is particularly simple to calculate. In earlier programs a rank 1 formula (the analogy of the one referred to below) was used to update $\underset{\sim}{G}$. In more recent programs however the more recent formula due to Powell was used, which is being reported in his paper at this conference.

Having decided upon a suitable approximation for the gradient, the choice of updating formula in the variable metric method will now be considered. There has been much interest in this subject, but we have confined ourselves to considering the two most well known formulae. One is the original Davidon rank 2 formula (see Fletcher and Powell (1963) for example), and the other the rank 1 formula which was suggested in 1968 or thereabouts by a number of authors. A list and details are given by Powell (1969b). It will now be shown that the rank 1 formula possesses an interesting property in respect of these penalty functions, which the rank 2 formula does not. Assume that Newton steps are being taken in the variable metric method, and let $\underset{\sim}{H}$ denote the matrix which approximates the inverse hessian of the penalty function ϕ being minimized. Then if $\underset{\sim}{x}$ and $\underset{\sim}{x}'$ denote successive points, whose difference is $\underset{\sim}{\delta} = -\underset{\sim}{H}\nabla\phi = \underset{\sim}{x} - \underset{\sim}{x}'$, then simple manipulation enables the predicted correction to $\underset{\sim}{x}'$ given by $\underset{\sim}{\delta}' = -\underset{\sim}{H}'\nabla\phi'$ to be simplified to $\underset{\sim}{\delta}' = -\Delta\underset{\sim}{H}\nabla\phi$ where $\Delta\underset{\sim}{H}$ is the correction to $\underset{\sim}{H}$ (rank 1 or rank 2 in the above). If all the constraints are linear and the initial $\underset{\sim}{H}$ matrix is chosen as recommended below, then it is simple to show that the first step from $\underset{\sim}{x}$ to $\underset{\sim}{x}'$ causes $\underset{\sim}{x}'$ to satisfy the linear constraints. The important property of the rank 1 formula is that it retains this property in the $\underset{\sim}{H}$ matrix, with the result that subsequent $\underset{\sim}{x}$ all satisfy the linear constraints. This result is not true for the rank 2 formula. The result is proved by stating that the condition for $\underset{\sim}{x}' + \underset{\sim}{\delta}'$ not to violate any constraints is that $\underset{\sim}{P}\underset{\sim}{\delta}' = \underset{\sim}{0}$, so substituting the above expression for $\underset{\sim}{\delta}'$, it is necessary that $\underset{\sim}{P}\Delta\underset{\sim}{H} = \underset{\sim}{0}$. For the rank 1 formula this reduces to requiring $\underset{\sim}{P}(\underset{\sim}{H}\underset{\sim}{\gamma} - \underset{\sim}{\delta})$ to be zero where $\underset{\sim}{\gamma} = \nabla\phi' - \nabla\phi$, or consequently that $\underset{\sim}{P}\underset{\sim}{H}\nabla\phi' = \underset{\sim}{0}$. Because $\underset{\sim}{c}'$ is zero, $\nabla\phi'$ turns out to be $\hat{\underset{\sim}{P}}\underset{\sim}{g}'$, and because $\underset{\sim}{P}\underset{\sim}{H}$ can be reduced to $[\underset{\sim}{P}(q\underset{\sim}{I} - \underset{\sim}{G})\underset{\sim}{P}]^{+}$, the scalar product $\underset{\sim}{P}\underset{\sim}{H}\nabla\phi'$ is zero. Thus the result is proved.

The results using the rank 1 formula are set out in Table 5. They were obtained using a more simple linear search to that used with the rank 2 formula as set out in Table 3. However the effect of this should be marginal, whereas a considerable improvement in rate of convergence is noticed on examining the results for the rank 1 formula.

A detailed examination of the computer results verifies the above theorem and suggests that it holds when there are a mixture of linear and nonlinear constraints. We feel quite certain that this is the feature which accounts for the improved rate of convergence.

It is usual in variable metric methods to set $\underset{\sim}{H} = \underset{\sim}{I}$ initially. However if the quadratic/linear approximation is made then an estimate of $\nabla^2 \phi_3$ is available and $\underset{\sim}{H}$ can be set as the inverse of this. It is most readily computed from the formula

$$\underset{\sim}{H} = (\underset{\sim}{G} - \underset{\sim}{P}\underset{\sim}{G} - \underset{\sim}{G}\underset{\sim}{P} + q\underset{\sim}{P})^{-1} \quad (3.3)$$

There is no guarantee of course that this is positive definite, and this should be taken into account. However some experiments, reported in Table 6, indicate that much more rapid convergence can be obtained. This can be ascribed to having taken account of the quadratic terms in ϕ_3, and also to the fact that use of (3.3) causes any linear constraints to be satisfied after the first iteration. Whilst on this subject, it will be noticed that if q is reset, then ϕ_3 is changed and so $\underset{\sim}{H}$ must be reset. It has again been found satisfactory to reset $\underset{\sim}{H}$ in accordance with (3.3).

It is also necessary to decide upon a strategy for carrying out the linear search. It is not appropriate to give all the details here, but they will be reported in detail by Lill (1971). The approach is based on using the estimated

Table 5
The effect of using a Rank 1 formula

Problem	Results with Approximation A to $\nabla \phi_3$		Comments
TP1	4*	6 $5 \cdot 10^{-6}$	$p = 2$, approximate $\tilde{G}$
TP2	48	51 $1 \cdot 10^{-5}$	$p = 10$, exact $\tilde{G}$
TP3	4	8 $9 \cdot 10^{-10}$	$p = 2$, exact $\tilde{G}$
PP1	4	5 $5 \cdot 10^{-5}$	$p = 10$, approximate $\tilde{G}$
Powell 1	5	9 $4 \cdot 10^{-7}$	$p = 10$, approximate $\tilde{G}$
Powell 2	12	28 $1 \cdot 10^{-6}$	$p = 10$, approximate $\tilde{G}$

*Entries as Table 2.

Table 6
Alternative initial settings for $\underset{\sim}{H}$

Problem	$\underset{\sim}{H} = \underset{\sim}{I}$			$\underset{\sim}{H} = (\underset{\sim}{G} - \underset{\sim}{P}\underset{\sim}{G} - \underset{\sim}{G}\underset{\sim}{P} + q\underset{\sim}{P})^{-1}$		
TP1	12*	15	3.10^{-6}	4	6	5.10^{-6}
TP3	7	10	1.10^{-5}	4	5	5.10^{-10}
Powell 1	8	12	2.10^{-7}	5	9	4.10^{-7}

*Entries as Table 2.

gradient at the initial point, and only function values elsewhere, so that the interpolations are based on as much reliable information as possible. Typically a trial step $\underset{\sim}{x} + \underset{\sim}{\delta}$ is taken and ϕ_3 is evaluated. A quadratic based on $\phi_3(\underset{\sim}{x})$, $\nabla\phi_3(\underset{\sim}{x})$ and $\phi_3(\underset{\sim}{x} + \underset{\sim}{\delta})$ is calculated and if its minimum lies at $\underset{\sim}{x} + \lambda\underset{\sim}{\delta}$ with $\lambda \varepsilon (0, 1)$, it is examined. If this interpolation fails then $\phi_3(\underset{\sim}{x} - \lambda\underset{\sim}{\delta})$ is calculated and a quadratic fitted to the three function values. Thus provision for a negative step is made in case $\nabla\phi$ is inaccurate or $\underset{\sim}{H}$ is not positive definite. The possibility of making a step with $\lambda > 1$ is also considered if ϕ_3 is improved on making the first step.

Finally, the results have only been examined with regard to making the best choice amongst several alternative strategies. However it is important to point out that on comparing these results with those obtained from conventional penalty function methods or those obtained by applying ϕ_3 without estimating $\nabla\phi_3$ as in Table 1, it will be seen that a considerable improvement in the rate of convergence has been obtained.

4. Inequality Constraints

The possible extension of these new penalty functions to deal with inequality constraints will now be discussed. In order not to increase the number of variables, an algorithm of exchange type is envisaged. In this a **basis** is kept of constraints (**active** constraints) which are being treated as equality constraints. Constraints are added to the basis if they prevent a minimum along a line being reached on an iteration. Decisions on removing constraints are taken according to the signs of the Lagrange multipliers. However it is necessary to prevent the phenomenon of "zigzagging" in which a constraint repeatedly enters and leaves the basis. Examples have been constructed for some optimization methods in which zigzagging causes convergence to a non-stationary point to take place, and it is certainly possible for a slow rate of linear convergence to occur. The former situation is of course disastrous, and in practice the latter can also prevent the solution being reached. Thus it is important to make the correct decision in this respect.

In earlier programs, the approximate Lagrange multipliers were examined after each iteration, and if any of the signs were positive then a constraint was removed from the basis. This was supplemented by an anti-zigzggging device in which a constraint was forced to stay in the basis if an oscillation had been observed. Although this was acceptable it was discarded in view of its ad-hoc nature. In an exchange type algorithm for quadratic programming given by Fletcher (1970) it is shown that finite convergence can be proved by only examining the signs of the Lagrange multipliers when $\underset{\sim}{x}$ is a minimum of the equality problem corresponding to the current basis. This is impracticable for non-quadratic functions, but a compromise can be reached by only examining the signs when an unconstrained minimum along a line is reached. In the quadratic/linear case then the same trajectory as given by Fletcher's quadratic programming algorithm would be obtained for strictly convex functions. Thus the new method terminates in a finite number of iterations for strictly convex quadratic/linear inequality problems.

The penalty function which has been used is ϕ_3 with approximation A to the gradient which neglects the curvature of the constraints. An advantage of using ϕ_3 is that a change of basis, although it causes a change in the penalty function being used, does not necessitate recomputation of q, which is based only on $\|G\|$. One effect of a change of basis is that the matrix $\tilde{H}$ now corresponds to a different penalty function. However it is possible to correct $\tilde{H}$ by making a change of rank 2 which would retain an exact value of $[\nabla^2 \phi_3]^{-1}$ in the quadratic/linear case. In this case the difference between the projection matrices $\tilde{P}_{(k)}$ and $\tilde{P}_{(k+1)}$, corresponding to k and k+1 constraints in the basis, is

$$\tilde{P}_{(k+1)} - \tilde{P}_{(k)} = \underline{vv}^T / \underline{v}^T \underline{v}$$

where $\underline{v} = \hat{\tilde{P}}_{(k)} \underline{n}$, $\underline{n}$ being the normal of the additional constraint. Hence the difference between $\nabla^2 \phi_3$ in the two cases can be written using (3.3) as

$$[\nabla^2 \phi]_{(k+1)} - [\nabla^2 \phi]_{(k)} = (q \underline{vv}^T - \underline{vv}^T G - G\underline{vv}^T)/\underline{v}^T \underline{v}$$

$$= \underline{rr}^T - \underline{ss}^T$$

where $\underline{s} = G\underline{v}/(q\underline{v}^T\underline{v})^{1/2}$ and $\underline{r} = \underline{s} - \underline{v}(q/\underline{v}^T\underline{v})^{1/2}$. It seems simplest to compute the new approximate inverse along these lines by considering the change to $\nabla^2 \phi$ as being two changes of rank 1. In this case the formula

$$(\underline{A} \pm \underline{bb}^T)^{-1} = \underline{A}^{-1} - \underline{A}^{-1}\underline{bb}^T\underline{A}^{-1}/(\underline{b}^T\underline{A}^{-1}\underline{b} \pm 1)$$

can be used to update the inverse matrix $\tilde{H}$.

Time has so far only permitted testing the algorithm on some of the smaller test problems. The results of this are given in Table 7, in which TP1, etc. refer to the original inequality problems and not to the modified versions used in sections 2 and 3. These results are very encouraging when compared with those in Table 1 for the basic method or for conventional penalty functions. The pattern which is observed is that a certain number of iterations are required to determine the correct basis (5 in the case of TP3) after which the rapid convergence to the solution of the corresponding equality problem is observed as in section 3. If this ability to isolate the correct basis in a reasonable number of iterations occurs in general, then the new method will indeed be a powerful tool for dealing with nonlinear programming.

Table 7

The method for inequality problems

TP3	9^*	25	2.10^{-6}
PP1	6	16	5.10^{-10}

*Entries as Table 2.
The rank 1 updating formula for $\underset{\sim}{H}$ was used with $p = 10$.

REFERENCES

1. Colville, A. R. (1968), "A comparative study on non-linear programming codes", IBM, New York Scientific Center Report No. 320-2949.

2. Fletcher, R. (1969a), "A class of methods for non-linear programming with termination and convergence properties", Harwell report T. P. 368, to be published in "Integer and Non-linear programming", J. Abadie, ed., (1970), North-Holland.

3. Fletcher R. (1969b), "A review of methods for unconstrained optimization", In "Optimization", R. Fletcher ed., Academic Press.

4. Fletcher, R. (1970), "A general quadratic programming algorithm", Harwell report T.P. 401.

5. Fletcher, R. and A. P. McCann (1969), "Acceleration techniques for non-linear programming", In "Optimization", R. Fletcher, ed., Academic Press.

6. Fletcher R. and M. J. D. Powell (1963), "A rapidly convergent descent method for minimization", Computer J., $\underline{6}$, p. 163.

7. Lill, S. A. (1971), Ph.D. thesis, University of Leeds (in preparation).

8. Powell, M. J. D. (1964), "An efficient method of finding the minimum of a function of several variables without calculating derivatives", Computer J., $\underline{7}$, p. 155.

9. Powell, M. J. D. (1969a), "A method for non-linear constraints in minimization problems", In "Optimization", R. Fletcher, ed., Academic Press.

10. Powell, M. J. D. (1969b), "Rank one methods for unconstrained optimization", Harwell report T.P. 372, to be published in "Integer and Non-linear programming", J. Abadie, ed., (1970), North-Holland.

11. Powell, M. J. D. (1969c), "On the convergence of the variable metric algorithm", Harwell Report T.P. 382.

12. Rosenbrock, H. H. (1960), "An automatic method for finding the greatest or least value of a function", Computer J., $\underline{3}$, p. 175.

Some Algorithms Based on the Principle of Feasible Directions

G. ZOUTENDIJK

ABSTRACT

A number of algorithms will be described which are based on the principle of feasible directions. Special problems like linear programming, unconstrained optimization, optimization subject to linear equality constraints, quadratic programming and linearly constrained nonlinear programming will be briefly dealt with.

1. Introduction

In this paper a number of algorithms, some well-known, some new, will be outlined which are based on the principle of feasible directions. Methods of feasible directions have originally been developed for nonlinear programming problems. However, the principle can be applied as well in other fields of numerical analysis such as approximation or solving systems of nonlinear equations. The principle entails that a sequence of trial solutions is obtained with improving value for some criterion function and that each new trial solution is obtained from the previous one by

(1) determining a direction in which progress can be made (a usable feasible direction) and

(2) determining the step length to be taken in that direction.

Several methods for generating directions will be outlined, while their efficiency will be demonstrated by means of an example. Methods of feasible directions have been described in many books and articles, e.g. in [13] and [14]. Several methods suggested by various authors for the unconmaximization or linearly constrained nonlinear programming problem are in fact methods of feasible directions.

2. Direction generators

To fix thoughts we will assume that the problem to be solved is a linearly constrained nonlinear programming problem:

FEASIBLE DIRECTIONS ALGORITHMS

$$\max\{f(x) \mid Ax \le b, \; x \ge 0\}, \qquad (1)$$

in which f is differentiable with continuous gradient vector $\nabla f(x) = \{\frac{\partial f}{\partial x_j}, \; j = 1, \ldots, n\}$, A an m by n matrix, $x \in E_n$ and $b \in E_m$. Let $\tilde{x} \in R = \{x \mid Ax \le b, \; x \ge 0\}$, $a_i^T \tilde{x} = b_i$ for $i \in I(\tilde{x}) \subset \{1, \ldots, m\}$ (a_i^T being the i·th row of the matrix A) and $\tilde{x}_j = 0$ for $j \in J(\tilde{x}) \subset \{1, \ldots, n\}$. Then a usable feasible direction $s \in E_n$ to be generated in $\tilde{x}$ must satisfy two requirements:

It should belong to the closed convex polyhedral cone of feasible directions:

$$s \in S(\tilde{x}) = \{s \mid a_i^T s \le 0 \text{ if } i \in I(\tilde{x}); \; s_j \ge 0 \text{ if } \qquad (2)$$
$$j \in J(\tilde{x})\} ;$$

It should make a sharp angle with the gradient in $\tilde{x}$:

$$\nabla f(\tilde{x})^T s > 0 \qquad (3)$$

A direction generator is a method generating a direction satisfying (2) and (3). We can distinguish between at least four classes of direction generators:

1. Methods which directly work with the trial solutions like the simplex method in linear programming. To solve (1) we could for instance linearize the objective function $f(x)$ at $\tilde{x} \in R$, so that its variable part becomes $\nabla f(\tilde{x})^T x$ and solve the linearized problem resulting in a solution x^*. The usable feasible direction will then be $s = x^* - \tilde{x}$.

2. Methods that determine the locally best direction according to some criterion, e.g. the one among all those with unit norm that maximizes the inner product $\nabla f(\tilde{x})^T s$:

$$\text{Max}\{p^T s \mid s \in S(\tilde{x}), \; \|s\| \le 1\}, \qquad (4)$$

where we have written p instead of $\nabla f(\tilde{x})$, or, equivalently (if $p^T s > 0$ in (4) and up to a proportionality factor)

$$\text{Min}\{\|s\| \mid s \in S(\tilde{x}), \ p^T s \geq 1\}. \quad (5)$$

For the norm we can for instance take:

the L_1 norm, $\sum_{j=1}^{n} |s_j| \leq 1,$ (2.1)

the L_2 norm, $(s^T s)^{1/2} \leq 1$, i.e. $s^T s \leq 1,$ (2.2)

the L_∞ (Chebyshev) norm (2.3)

$\max_j |s_j| \leq 1,$ i.e. $\forall_j \ -1 \leq s_j \leq 1,$

a metricized norm $s^T P s \leq 1$ with P being a symmetric positive definite matrix (2.4)

Since the only reasons to introduce the norm have been to avoid infinite solutions in the direction problem and to obtain a measure of comparison we could as well consider other normalizations which are no norms in the strict sense such as

$s_j \leq 1$ if $p_j > 0$; $s_j \geq -1$ if $p_j \leq 0,$ (2.5)

$s_j \leq p_j$ if $p_j > 0$; $s_j \geq -p_j$ if $p_j < 0,$ (2.6)

$\sum_{p_j > 0} s_j \leq 1; \quad \sum_{p_j < 0} s_j \geq -1$ (2.7)

3. Methods in which we do not normalize but in which the direction is fixed by a division of the variables into basic and non-basic variables. Let $A(\tilde{x})$ be the $|I(\tilde{x})|$ by n matrix consisting of the rows a_i^T with $i \in I(\tilde{x})$. Then we can consider the system:

$$A(\tilde{x})s + t = 0, \quad t \geq 0$$

$$s_j \geq 0, \quad j \in J(\tilde{x})$$

$$p^T s = 1$$

Let B be an $|I(\tilde{x})|$ by $|I(\tilde{x})|$ square non-singular submatrix of $(A(\tilde{x}), I)$. In an obvious notation we can write:

$$D\bar{s}_D + B\bar{s}_B = 0,$$

$$\bar{p}_D^T \bar{s}_D + \bar{p}_B^T \bar{s}_B = 1,$$

in which (after rearranging of columns or components)

$$(A(\tilde{x}), I) = (D, B); \quad \bar{s} = \binom{s}{t} = \binom{\bar{s}_D}{\bar{s}_B}; \quad \bar{p} = \binom{p}{0} = \binom{\bar{p}_D}{\bar{p}_B}.$$

It follows:

$$B^{-1}D\bar{s}_D + \bar{s}_B = 0,$$

$$-(\bar{p}_B^T B^{-1} D - \bar{p}_D^T)\bar{s}_D = 1 \quad \text{or} \quad -\bar{u}_D^T \bar{s}_D = 1 \tag{6}$$

We can now act in one of the following ways:

3.1. Select the most attractive $\bar{s}_D$ variable (in terms of its reduced "price") and give it a value α (> 0 (< 0) if the corresponding $(\bar{u}_D)_j < 0 (> 0)$):

$$(\bar{s}_D)_{k'} = \alpha; \quad (\bar{s}_D)_k = 0, \quad k \neq k'; \quad \bar{s}_B = -\alpha B^{-1} \bar{a}_{.k'}$$

with α such that $p^T s = 1$ (not really necessary; we can as well take $\alpha = \pm 1$).

3.2. Take

$$(\bar{s}_D)_j = \max\{-(\bar{u}_D)_j, 0\} \quad \text{if} \quad (\bar{x}_D)_j = 0,$$

$$= -(\bar{u}_D)_j \quad \text{if} \quad (\bar{x}_D)_j > 0 .$$

3.3. Any combination of 3.1 and 3.2, e.g. we only give $(\bar{s}_D)_j$ a non-zero value if $|(\bar{u}_D)_j| \geq \epsilon$ for some adjustable $\epsilon > 0$.

In order that a non-zero step can be made the basis B to be chosen should be such that either

$$(\bar{x}_B)_i > 0$$

or

$$(\bar{x}_B)_i = 0 \quad \text{and} \quad (-B^{-1} D \bar{s}_D)_i \geq 0$$

To arrive at this situation in the case of method 3.1 we pivot in the matrix using the normal simplex method criterion to select pivot columns and rows until no pivot element can be found anymore in the matrix. In the case of methods 3.2 or 3.3 we could then give the corresponding variable its proper value in the direction vector, update the right-hand side, and continue pivoting, if necessary, with a matrix in which one column will be left out of consideration.

In practice the starting basis for this pivoting operation will be the final basis of the previous direction problem, so that except in the first direction problem only a few changes of basis (0, 1 or 2) will be required.

Notes

a. Equality constraints ($a_i^T . x = b_i$ instead of $\leq b_i$) will result in relations of the type $a_i^T . s = 0$ in the direction problem which do not present any problem.

b. In the direction problems, previously described, we could replace $S(\tilde{x})$ by $S_0(\tilde{x}) = \{s \mid a_i^T s = 0, \ i \in I(\tilde{x}); s_j = 0, \ j \in J(\tilde{x})\}$. If this requirement turns out to be too restrictive, so that $p^T s/|s|$ becomes too small in (4) or (6) becomes inconsistent, then we must replace one of the equality signs by the weaker $\leq$ sign. The philosophy behind replacing $S(\tilde{x})$ by $S_0(\tilde{x})$ is that it might be computationally more advantageous to stay as long as possible in a certain combination of hyperplanes and, if necessary, to leave only one of them. This philosophy which seems to make sense in the (nearly)linear case can be found in the simplex method and in the gradient projection method [10].

4. Methods in which the direction is obtained by premultiplying the gradient by a matrix which is obtained from an approximation to the matrix of second partial derivatives by projecting this matrix onto a suitable subset of the set of binding hyperplanes. Although one of these methods is outlined in section 4 (Goldfarb's method) this paper is primarily devoted to the other classes of methods. It should be said, however, that most computational experience has been obtained with class 4 methods, in particular for unconstrained optimization (see section 3).

To compare some of the different direction generators we shall determine a usable feasible direction for the following problem:

$$s_1 - 2s_2 + s_3 + 3s_4 - s_5 \leq 0$$

$$-s_1 + 3s_2 - 2s_3 - s_4 + s_5 \leq 0 \qquad (7)$$

$$s_1 \geq 0 \qquad\qquad s_5 \geq 0$$

$$-3s_1 + 2s_2 - s_3 + 3s_4 + s_5 > 0$$

As measure of comparison we shall use the criterion
$(\frac{p^T s}{|s|})^2 = \frac{(p^T s)^2}{s^T s}$. This obviously is the criterion to compare directions locally. It should be observed, however, that a method generating directions closer to the gradient is not necessarily a more efficient one. The amount of work per step as well as numerical stability are other important factors to be considered.

This is to a very large extent problem dependent. As an example we may mention the simplex method for linear programming which is very efficient but which certainly does not lead to very good local directions.

Method 2.2. The optimality conditions for the problem $\max\{p^T s \mid As \leq 0, \; s^T s \leq 1\}$ are $p = A^T u + \beta s$, $u \geq 0$, $\beta \geq 0$ (and > 0 if $p^T s > 0$), $u^T As = 0$. Multiplying from the left by A and introducing $v = -\beta As \geq 0$ we obtain

$$-Ap = -AA^T u + v, \quad u \geq 0, \; v \geq 0, \; u^T v = 0 \qquad (8)$$

If, in the final solution, it turns out that $A_1 s = 0$ and $A_2 s < 0$ holds - A_1 and A_2 being a horizontal partitioning of the matrix A -, then it follows that $u = (A_1 A_1^T)^{-1} A_1 p$ and $s = \{I - A_1^T (A_1 A_1^T)^{-1}\} p$. We have taken $\beta = 1$, so that $s = p - A^T u$ is proportional to the solution of the direction problem. It is the projection of the vector p onto the linear subspace $A_1 s = 0$; it is also the projection of p onto the cone $\{s \mid As \leq 0\}$. To solve (8) we select a $v_i < 0$ and interchange v_i and u_i by taking the diagonal element of the matrix as pivot element (this element will always be < 0 when $v_i < 0$, see [13], pp. 82-83). These interchanges of complementary variables will be continued until the right-hand side is completely nonnegative. The choice of the pivot rows can be made in such a way that finiteness is guaranteed (see [13], p. 84). In practice that row will be selected among the candidates for which the quotient of right-hand side and diagonal element is largest.

In our example we have:

$$\begin{pmatrix} 1 & -2 & 1 & 3 & -1 \\ -1 & 3 & -2 & -1 & 1 \\ -1 & & & & \\ & & & & -1 \end{pmatrix}, \; p^T = (-3, 2, -1, 3, 1).$$

The tableau becomes:

FEASIBLE DIRECTIONS ALGORITHMS

	u_1	u_2	u_3	u_4			u_1	u_2	v_3	u_4
v_1	0	-16	13	1	-1	v_1	-3	-15	12	1	-1
v_2	-9	13	-16	-1	1	$\rightarrow$ v_2	-6	12	$\boxed{-15}$	-1	1	$\rightarrow$
v_3	-3	1	-1	$\boxed{-1}$	0	u_3	3	-1	1	-1	0
v_4	1	-1	1	0	-1	v_4	1	-1	1	0	-1

	u_1	v_2	v_3	u_4			v_1	v_2	v_3	u_4
v_1	$-\dfrac{39}{5}$	$\boxed{\dfrac{27}{5}}$	$\dfrac{4}{5}$	$\dfrac{1}{5}$	$-\dfrac{1}{5}$	u_1	$\dfrac{13}{9}$	$-\dfrac{5}{27}$	$-\dfrac{4}{27}$	$-\dfrac{1}{27}$	$\dfrac{1}{27}$
u_2	$\dfrac{2}{5}$	$-\dfrac{4}{5}$	$-\dfrac{1}{15}$	$\dfrac{1}{15}$	$-\dfrac{1}{15}$	u_2	$\dfrac{14}{9}$	$-\dfrac{4}{27}$	$-\dfrac{5}{27}$	$\dfrac{1}{27}$	$-\dfrac{1}{27}$
u_3	$\dfrac{13}{5}$	$-\dfrac{1}{5}$	$\dfrac{1}{15}$	$-\dfrac{16}{15}$	$\dfrac{1}{15}$	$\rightarrow$ u_3	$\dfrac{26}{9}$	$-\dfrac{1}{27}$	$\dfrac{1}{27}$	$-\dfrac{29}{27}$	$\dfrac{2}{27}$
v_4	$\dfrac{3}{5}$	$-\dfrac{1}{5}$	$\dfrac{1}{15}$	$-\dfrac{1}{15}$	$-\dfrac{14}{15}$	v_4	$\dfrac{8}{9}$	$-\dfrac{1}{27}$	$\dfrac{1}{27}$	$-\dfrac{2}{27}$	$-\dfrac{25}{27}$

Optimal. It follows that $s = p - A^T u = (0, \dfrac{2}{9}, \dfrac{2}{3}, \dfrac{2}{9}, \dfrac{8}{9})^T$,
$\dfrac{(p^T s)^2}{s^T s} = \dfrac{4}{3} = 1.33$. The same direction is also given by Rosen's gradient projection method.

Advantages of method 2.1: it gives the direction closest to the gradient; the solution is unique. Disadvantages: non-negative constraints increase the size of AA^T; this matrix is dense and does not show structure (a special structure of A may be partially or completely destroyed); the condition of AA^T will usually be worse than the condition of A; an additional calculation $s = p - A^T u$ has to be made. Conclusion: method 2.1 should only be considered in the case of a highly nonlinear objective function and a matrix A without special structure.

Notes 1. In the case of equalities among the constraints the corresponding u_i are unrestricted. When solving (8) they can be put into the basis one after the other.

2. Method 2.4 can be dealt with in exactly the same way: The optimality conditions for the problem $\max\{p^T s \mid As \leq 0, s^T Ps \leq 1\}$ are $p = A^T u + \beta Ps$, $u \geq 0$, $\beta \geq 0$, $u^T As = 0$. Hence $P^{-1}p = P^{-1}A^T u + \beta s$ and

$$-AP^{-1}p = -AP^{-1}A^T u + v, \quad u \geq 0, \quad v \geq 0, \quad u^T v = 0 \qquad (9)$$

which can be solved in the same way as (8):

$$u = (A_1 P^{-1} A_1^T)^{-1} A_1 P^{-1} p, \quad s = P^{-1}p - P^{-1}A_1^T (A_1 P^{-1} A_1^T)^{-1} A_1 P^{-1} p.$$

Method 2.4 might be of use when an estimate of the matrix of second partial derivatives of $f(x)$ can easily be made. This is for instance the case when the objective function is separable: $f(x) = \sum_{j=1}^{n} f_j(x_j)$. Another application will be given in sections 3 and 4.

Method 2.3. We can use the simplex method to solve the direction problem taking account of the fact that all variables have upper bounds $+1$ and lower bounds -1 or 0 (s_1 and s_5, to be indicated by a $+$ index):

	s_1^+	s_2	s_3	s_4	s_5^+
t_1	0	1	-2	1	⌐3⌐	-1
t_2	0	-1	3	-2	-1	1
	3	-2	1	-3	-1

	s_1^+	s_2	s_3	t_1	s_5^+
s_4	0	$\frac{1}{3}$	$-\frac{2}{3}$	$\frac{1}{3}$	$\frac{1}{3}$	$-\frac{1}{3}$
t_2	0	$-\frac{2}{3}$	⌐$\frac{7}{3}$⌐	$-\frac{5}{3}$	$\frac{1}{3}$	$\frac{2}{3}$ →
	4	-4	2	1	-2

FEASIBLE DIRECTIONS ALGORITHMS

		s_1^+	t_2	s_3	t_1	s_5^+
s_4	0	$\frac{1}{7}$	$\frac{2}{7}$	$-\frac{1}{7}$	$\frac{3}{7}$	$-\frac{1}{7}$
s_2	0	$-\frac{2}{7}$	$\frac{3}{7}$	$-\frac{5}{7}$	$\frac{1}{7}$	$\frac{2}{7}$
		$\frac{20}{7}$	$\frac{12}{7}$	$-\frac{6}{7}$	$\frac{11}{7}$	$-\frac{6}{7}$

$\xrightarrow{(s_3=1)}$

		s_1^+	t_2	$\bar{s}_3$	t_1	s_5^+
s_4		$\frac{1}{7}$	$\frac{1}{7}$	$\frac{2}{7}$	$\frac{1}{7}$	$\frac{3}{7}$	$-\frac{1}{7}$
s_2		$\frac{5}{7}$	$-\frac{2}{7}$	$\frac{3}{7}$	$\frac{5}{7}$	$\frac{1}{7}$	$\frac{2}{7}$
		$\frac{6}{7}$	$\frac{20}{7}$	$\frac{12}{7}$	$\frac{6}{7}$	$\frac{11}{7}$	$-\frac{6}{7}$

$\xrightarrow{(s_5=1)}$

	s_1^+	t_2	$\bar{s}_3$	t_1	$\bar{s}_5^+$
s_4	$\frac{2}{7}$	$\frac{1}{7}$	$\frac{2}{7}$	$\frac{1}{7}$	$\frac{3}{7}$	$\frac{1}{7}$
s_2	$\frac{3}{7}$	$-\frac{2}{7}$	$\frac{3}{7}$	$\frac{5}{7}$	$\frac{1}{7}$	$-\frac{2}{7}$
	$\frac{12}{7}$	$\frac{20}{7}$	$\frac{12}{7}$	$\frac{6}{7}$	$\frac{11}{7}$	$\frac{6}{7}$

, optimal; $s = (0, \frac{3}{7}, 1, \frac{2}{7}, 1)^T$,

$$\frac{(p^T s)^2}{s^T s} = \frac{48}{37} = 1.30$$

optimum reached with two changes of basis.

Notes

1. We could also have applied the dual bounded variables method with rejection of pivot elements as explained in [13], pp. 57-58. This method has some advantages over the primal method.

2. From the third tableau above we see that method 3.1 would have given the solution $s = (0, \frac{5}{7}, 1, \frac{1}{7}, 0)^T$ with $\frac{(p^T s)^2}{s^T s} = \frac{12}{25} = 0.48$. Method 3.2 would have given the same solution as 2.3.

3. Using method 2.5 or 2.6, s_3 would not have had an upper bound. Therefore we would have pivoted in the third tableau, the pivot element being determined by s_2 and $s_3(-\frac{5}{7})$. After this s_2 would have been put at its upper bound (1 in 2.5, 2 in 2.6). In both cases s_5 could then also be put at its upper bound. We would have obtained:

in 2.5: $s = (0, 1, \frac{9}{5}, \frac{2}{5}, 1)^T$, $\frac{(p^T s)^2}{s^T s} = \frac{16}{15} = 1.07$;

in 2.6: $s = (0, 2, \frac{16}{5}, \frac{3}{5}, 1)^T$, $\frac{(p^T s)^2}{s^T s} = \frac{54}{65} \doteq 0.83$.

Method 2.1. A special method can be developed for the problem $\text{Max}\{p^T s \mid As \leq 0, \Sigma |s_j| \leq 1\}$ which will not be explained here. After four changes of basis we would have obtained: $s^T = (0, 0, \frac{1}{4}, \frac{1}{8}, \frac{5}{8})$ with $\frac{(p^T s)^2}{s^T s} = \frac{6}{5} = 1.20$.
The solution of the direction problem is slightly more complicated and the resulting direction vector is not better. There does not appear to be any advantage in 2.1 compared to 2.3 or 3.2. Finally method 2.7 would have given the same solution as 3.1.

Although just one example is obviously not sufficient other experiments (with larger problems) have supported the conclusion that methods 1, 3.1, 2.3, 3.2, 2.2 and 2.4 (special version to be explained later on) are the ones to consider and that in this order they are more suited for problems that are more highly nonlinear and have less matrix structure.

3. Unconstrained Optimization

If we have a direction generator and a method for step length determination, then we have a method for solving the unconstrained maximization problem:

$$\max\{f(x)\}$$

Such a method will be (weakly)convergent if at least one point of accumulation of the sequence x^k, say $\bar{x}$, is a stationary point, hence satisfies $\nabla f(\bar{x}) = 0$. The simplest direction generator is

FEASIBLE DIRECTIONS ALGORITHMS

$$s^k = g^k = g(x^k),$$

where we have written g^k for $\nabla f(x^k)$. Together with one dimensional maximization along the direction chosen, $f(x^{k+1}) = \max(f(x^k + \lambda s^k))$, we would have obtained the optimal gradient method (method of steepest ascent) which is strongly convergent (any point of accumulation is stationary) but – due to its slow rate of convergence and erratic behavior in some cases – not very useful in practice.

For a quadratic objective function, $f(x) = p^T x - \frac{1}{2} x^T P x$ with P symmetric and positive definite, we can obtain a finite method by adding the conjugacy relation

$$(s^{k'})^T P s = 0 \qquad (11)$$

to the next direction problem, so that these problems will contain more and more linear relations of type (11) until after a finite number $\leq n$ of steps no usuable s can be found anymore which means that we have arrived at the maximum. The relation (11) is equivalent to

$$(\Delta g^k)^T s = 0, \qquad (12)$$

since $\Delta g^k = g^{k+1} - g^k = -\lambda_k P s^k$.

Relations of type (12) can be used as well in the case of a general function. If this function can "reasonably well" be approximated by a quadratic function adding of relations of type (12) will make sense.

This leads to the following general method for the unconstrained optimization problem:
1. Start with any x^0, $h_0 = 0$, $k = 0$.
2. Solve the direction problem

$$(\Delta g^h)^T s = 0, \quad h = h_0, \; h_0 + 1, \ldots, k-1$$

(for $k = 0$ no relations),

$$(g^k)^T s > 0,$$

by using any direction generator.

3. If 2 gives a usuable direction (as defined below), go to 4; else

if $h_0 \neq k$, then $h_0 := h_0 + 1$, go to 2;

if $h_0 = k$, stop (maximum obtained).

4. Determine $\lambda_k = \max\{\lambda \mid \nabla f(x^k + \lambda s^k)^T s^k \geq 0\}$,
5. $x^{k+1} = x^k + \lambda_k s^k$; calculate g^{k+1} and $f(x^{k+1})$
6. $k := k+1$; go to 2.

A usable direction in the algorithm will be determined by the requirement that $(g^k)^T s^k > \epsilon_k |s^k|$ should hold for some adjustable $\epsilon_k \geq \epsilon > 0$ (ϵ being the near optimality criterion).

If $\theta_k = \dfrac{(g^k)^T s^k}{|g^k| |s^k|}$, the cosine of the angle between g^k and s^k, then it can be proved (see [14]) that the method is weakly convergent provided $\Sigma \theta_k^2$ is divergent. We can adjust ϵ_k in such a way that this is the case. The method obviously has the quadratic termination property (i.e. is finite for $f(x)$ quadratic). Alternatively, instead of increasing h_0 by 1 (i.e. dropping the oldest conjugacy relation) we could restart completely when the direction generator does not supply a usable direction. The last obtained trial solution x^k would be the new starting solution. This would considerably reduce the amount of work per step but it would increase the number of steps. We do not recommend it.

Closely related to the methods of conjugate feasible directions is the variable metric method [4] in which

1. $s^k = H_k g^k$
2. $H_{k+1} = H_k + P_k + Q_k$ ($H_0 = I$ or any other symmetric positive definite matrix) with

$$P_k = -\dfrac{\lambda_k s^k (s^k)^T}{(\Delta g^k)^T s^k} = \dfrac{\lambda_k s^k (s^k)^T}{(g^k)^T s^k}, \quad (13)$$

$$Q_k = - \frac{H_k \Delta g^k (\Delta g^k)^T H_k}{(\Delta g^k)^T H_k \Delta g^k} \qquad (14)$$

It is not difficult to prove that the matrices H_k are positive definite, so that $(g^k)^T s^k > 0$ will always hold. If $f(x) = p^T x - \frac{1}{2} x^T P x$, then

    a. the directions s^k are mutually conjugate (this is caused by adding Q_k at each step);

    b.
$$\sum_{h=0}^{n-1} P_h = P^{-1} \qquad (15)$$

    c. $H_k P s^h = s^h$ for $h = 0, 1, \ldots, k-1$,

so that $H_n = P^{-1}$

Hence in the case of a general function the correction P_k is added to obtain an approximation to the inverse of the matrix of second partial derivatives and Q_k is added to obtain generalized conjugacy.

    Computational experience with the variable metric method has been quite good, although there have been examples in which the method did not appear to converge. McCormick and Pearson [8] have reported that according to their experience a periodic restart is also recommended from a computational point of view. They compared the variable metric method without and with restart with our method 2.2 with restart after each n steps. The latter two behaved better than the variable metric without restart; variable metric with restart needed about 20% fewer steps than 2.2 with restart. They also observed that real progress was only made during the last step of each cycle of n steps. This is a strong argument for our methods without restart but with a moving tableau as explained above. In that case the past is gradually, not suddenly forgotten which makes it different from the variable metric method without restart in which the past is never forgotten. From the reported experience it could be concluded that it is the "conjugacy" of the directions which is of real importance, rather than the approximation of the matrix of second partial derivatives.

However, the idea of making use of information on this matrix could be used as well in our method 2.4. Starting with $P = I$ (so that 2.4 becomes equivalent to 2.2) we perform a cycle of n steps. At the same time we store the vectors s^k and the scalars $-\lambda_k/(\Delta g^k)^T s^k = \lambda_k/(g^k)^T s^k$. Next we recalculate P^{-1} using (15). During the next cycle of n steps the direction problems will be of the form

$$\text{Max}\{(g^k)^T s \mid (\Delta g^h)^T s = 0, \quad h = 0, 1, \ldots, k-1;$$

$$s^T P s \leq 1\},$$

after which P^{-1} is again recalculated, etc.

Recently it has been tried (see [9] and [3]) to get rid of the need to solve the time consuming one-dimensional maximization problems, even at the cost of giving up quadratic termination. Computational experience, although encouraging, is still limited, so that no definite conclusion can be drawn.

It should be emphasized that it is also possible in our method to abandon linear searches if that proves worthwhile. Having a trial solution $\tilde{x}$ with gradient $\tilde{g}$ and a direction $\tilde{s}$ such that $\tilde{g}^T \tilde{s} > 0$ we select a $\lambda' > 0$, $x' = \tilde{x} + \lambda' \tilde{s}$ and calculate g' and $(g')^T \tilde{s}$. The new trial solution $\bar{x}$ to be determined will satisfy $\bar{x} = \tilde{x} + \alpha(x' - \tilde{x})$. Assuming (near) linearity of the gradient we have:

$$0 = \bar{g}^T \tilde{s} \approx (1-\alpha)\tilde{g}^T \tilde{s} + \alpha(g')^T \tilde{s}, \quad \text{so that} \quad \alpha = \frac{\tilde{g}^T \tilde{s}}{-(g'-\tilde{g})^T \tilde{s}}$$

approximation. This way $\bar{x}$ is determined. We then calculate $\bar{g}$ and add $(\bar{g} - \tilde{g})^T s = 0$ to the tableau. Note that quadratic termination will still hold with this method. A good choice of λ' is of course of importance.

4. Linearly Constrained Nonlinear Programming

After what has been said in section 2 and 3 a general method of solution for the linearly constrained nonlinear programming problem can be easily developed. In such a method we start with a feasible trial solution and assume that it lies in the correct set of hyperplanes. We then add and drop conjugacy relations in the way described in section 3 until we are either (near) optimal or we find out during the step length determination that another hyperplane has to be added to the set of binding constraints. In that case the conjugation process is restarted, i.e. old conjugacy relations, if any, are removed from the direction problem. A minor complication is the possibility of zigzagging to a nonoptimal point. This will be prevented by a so-called anti-zigzagging precaution for which we choose that after arrival in a new hyperplane we will require $a_i^T . s = 0$ instead of ≤ 0. This additional requirement will hold for all consequent direction problems and will only be relaxed if otherwise no usable direction can be found. A usable direction in $\tilde{x}$ will then be defined as a direction $s \in S(\tilde{x})$ (see (2)) satisfying $\nabla f(\tilde{x})^T \frac{s}{\|s\|} > \epsilon > 0$ in which ϵ is a predetermined small nonnegative number chosen in such a way that near optimality will follow from $\nabla f(\tilde{x})^T \frac{s}{\|s\|} \leq \epsilon$ for all $s \in S(\tilde{x})$. If $f(x)$ is quadratic we will take $\epsilon = 0$. Given a problem of type (1) the method is as follows:
1. Choose $x^0 \in R$; calculate g^0 and $f(x^0)$; Define the sets $I_0 = J_0 = K_0 = \emptyset$, set $k = 0$.
2. Use any direction generator to find an s^k satisfying:
   (1) $a_i^T . s = 0$ if $i \in I_k$ and ≤ 0 if $i \in I(x^k) - I_k$,

(2) $s_j = 0$ if $j \in J_k$ and ≥ 0 if $j \in J(x^k) - J_k$,
(3) $(\Delta g^h)^T s = 0$ if $h \in K_k$,
(4) $(g^k)^T s > \epsilon \|s\|$.

3. If the direction problem is infeasible, then
   if $I_k \cup J_k \cup H_k \neq \emptyset$ drop the oldest index from either I_k, J_k or H_k; go to 2;
   if $I_k \cup J_k \cup H_k = \emptyset$ stop, the optimal solution has been obtained.

4. Determine $\lambda_k = \min(\lambda'_k, \lambda''_k)$ with

$$\lambda'_k = \max\{\lambda \mid \nabla f(x^k + \lambda s^k)^T s^k \geq 0\},$$

$$\lambda''_k = \max\{\lambda \mid x^k + \lambda s^k \in R\} =$$

$$= \min\{\min(\frac{x_j^k}{-s_j^k} \mid s_j^k < 0), \min(\frac{y_i^k}{-t_i^k} \mid t_i^k < 0)\}.$$

5. If $\lambda_k = \infty$, stop (infinite solution).
6. Calculate $x^{k+1} = x^k + \lambda_k s^k$, g^{k+1} and $f(x^{k+1})$.
7. If $\lambda_k = \lambda'_k$, then $K_{k+1} = K_k \cup \{k\}$; $I_{k+1} = I_k$, $J_{k+1} = J_k$;
   If $\lambda_k = \lambda''_k$, then $I_{k+1} = I_k \cup \{i \mid y_i^{k+1} = 0$ and $y_i^k > 0\}$,
   $J_{k+1} = J_k \cup \{j \mid x_j^{k+1} = 0$ and $x_j^k > 0\}$
   $K_{k+1} = \emptyset$;
   If $\lambda_k = \lambda'_k = \lambda''_k$, then $K_{k+1} = \{k\}$.
8. $k := k+1$, go to 2.

In practice ϵ will be gradually reduced to a lower limit ϵ' corresponding with the near-optimality. The direction generator should be such that for all k $(g^k)^T \frac{s^k}{\|s^k\|} \geq \delta (g^k)^T \frac{s_0^k}{\|s_0^k\|}$, $0 < \delta \leq 1$, independent of k, in which s_0^k is the solution of the direction problem with method 2.2 (gradient projection). A variant with little theoretical justification which has been successful computationally is obtained by not changing K when in step 7 $\lambda_k = \lambda_k''$ holds (hence by putting $K_{k+1} = K_k$ instead of $K_{k+1} = \emptyset$).

It should be noted that we have actually described a class of methods the members of which can vary considerably depending on the direction generator used.

A special method in this class which uses a type 1 direction generator is the linear approximation method which was suggested by Frank and Wolfe [5], however without conjugacy requirements. The latter appear to be essential for a reasonable rate of convergence. A description of the method follows:

1. Choose $x^0 \in R$; calculate g^0;
   Define $K_0 = \emptyset$, set $k = 0$.
2. Solve the linear subproblem:

$$\text{Max}\{(g^k)^T x \mid Ax \leq b, \ x \geq 0, \ \Sigma x_j \leq \mu,$$

$$(\Delta g^h)^T (x - x^h) = 0, \ h \in K_k \} .$$

   Let the solution be $\bar{x}^k$.
3. If $(g^k)^T \bar{x}^k = (g^k)^T x^k$, then
   if $K_k = \emptyset$, stop (optimal solution),
   if $K_k \neq \emptyset$, drop the oldest index from K_k,
   go to 2;
4. $s^k = \bar{x}^k - x^k$, determine
   $\lambda_k = \max \{\lambda \mid \nabla f(x^k + \lambda s^k)^T s^k \geq 0, \ \lambda_k \leq 1\}$
5. $x^{k+1} = x^k + \lambda_k s^k$; calculate g^{k+1}
6. If $\lambda_k < 1$, then $H_{k+1} = H_k \cup \{k\}$;
   if $\lambda_k = 1$, then $H_{k+1} = \emptyset$.
7. $k := k+1$, go to 2 unless some optimality criterion is met.

The number μ in step 2 should be chosen so large that the artificial constraint $\Sigma x_j \leq \mu$ which has been added to prevent infinite solutions in the linear subproblems is redundant for the optimal solution.

A method of feasible directions which is a direct extension of the variable metric method to linearly constrained problems has been suggested by Goldfarb [6]. In this method the matrix H (see section 3) is projected onto the linear subspace $\{s \mid a_{i.}^T s = 0, \; i \in I(\tilde{x}); \; s_j = 0, \; j \in J(\tilde{x})\}$. Note the equality instead of the inequality signs (see note b, p. 99). To simplify the description of the method we will assume that the nonnegativity constraints are contained in the matrix A, so that the problem is

$$\text{Max}\{f(x) \mid Ax \leq b\} .$$

The submatrix of A consisting of those rows which are binding in x^k will be denoted by A_k; with row $a_{i.}^T$ omitted we will obtain the matrix $A_{k,i}$. The method is as follows:

1. Choose $x^0 \in R$ and determine A_0 (m_0 by n) and $I(x^0) = \{i_0, i_1, \ldots, i_{m_0}\}$. Take H_0 symmetric and positive definite, n by n. Calculate for $h = 1, \ldots, m_0$

$$H_0^h = H_0^{h-1} - \frac{H_0^{h-1} a_{i_h} a_{i_h}^T H_0^{h-1}}{a_{i_h}^T H_0^{h-1} a_{i_h}}$$

Set $k = 0$. Calculate g^0.

2. Compute $s^k = H_k^{m_k} g^k$.
3. If $(g^k)^T s^k \leq \epsilon$, then
   if $u^k = \{A_k A_k^T\}^{-1} A_k g^k \geq 0$, go to 8; else
   if $\exists_i (u_i^k < 0)$, then take r such that u_r most negative and drop a_r. by

$$H_k^{m_k-1} = H_k^{m_k} + \frac{P_{A_{k,r}} a_{r.} a_{r.}^T P_{A_{k,r}}}{a_{r.}^T P_{A_{k,r}} a_{r.}}$$

in which $P_A = I - A^T(AA^T)^{-1}A$;
set $m_k := m_{k-1}$, $A_k = A_{k,r}$, go to 2;

4. If $(g^k)^T s^k > \epsilon$, then calculate λ_k.
5. If $\lambda_k = \infty$, stop (infinite solution);
   if $\lambda_k < \infty$, then $x^{k+1} = x^k + \lambda_k s^k$, calculate g^k.
6. If $\lambda_k = \lambda_k''$, then

$$H_{k+1}^{m_k+h} = H_{k+1}^{m_k+h-1} - \frac{H_{k+1}^{m_k+h-1} a_{i_h} \cdot a_{i_h}^T H_{k+1}^{m_k+h-1}}{a_{i_h}^T H_{k+1}^{m_k+h-1} a_{i_h}}$$

for $i_h \in I(x^{k+1}) - I(x^k)$ (say $h = 1, \ldots, \ell$);
$H_{k+1}^{m_k} = H_k^{m_k}$;
$k := k+1$, $m_k := m_k + \ell$, go to 2.

7. If $\lambda_k = \lambda_k'$, then

$$H_{k+1}^{m_k} = H_k^{m_k} - \lambda_k \frac{s^k (s^k)^T}{(\Delta g^k)^T s^k} - \frac{H_k^{m_k} \Delta g^k (\Delta g^k)^T H_k^{m_k}}{(\Delta g^k)^T H_k^{m_k} \Delta g^k},$$

$k := k+1$, go to 2.

8. If an optimality criterion is met, stop; else calculate λ_k, go to 5.

This method has a number of unpleasant properties:

1. Not only does the matrix H have to be updated but the matrix $(A_k A_k^T)^{-1}$ as well.
2. When we have to leave a hyperplane the updating of H is such (see step 3) that new directions will no longer be (near) conjugate to those obtained since the last time a hyperplane was added to the set of active constraints.
3. We can make many small steps in the wrong combination of hyperplanes before step 3 will be executed. This might also be the case in the other methods but to a lesser extent since the way

constraining and conjugacy relations are added ensures that we will leave a wrong combination of hyperplanes after about n steps.
4. Linear dependency of constraints (which often happens in practice) may lead to steps 2 and 3 that do not result in a usable direction. The problem can be overcome at the cost of further complicating the method.
5. Little use can be made of a special structure of the matrix A, if any, and of the simple form of the nonnegativity constraints.

Although some of these disadvantages might be overcome by modifying the method it is to be expected that the other methods of conjugate feasible directions are to be preferred in most cases.
If it turns out to be worthwhile to approximate the matrix of second partial derivatives, then just as in the unconstrained case we can store the vectors s^k and scalars $\lambda_k/(g^k)^T s^k$. After $n - m'$ interior steps in the same combination of hyperplanes (m' being the number of active constraints) we can calculate ΣP_h (see p.108) and use this as a new estimate for P^{-1} in direction generator 2.4. Each time $\lambda_k = \lambda_k'''$ the vectors s^k, already stored, must be left away.

All the methods outlined above are finite in the case of a quadratic objective function ($\epsilon = 0$ in this case).

Another way of using the principle of (pseudo) conjugate directions has been explained by Zangwill [12]. Our method of adding linear relations to the direction problem looks more natural and simpler.

Whether recent developments of methods not requiring line searches will lead to real improvements remains subject to doubt. For the nonlinear programming problems of the economic planning type (large, sparse and structured matrices few and well-bahaved nonlinearities) this is probably not the case.

With regards to the efficiency of the various methods of conjugate feasible directions the reader is referred to the last paragraph of section 2.

5. A partitioning method

An interesting special problem of considerable practical importance is:

$$\max\{p_0(z) + p(z)^T x \mid A(z)x \leq b(z),\ x \geq 0,\ z \in Z \subset E_r\}, \quad (16)$$

which is a linear program if we fix the "decision" variables z. The set Z is supposed to be convex.

This problem can be solved along the following lines:
1. Start with some $z^0 \in Z$ for which the linear program is feasible (if necessary we can modify the problem by adding one additional variable which should ultimately be 0, so that a given z^0 is feasible in the modified problem).
2. At step k solve the linear program for $z = z^k$, resulting in the solution x^k.
3. Find a usable feasible direction in (x^k, z^k).
4. Determine the step length λ_k as well as z^{k+1}.
5. Repeat 2 - 4 for k: = k +1 unless the optimality test is satisfied.

The direction to be found must satisfy the following requirements:

1. $$\sum_{j=1}^{n} \sum_{\ell=1}^{r} \frac{\partial a_{ij}(\tilde{z})}{\partial z_\ell} \tilde{x}_j (s_z)_\ell + \sum_{j=1}^{n} a_{ij}(\tilde{z})(s_x)_j - \sum_{\ell=1}^{r} \frac{\partial b_i(\tilde{z})}{\partial z_\ell}(s_z)_\ell <$$

$$< 0 \ (\text{e.g.} \leq -1) \ \text{if} \ a_{i\cdot}(\tilde{z})^T \tilde{x} = b_i(\tilde{z});$$

2. s_z feasible in $\tilde{z}$ with regards to Z;
3. $(s_x)_j \geq 0$ if $\tilde{x}_j = 0$;

4. $$\sum_{\ell=1}^{r} \frac{\partial p_0(\tilde{z})}{\partial z_\ell}(s_z)_\ell + \sum_{j=1}^{n}\sum_{\ell=1}^{r} \frac{\partial p_j(\tilde{z})}{\partial z_\ell} \tilde{x}_j (s_z)_\ell +$$

$$\sum_{j=1}^{n} p_j(\tilde{z})(s_x)_j > 0.$$

Any direction generator can be used.

The step length λ to be determined next will be nonzero. A golden section search might be applied here. The efficiency of the method will depend heavily on the complexity of the step length calculation. Very little is known about the convergence properties of this method, except in very special circumstances.

If $A(z)$ does not depend on z the problem simplifies considerably. We can then generate the directions using the information of the previous final LP tableau which has the advantage that many of the rows will be non-binding. It is then also possible to require $s_x = 0$ in the direction problem (i.e. assume that the final non-basic variables are correctly 0) as long as a usable direction can be obtained in that way. Note that our method of determining a usable feasible direction guarantees a nonzero step length, so that we do not run into the problem of alternative bases which Rosen encounters in his convex partitioning programming [11]. If, in addition, $p(z)$ does not depend on z, if $b(z)$ and $p_0(z)$ are linear functions of z and if Z is a convex polyhedron, then our method specializes to a partitioning method for linear programming which is closely related but not equivalent to the one, suggested by Beale [1]. Due to the linearity we will now have ≤ 0 instead of < 0 in requirement 1 of the direction problem. The partitioning method will be summarized for the linear program:

$$\max\{p^T x + q^T z \mid Ax + Qz \leq b, \; x \geq 0, \; z \geq 0\} \quad (17)$$

1. Start with $z = \tilde{z}$ and solve the linear program:

$$\max\{p^T x \mid Ax \leq b - Q\tilde{z}, \; x \geq 0\}$$

Indicating the data of the final tableau with an asterisk we have the equivalent problem:

$$\max\{(p^*)^T x^* + (q^*)^T z \mid A^* x^* + Q^* z \leq b^*, \; x^* \geq 0, \quad (18)$$
$$z \geq 0\},$$

which has the solution $x^* = 0$ for $z = \tilde{z}$.

FEASIBLE DIRECTIONS ALGORITHMS

2. Adapt the variables z under the assumption $x^* = 0$ by applying any method of feasible directions starting in $\tilde{z}$ to the problem:

$$\max\{(q^*)^T z \mid Q^* z \leq b^*, \; z \geq 0\} \tag{19}$$

This gives a new estimate $\bar{z}$ ($\bar{z} = \tilde{z}$ is possible). Suppose $y_i^* = 0$ for $i \in I(\bar{z})$.

3. Find a feasible direction satisfying

$$(a_{i\cdot}^*)^T s(x^*) + (q_{i\cdot}^*)^T s(z) \leq 0 \quad \text{if} \quad i \in I(\bar{z})$$

$$s(x^*) \geq 0$$

$$(s(z))_\ell \geq 0 \quad \text{if} \quad z_\ell = 0$$

$$(p^*)^T s(x^*) + (q^*)^T s(z) > 0$$

by applying any direction generator.

4. If the direction problem is infeasible we have obtained the optimum solution (this occurs after a finite number of steps).

5. Determine the step length $\bar{\lambda}$ ($= \max\{\lambda \mid y_1^*(\lambda) \geq 0, z(\lambda) \geq 0\}$).

6. If the step length is determined by one of the z-variables becoming 0 (and $y_i^*(\bar{\lambda}) > 0$ for $i \in I - I(\bar{z})$, then $\bar{z}: = \bar{z} + \bar{\lambda} s(\tilde{z})$, $\tilde{x}^*: = x^* + \bar{\lambda} s(x^*)$, go to 3 ($\{s(x^*)\}_j \geq 0$ need not hold for all j anymore);
If the step length is determined by one of the y^* variables becoming 0, $i \in I - I(\bar{z})$, then $\tilde{z}: = \bar{z} + \bar{\lambda} s(z)$, go to 1 (starting with a reinversion).

Like the other partitioning or decomposition methods this method is particularly useful when A can be decomposed and/or has a special structure (e.g. network structure).

As an example we consider the problem:

		x_1	x_2	x_3	x_4	x_5	x_6	x_7	x_8	z_1	z_2
y_1	4	1		2	1					2	-1
y_2	3		-1	1						-1	2
y_3	6	2	-3		1						2
y_4	3					-1		2		-2	3
y_5	5						1	2	-2		-2
y_6	6						1	-1	3	1	-1
		-2	$\frac{3}{2}$	-2	1	1	-2	-1	0	-1	$-\frac{9}{4}$

Fixing $z_1 = 1$, $z_2 = 1$ we obtain the tableau:

		y_3	x_2	y_1	x_4	x_5	y_5	y_4	y_6	z_1	z_2
x_3	$\frac{1}{2}$	$-\frac{1}{4}$	$\frac{3}{4}$	$\frac{1}{2}$	$\frac{1}{4}$					$\frac{1}{2}$	$-\frac{1}{2}$
y_2	$\frac{3}{2}$	$\frac{1}{4}$	$-\frac{7}{4}$	$-\frac{1}{2}$	$-\frac{1}{4}$					$-\frac{3}{2}$	$\frac{5}{2}$
x_1	2	$\frac{1}{2}$	$-\frac{3}{2}$		$\frac{1}{2}$					1	
x_7	1				$-\frac{1}{2}$		$\frac{1}{2}$			-1	$\frac{3}{2}$
x_6	$\frac{35}{8}$				$\frac{1}{4}$		$\frac{3}{8}$	$\frac{1}{8}$	$\frac{1}{4}$		$-\frac{5}{8}$
x_8	$\frac{7}{8}$				$-\frac{1}{4}$		$-\frac{1}{8}$	$\frac{1}{8}$	$\frac{1}{4}$		$\frac{3}{8}$
	18	$\frac{1}{2}$	0	1	$\frac{5}{2}$	1	$\frac{3}{4}$	$\frac{3}{4}$	$\frac{1}{2}$	1	-3

Next we try to adjust the z_ℓ under the assumption $x^* = 0$. We will use method 3.1. First direction problem: $-s_{z_1} + 3s_{z_2} > 0$. Hence $s_{z_1} = 0$, $s_{z_2} = 1$, $\lambda = \frac{3}{5}$ ($y_2 = 0$), $z^T = (1, \frac{8}{5})$, $y^* = (\frac{4}{5}, 0, 2, \frac{1}{10}, \frac{19}{4}, \frac{13}{20})^T$.

Second direction problem: $-\frac{3}{2}s_{z_1} + \frac{5}{2}s_{z_2} \leq 0$, $-s_{z_1} + 3s_{z_2} > 0$. This results in $s_{z_1} = \frac{5}{3}$, $s_{z_2} = 1$, $\lambda = \frac{6}{5}$ $(x_1 = 0)$; $z^T = (3, \frac{14}{5})$, $y^* = (\frac{2}{5}, 0, 0, \frac{3}{10}, \frac{11}{2}, \frac{1}{5})^T$.

New direction problem: $-\frac{3}{2}s_{z_1} + \frac{5}{2}s_{z_2} \leq 0$, $s_{z_1} \leq 0$, $-s_{z_1} + \frac{5}{2}s_{z_2} > 0$ No solution. We now also take the x^* variables into account. Direction problem:

$\frac{1}{4}s_{y_3}^+ - \frac{7}{4}s_{x_2}^+ - \frac{1}{2}s_{y_1}^+ - \frac{1}{4}s_{x_4}^+ - \frac{3}{2}s_{z_1}^+ + \frac{5}{2}s_{z_2} \leq 0$, $\frac{1}{2}s_{y_3}^+ -$

$-\frac{3}{2}s_{x_2}^+ + \frac{1}{2}s_{x_4}^+ + s_{z_1} \leq 0$, $-\frac{1}{2}s_{y_3}^+ - s_{y_1}^+ - \frac{5}{2}s_{x_4}^+ +$

$(-s_{x_5}^+ - \frac{3}{4}s_{y_5}^+ - \frac{3}{4}s_{y_4}^+ - \frac{1}{2}s_{y_6}^+) - s_{z_1} + 3s_{z_2} > 0$.

Solution: $s_{x_2} = 1$, $s_{w_1} = \frac{3}{2}$, $s_{w_2} = \frac{8}{5}$, $s_j = 0$ for $x_j^* \neq x_2^*$; $\lambda = \frac{1}{3}$ $(x_7 = 0)$, $z^T = (\frac{7}{2}, \frac{10}{3})$.

Although we could calculate y^* and solve a second direction problem we prefer to substitute $z = (\frac{7}{2}, \frac{10}{3})^T$, i.e. $\Delta z = (\frac{5}{2}, \frac{7}{3})^T$ in the last tableau, so that $y^* = (\frac{5}{12}, -\frac{7}{12}, -\frac{1}{2}, 0, \frac{35}{6}, 0)$ and to retain primal feasibility by dual pivot operations. The first pivot is determined by y_2 and x_2, so that we obtain the tableau:

	y_3	y_2	y_1	x_4	x_5	y_5	y_4	y_6	z_1	z_2
x_3	$\frac{1}{6}$	$-\frac{1}{7}$	$\frac{3}{7}$	$\frac{2}{7}$	$\frac{1}{7}$				$-\frac{1}{7}$	$\frac{4}{7}$
x_2	$\frac{1}{3}$	$-\frac{1}{7}$	$-\frac{4}{7}$	$\frac{2}{7}$	$\frac{1}{7}$				$\frac{6}{7}$	$-\frac{10}{7}$
x_1	0	$\frac{2}{7}$	$-\frac{6}{7}$	$\frac{3}{7}$	$\frac{2}{7}$				$+\frac{6}{7}$	$-\frac{15}{7}$
x_7	0					$-\frac{1}{2}$	$\frac{1}{2}$		-1	$\frac{3}{2}$
x_6	$\frac{35}{6}$					$\frac{1}{4}$	$\frac{3}{8}$	$\frac{1}{8}$	$\frac{1}{4}$	$-\frac{5}{8}$
x_8	0					$-\frac{1}{4}$	$-\frac{1}{8}$	$\frac{1}{8}$	$\frac{1}{4}$	$\frac{3}{8}$
	$22\frac{1}{2}$	$\frac{1}{2}$	0	1	$\frac{5}{2}$	1	$\frac{3}{4}$	$\frac{3}{4}$	$\frac{1}{2}$	1	-3

Neither the direction problem in the z-variables only, nor the extended direction problem has a solution, so that we have solved the problem.

REFERENCES

1. E. M. L. Beale, The Simplex Method using Pseudo-basic Variables, pp. 133-148 of R. L. Graves and P. Wolfe (eds.), Recent Advances in Mathematical Programming, McGraw-Hill, 1963.

2. E. W. Cheney and A. A. Goldstein, Newton's Method for Convex Programming and Tchebycheff Approximation, Numerische Mathematik 1 (1959), 253-268.

3. R. Fletcher, A new Approach to Variable Metric Algorithms, report TP 383 of the Atomic Energy Research Establishment, Harwell, U.K., October, 1969.

4. R. Fletcher and M. J. D. Powell, A Rapidly Convergent Descent Method for Minimization, Computer Journal 6 (1963), 163-168.

5. M. Frank and P. Wolfe, An Algorithm for Quadratic Programming, Naval Research Logistics Quarterly 3 (1956), 95-110.

6. D. Goldfarb, Extension of Davidon's Variable Metric Method to Maximization under Linear Inequality and Equality Constraints, SIAM J. Appl. Math. 17 (1969) 739-764.

7. J. E. Kelly, The Cutting Plane Method for Solving Convex Programs, J. SIAM 8 (1960), 703-712.

8. G. P. McCormick and J. D. Pearson, Variable Metric Methods and Unconstrained Optimization, Proceedings of the Keele Conference on Optimization, March 1968, Acad. Press, pp. 307-325.

9. B. A. Murtagh and R. W. H. Sargent, A Constrained Minimization Method with Quadratic Convergence, Proceedings of the Keele Conference on Optimization, March 1968, Acad. Press, pp. 215-346.

10. J. B. Rosen, The Gradient Projection Method for Nonlinear Programming, Part I: Linear Constraints, J. SIAM 8 (1960), 181-217.

11. J. B. Rosen, Convex Partitioning Programming, pp. 159-176 of R. L. Graves and P. Wolfe (eds.), Recent Advances in Mathematical Programming, McGraw-Hill, 1963.

12. W. I. Zangwill, Nonlinear Programming, Prentice-Hall, 1969.

13. G. Zoutendijk, Methods of Feasible Directions, Elsevier, 1960.

14. G. Zoutendijk, Nonlinear Programming, Computational Methods, in J. Abadie (ed.), Nonlinear and Integer Programming, North-Holland Publ. Co., 1970, pp. 37-86.

Numerical Techniques in Mathematical Programming

R. H. BARTELS, G. H. GOLUB, AND M. A. SAUNDERS

ABSTRACT

The application of numerically stable matrix decompositions to minimization problems involving linear constraints is discussed and shown to be feasible without undue loss of efficiency.

Part A describes computation and updating of the product-form of the LU decomposition of a matrix and shows it can be applied to solving linear systems at least as efficiently as standard techniques using the product-form of the inverse.

Part B discusses orthogonalization via Householder transformations, with applications to least squares and quadratic programming algorithms based on the principal pivoting method of Cottle and Dantzig.

Part C applies the singular value decomposition to the nonlinear least squares problem and discusses related eigenvalue problems.

Table of Contents

Introduction

    A. THE USE OF LU DECOMPOSITION IN EXCHANGE ALGORITHMS

1.   LU Decomposition
2.   Exchange Algorithms
3.   Updating the LU Decomposition
4.   Round-off Considerations
5.   Efficiency Considerations
6.   Storage Considerations
7.   Accuracy Considerations

    B. THE QR DECOMPOSITION AND QUADRATIC PROGRAMMING

8.   Householder Triangularization
9.   Projections
10.   Orthogonalization with Respect to Positive Definite Forms
11.   Linear Least Squares and Quadratic Programming
12.   Positive Definite Programming
13.   Semi-Definite Programming

C. THE SVD AND NONLINEAR LEAST SQUARES

14. The Singular Value Decomposition
15. Nonlinear Least Squares
16. Modified Eigensystems

References

Introduction

This paper describes the application of numerically stable matrix decompositions to minimization problems involving linear constraints. Algorithms for solving such problems are fundamentally techniques for the solution of selected systems of linear equations, and during the last fifteen years there has been a major improvement in the understanding of these and other linear algebraic problems. We show here that methods which have been analysed by various workers and proven to be numerically stable may be employed in mathematical programming algorithms without undue loss of efficiency.

Part A describes means for computing and updating the product-form of the LU decomposition of a matrix. The solution of systems of equations by this method is shown to be stable and to be at least as efficient as standard techniques which use the product-form of the inverse.

In Part B we discuss orthogonalization via Householder transformations. Applications are given to least squares and quadratic programming algorithms based on the principal pivoting method of Cottle and Dantzig [5]. For further applications of stable methods to least squares and quadratic programming, reference should be made to the recent work of R. J. Hanson [13] and of J. Stoer [26] whose algorithms are based on the gradient projection method of J. B. Rosen [24].

In Part C the application of the singular value decomposition to the nonlinear least squares problem is discussed, along with related eigenvalue problems.

A. THE USE OF LU DECOMPOSITION IN EXCHANGE ALGORITHMS

1. LU Decomposition

If B is an $n \times n$, nonsingular matrix, there exists a permutation matrix π, a lower-triangular matrix L with ones on the diagonal, and an upper-triangular matrix U such that

$$\pi B = LU . \qquad (1.1)$$

It is possible to choose π, L, and U so that all elements of L are bounded in magnitude by unity.

A frequently-used algorithm for computing this decomposition is built around Gaussian elimination with row interchanges. It produces the matrices π and L in an implicit form as shown:

For $k = 1, 2, \ldots, n-1$ in order carry out the following two steps:

> Find an element in the k-th column of B, on (1.2)
> or below the diagonal, which has maximal
> magnitude. Interchange the k-th row with the
> row of the element found.

> Add an appropriate multiple of the resulting (1.3)
> k-th row to each row below the k-th in order
> to create zeros below the diagonal in the k-th
> column.

Each execution of the first step (1.2), in matrix notation, amounts to the premultiplication of B by a suitable permutation matrix π_k. The following step (1.3) may be regarded as the premultiplication of B by a matrix Γ_k of the form

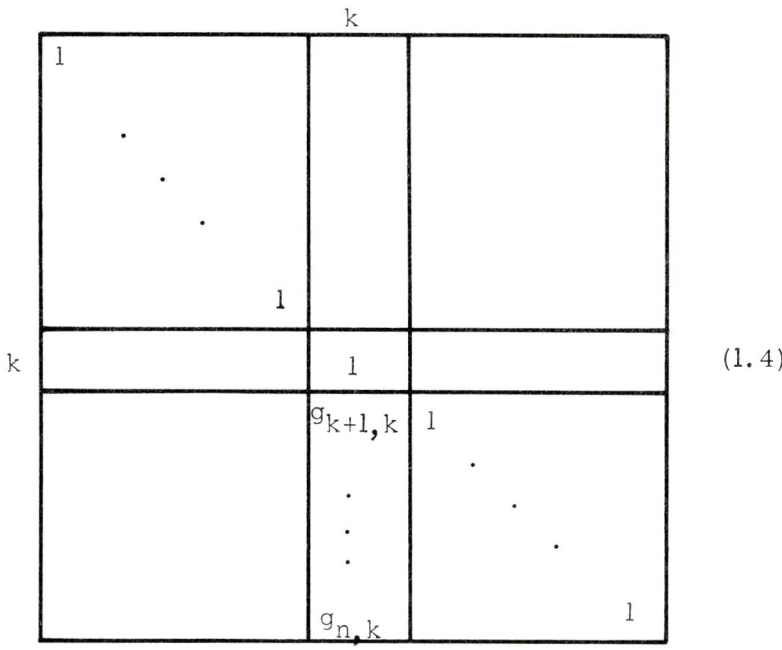

(1.4)

where $|g_{i,k}| \leq 1$ for each $i = k+1, \ldots, n$.

By repeating the two steps n-1 times, B is transformed into U. And at the same time the matrix $(L^{-1}\pi)$ is collected in product form

$$L^{-1}\pi = \Gamma_{n-1}\pi_{n-1} \cdots \Gamma_1 \pi_1 . \tag{1.5}$$

This algorithm requires $n^3/3 + O(n^2)$ multiplication/division operations and again this many addition/subtraction operations. Both U and all of the $g_{i,j}$ can be stored in the space which was originally occupied by B. An additional n locations are required for the essential information contained in the π_k.

2. Exchange Algorithms

Many algorithms require the solving of a sequence of linear equations

$$B^{(i)} x = v^{(i)} \qquad (2.1)$$

for which each $B^{(i)}$ differs from its predecessor in only one column. Examples of such algorithms are: the simplex method, Stiefels' exchange method for finding a Chebyshev solution to an overdetermined linear equation system, and adjacent-path methods for solving the complementary-pivot programming problem.

Given that $B^{(0)}$ has a decomposition of the form

$$B^{(0)} = L^{(0)} U^{(0)}, \qquad (2.2)$$

where $U^{(0)}$ is upper-triangular, and given that ${L^{(0)}}^{-1}$ has been stored as a product

$${L^{(0)}}^{-1} = \Gamma_{n-1}^{(0)} \pi_{n-1}^{(0)} \cdots \Gamma_{1}^{(0)} \pi_{1}^{(0)}, \qquad (2.3)$$

the initial system of the sequence is readily solved: Set

$$y = {L^{(0)}}^{-1} v^{(0)}, \qquad (2.4)$$

and then back-solve the triangular system

$$U^{(0)} x = y. \qquad (2.5)$$

3. Updating the LU Decomposition

Let the column r_0 of $B^{(0)}$ be replaced by the column vector $a^{(0)}$. So long as we revise the ordering of the unknowns accordingly, we may insert $a^{(0)}$ into the last column position, shifting columns $r_0 + 1$ through n of $B^{(0)}$ one position to the left to make room. We will call the result $B^{(1)}$, and we can easily check that it has the decomposition

$$B^{(1)} = L^{(0)} H^{(1)} , \qquad (3.1)$$

where $H^{(1)}$ is a matrix which is upper-Hessenberg in its last $n - r_0 + 1$ columns and upper-triangular in its first $r_0 - 1$ columns. That is, $H^{(1)}$ has the form

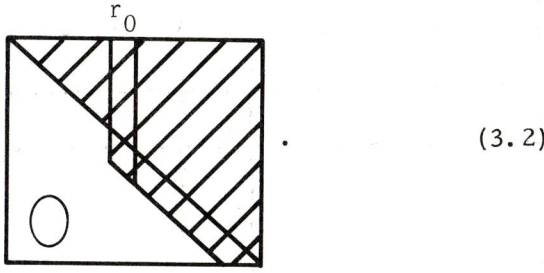

$$(3.2)$$

The first $r_0 - 1$ columns of $H^{(1)}$ are identical with those of $U^{(0)}$. The next $n - r_0$ are identical with the last $n - r_0$ columns of $U^{(0)}$. And the last column of $H^{(1)}$ is the vector $L^{(0)^{-1}} a^{(0)}$.

$H^{(1)}$ can be reduced to upper-triangular form by Gaussian elimination with row interchanges. Here, however, we need only concern ourselves with the interchanges of pairs of adjacent rows. Thus $U^{(1)}$ is gotten from $H^{(1)}$ by applying a sequence of simple transformations:

$$U^{(1)} = \Gamma^{(1)}_{n-1} \pi^{(1)}_{n-1} \cdots \Gamma^{(1)}_{r_0} \pi^{(1)}_{r_0} H^{(1)} , \qquad (3.3)$$

where each $\Gamma^{(1)}_i$ has the form

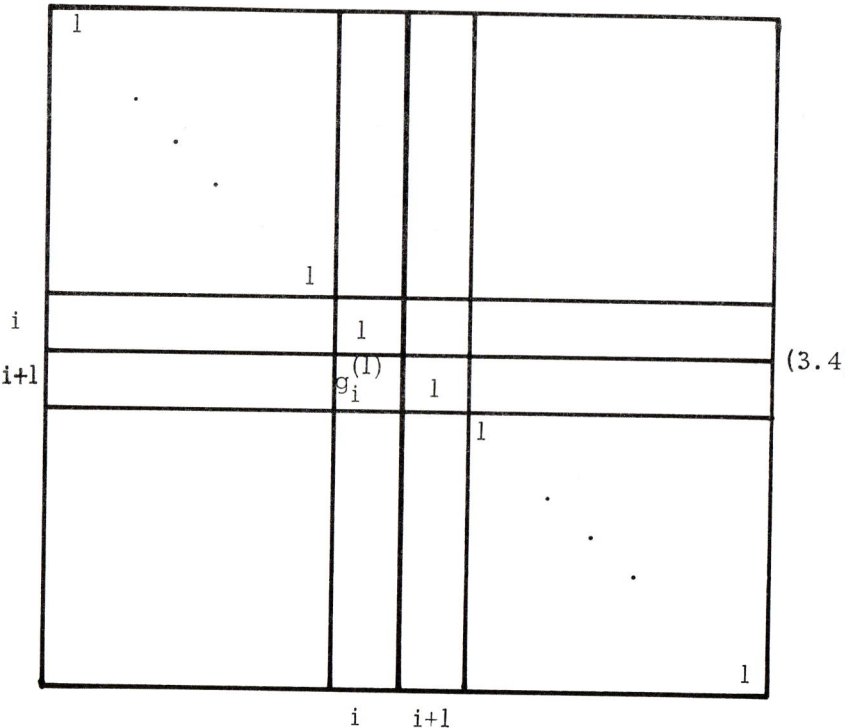
(3.4)

and each $\pi_i^{(1)}$ is either the identity matrix or the identity with the i-th and i+1-st rows exchanged, the choice being made so that $|g_i^{(1)}| \leq 1$.

The essential information in all of these transformations can be stored in $n - r_0$ locations plus an additional $n - r_0$ bits (to indicate the interchanges). If we let

$$L^{(1)^{-1}} = \Gamma^{(1)}_{n-1}\pi^{(1)}_{n-1} \cdots \Gamma^{(1)}_{r_0}\pi^{(1)}_{r_0} L^{(0)^{-1}}, \qquad (3.5)$$

then we have achieved the decomposition

$$B^{(1)} = L^{(1)} U^{(1)}. \qquad (3.6)$$

The transition from $B^{(i)}$ to $B^{(i+1)}$ for any i is to be made exactly as was the transition from $B^{(0)}$ to $B^{(1)}$. Any system of linear equations involving the matrix $B^{(i)}$ for any i is to be solved according to the steps given in (2.4) and (2.5).

4. Round-off Considerations

For most standard computing machines the errors in the basic arithmetic operations can be expressed as follows:

$$f\ell(a \pm b) = a(1 + \varepsilon_1) \pm b(1 + \varepsilon_2)$$

$$f\ell(a \times b) = ab(1 + \varepsilon_3) \qquad (4.1)$$

$$f\ell(a/b) = (a/b)(1 + \varepsilon_4) ,$$

where $|\varepsilon_i| < \beta^{1-t}$. Here β stands for the base of the number system in which machine arithmetic is carried out and t is the number of significant figures which the machine retains after each operation. The notation $f\ell(a \text{ "op" } b)$ stands for the result of the operation "op" upon the two, normal-precision floating-point numbers a and b when standard floating-point arithmetic is used.

The choice of an LU decomposition for each $B^{(i)}$ and the particular way in which this decomposition is updated were motivated by the desire to find a way of solving a sequence of linear equations (2.1) which would retain a maximum of information from one stage to the next in the sequence and which would be as little affected by round-off errors as possible. Under the assumption that machine arithmetic behaves as given in (4.1), the processes described in Sections 2 and 3 are little affected by round-off errors. The efficiency of the processes will vary from algorithm to algorithm, but we will argue in a subsequent section that the processes should cost roughly as much as those based upon product-form inverses of the $B^{(i)}$.

We will now consider the round-off properties of the basic steps described in Sections 2 and 3.

The computed solution to the triangular system of linear equations

$$U^{(i)} x = y \qquad (4.1)$$

can be shown, owing to round-off errors, to satisfy a perturbed system

$$(U^{(i)} + \delta U^{(i)}) x = y . \qquad (4.3)$$

It is shown in Forsythe and Moler [9] that

$$\frac{\|\delta U^{(i)}\|}{\|U^{(i)}\|} \leq \frac{n(n+1)}{2} (1.01) \beta^{1-t} \qquad (4.4)$$

where $\|\ldots\|$ denotes the infinity norm of a matrix, and thus round-off errors in the back-solution of a triangular system of linear equations may be regarded as equivalent to relatively small perturbations in the original system.

Similarly, the computed L and U obtained by Gaussian elimination with row interchanges from an upper-Hessenberg matrix H satisfy the perturbed equation

$$H + \delta H = LU , \qquad (4.5)$$

where Forsythe and Moler show that

$$\frac{\|\delta H\|}{\|H\|} \leq n^2 \rho \beta^{1-t} , \qquad (4.6)$$

and Wilkinson [28] establishes that $\rho \leq n$. Thus, the computational process indicated in (3.3) can be regarded as introducing only relatively small perturbations in each of the $H^{(i)}$.

Similar results hold for the initial LU decomposition (2.2) with a different bound for ρ. The reader is referred again to Forsythe and Moler.

The most frequent computational step in the processes which we have described is the application of one Gaussian elimination step Γ to a column vector v:

$$w = \Gamma v = \begin{pmatrix} 1 & & & & & \\ & \ddots & & & & \\ & & 1 & & & \\ & & & 1 & & \\ & & & & \ddots & \\ & & & & & 1 \\ & & g & & 1 & \\ & & & & & 1 \\ & & & & & & \ddots \\ & & & & & & & 1 \end{pmatrix} \begin{pmatrix} v_1 \\ \vdots \\ v_{i-1} \\ v_i \\ v_{i+1} \\ \vdots \\ v_{j-1} \\ v_j \\ v_{j+1} \\ \vdots \\ v_n \end{pmatrix} \quad (4.7)$$

The computed vector w satisfies

$$w_k = v_k \quad \text{for } k \neq j$$

$$\begin{aligned} w_j &= fl\,(fl\,(gv_i) + v_j) \\ &= gv_i(1 + \varepsilon_3)(1 + \varepsilon_1) + v_j(1 + \varepsilon_2) \\ &= gv_i + v_j + gv_i(\varepsilon_1 + \varepsilon_3 + \varepsilon_1\varepsilon_3) + v_j\varepsilon_2 \,. \end{aligned} \quad (4.8)$$

Thus we may regard the computed vector w as the exact result of a perturbed transformation

$$w = (\Gamma + \delta\Gamma)v, \qquad (4.9)$$

where

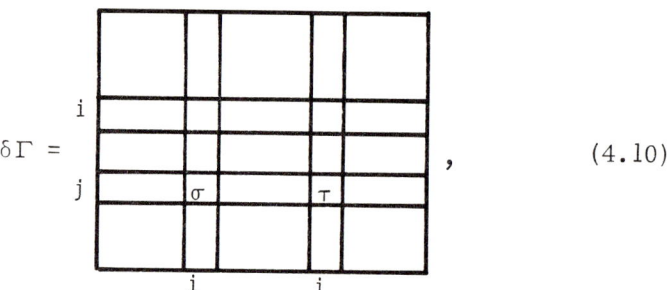
$$\delta\Gamma = \quad , \qquad (4.10)$$

and

$$\sigma = g(\varepsilon_1 + \varepsilon_3 + \varepsilon_1\varepsilon_3)$$
$$\tau = \varepsilon_2. \qquad (4.11)$$

Therefore we have

$$\frac{\|\delta\Gamma\|}{\|\Gamma\|} \leq \frac{|g| \, |\varepsilon_1 + \varepsilon_3 + \varepsilon_1\varepsilon_3| + |\varepsilon_2|}{1 + |g|}, \qquad (4.12)$$

where the right-hand side is bounded, since $|g| \leq 1$, according to

$$\frac{\|\delta\Gamma\|}{\|\Gamma\|} \leq \beta^{1-t}[3 + \beta^{1-t}] < 3.01\beta^{1-t} \text{ (say)}. \qquad (4.13)$$

Hence, the computations which we perform using transformations (4.7) also introduce relatively small perturbations into the quantities which we manipulate.

It is precisely with regard to such transformations that we feel our method of computation, based upon LU decompositions, is superior to methods based upon the inverses of the matrices $B^{(i)}$. Such methods use transformations of the form

$$\begin{array}{c} \begin{pmatrix} 1 & & & \eta_1 & & & \\ & \ddots & & \vdots & & & \\ & & 1 & \eta_{k-1} & & & \\ & & & \eta_k & & & \\ & & & \eta_{k+1} & 1 & & \\ & & & \vdots & & \ddots & \\ & & & \eta_K & & & 1 \end{pmatrix} \quad k \end{array} \quad (4.14)$$

These are applied to each column in $B^{(i-1)^{-1}}$ to produce $B^{(i)^{-1}}$; or alternatively, in product-form methods, they are applied to the vector $v^{(i)}$ to produce the solution to system (2.1). As such, they involve successive computations of the form (4.7). Each such computation may be regarded as satisfying (4.9). But, since the η_j may be unrestricted in magnitude, no bound such as (4.13) can be fixed.

5. Efficiency Considerations

As we have already pointed out, it requires

$$n^3/3 + O(n^2) \qquad (5.1)$$

multiplication-type operations to produce an initial LU decomposition (2.2) To produce the product-form inverse of an $n \times n$ matrix, on the other hand, requires

$$n^3/2 + O(n^2) \qquad (5.2)$$

operations.

The solution for any system (2.1) must be found according to the LU-decomposition method by computing

$$y = L^{(i)^{-1}} v^{(i)} \qquad (5.3)$$

followed by solving

$$U^{(i)} x = y . \qquad (5.4)$$

The application of $L^{(0)^{-1}}$ to $v^{(i)}$ in (5.3) will require

$$\frac{n(n-1)}{2} \qquad (5.5)$$

operations. The application of the remaining transformations in $L^{(i)^{-1}}$ will require at most

$$i(n-1) \qquad (5.6)$$

operations. Solving (5.4) costs

NUMERICAL TECHNIQUES IN PROGRAMMING

$$\frac{n(n+1)}{2} \tag{5.7}$$

operations. Hence, the cost of (5.3) and (5.4) together is not greater than

$$n^2 + i(n-1) \tag{5.8}$$

operations, and a reasonable expected figure would be $n^2 + \frac{i}{2}(n-1)$.

On the other hand, computing the solution to (2.1) using the usual product form of $B^{(i)^{-1}}$ requires the application of n+i transformations of type (4.12) to $v^{(i)}$ at a cost of

$$n^2 + in \tag{5.9}$$

operations.

If a vector $a^{(i)}$ replaces column r_i in $B^{(i)}$, then the updating of $B^{(i)^{-1}}$ requires that the vector

$$z = B^{(i)^{-1}} a^{(i)} \tag{5.10}$$

be computed. This will cost $n^2 + in$ operations, as shown in (5.9). Then a transformation of form (4.14) must be produced from z, and this will bring the total updating cost to

$$n^2 + (i+1)n . \tag{5.11}$$

The corresponding cost for updating the LU decomposition will be not more than

$$\frac{n(n-1)}{2} + i(n-1) \tag{5.12}$$

137

operations to find $L^{(i)^{-1}} a^{(i)}$, followed by at most

$$\frac{n(n+1)}{2} \qquad (5.13)$$

operations to reduce $H^{(i+1)}$ to $U^{(i+1)}$ and generate the transformations of type (3.4) which effect this reduction. This gives a total of at most

$$n^2 + i(n-1) \qquad (5.14)$$

operations, with an expected figure closer to $n^2 + \frac{i}{2}(n-1)$.

Hence, in every case the figures for the LU decomposition: (5.14), (5.8), and (5.1) are smaller than the corresponding figures (5.11), (5.9), and (5.2) for the product-form inverse method.

6. Storage Considerations

All computational steps for the LU-decomposition method may be organized according to the columns of the matrices $B^{(i)}$. For large systems of data this permits a two-level memory to be used, with the high-speed memory reserved for those columns being actively processed.

The organization of Gaussian elimination by columns is well-known, and it is clear how the processes (5.3) may be similarly arranged. Finally, the upper-triangular systems (5.4) can be solved columnwise as indicated below in the 4 × 4 case:

$$\begin{pmatrix} u_{11} & u_{12} & u_{13} & u_{14} \\ 0 & u_{22} & u_{23} & u_{24} \\ 0 & 0 & u_{33} & u_{34} \\ 0 & 0 & 0 & u_{44} \end{pmatrix} \begin{pmatrix} x_1 \\ x_2 \\ x_3 \\ x_4 \end{pmatrix} = \begin{pmatrix} y_1 \\ y_2 \\ y_3 \\ y_4 \end{pmatrix}. \qquad (6.1)$$

Bring the y vector and the last column of U into high-speed memory. Set $x_4 = y_4/u_{44}$. Set $y_i' = y_i - u_{i4}x_4$ for $i = 3, 2, \ldots$ This leaves us with the following 3×3 system:

$$\begin{pmatrix} u_{11} & u_{12} & u_{13} \\ 0 & u_{22} & u_{23} \\ 0 & 0 & u_{33} \end{pmatrix} \begin{pmatrix} x_1 \\ x_2 \\ x_3 \end{pmatrix} = \begin{pmatrix} y_1' \\ y_2' \\ y_3' \end{pmatrix}. \quad (6.2)$$

We process it as suggested in the 4×4 case, using now the third column of U to produce x_3. Repeat as often as necessary.

In the event that the matrices $B^{(i)}$ are sparse as well as large, we wish to organize computations additionally in such a way that this sparseness is preserved as much as possible in the decompositions. For the initial decomposition (2.2), for example, we would wish to order the columns of $B^{(0)}$ in such a way that the production of $L^{(0)^{-1}}$ and $U^{(0)}$ introduce as few new nonzero elements as possible. And at subsequent stages, if there is a choice in the vector $a^{(i)}$ which is to be introduced as a new column into the matrix $B^{(i)}$ to produce $B^{(i+1)}$, it may be desirable to make this choice to some extent on sparseness considerations.

It is not generally practical to demand a minimum growth of nonzero elements over the entire process of computing the initial decomposition. However, one can easily demand that, having processed the first k-1 columns according to (1.2) and (1.3), the next column be chosen from those remaining in such a way as to minimize the number of nonzero elements generated in the next execution of steps (1.2) and (1.3). See, for example, Tewarson [27]. Choice of the next column may also be made according to various schemes of "merit"; e.g., see Dantzig et al. [6].

The introduction of new nonzero elements during the process of updating the i-th decomposition to the i+1-st depends upon the nonzero elements in $L^{(i)^{-1}} a^{(i)}$ over those in $a^{(i)}$, (6.3)

and

the number r_i of the column to be removed from $B^{(i)}$. (6.4)

No freedom is possible in the reduction of $H^{(i+1)}$ to $U^{(i+1)}$ once $a^{(i)}$ has been chosen and the corresponding r_i has been determined.

The growth (6.3) can be determined according to the techniques outlined in Tewarson's paper, at a cost for each value of i, however, which is probably unacceptable. The more important consideration is (6.4). The larger the value of r_i, the fewer elimination steps must be carried out on $H^{(i+1)}$ and the less chance there is for nonzero elements to be generated. Again, however, the determination of the value of r_i corresponding to each possible choice of $a^{(i)}$ may prove for most algorithms to be unreasonably expensive.

7. Accuracy Considerations

During the execution of an exchange algorithm it sometimes becomes necessary to ensure the highest possible accuracy for a solution to one of the systems (2.1). High accuracy is generally required of the last solution in the sequence, and it may be required at other points in the sequence when components of the solution, or numbers computed from them, approach critical values. For example, in the simplex method inner products are taken with the vector of simplex multipliers, obtained by solving a system involving $B^{(i)}$, and each of the non-basic vectors. The computed values are then subtracted from appropriate components of the cost vector, and the results are compared to zero. Those which are of one sign have importance in determining how the matrix $B^{(i+1)}$ is to be obtained from $B^{(i)}$. The value zero, of course, is critical.

The easiest way of ensuring that the computed solution to a system

$$Bx = v \qquad (7.1)$$

has high accuracy is by employing the technique of iterative refinement [9, Chapter 13]. According to this technique, if $x^{(0)}$ is any sufficiently good approximation to the solution of (7.1) (for example, a solution produced directly via the LU-decomposition of B) then improvements may be made by computing

$$r^{(j)} = v - Bx^{(j)}, \qquad (7.2)$$

solving

$$Bz^{(j)} = r^{(j)}, \qquad (7.3)$$

and setting

$$x^{(j+1)} = x^{(j)} + z^{(j)} \qquad (7.4)$$

for $j = 0, 1, 2, \ldots$ until $\|z^{(j)}\|$ is sufficiently small. The inner products necessary to form the residuals (7.2) <u>must</u> be computed in double-precision arithmetic. If this rule is observed, however, and if the condition of the system, measured as

$$\text{cond}(B) = \|B\| \, \|B^{-1}\|, \qquad (7.5)$$

is not close to β^{t-1}, the refinement process can be counted on to terminate in a few iterations. The final vector $x^{(j)}$ will then be as accurate a solution to (7.1) as the significance of the data in B and v warrant.

Step (7.3) is most economically carried out, of course, via the same LU-decomposition which was used to produce $x^{(0)}$. If this is done, each repetition of steps (7.2) through (7.4) will cost only $O(n^2)$ operations. The alternative approach of producing a highly accurate solution to (7.1) by solving the system entirely in double-precision arithmetic is generally more expensive than iterative refinement by a factor of n.

B. THE QR DECOMPOSITION AND QUADRATIC PROGRAMMING

8. Householder Triangularization

Householder transformations have been widely discussed in the literature. In this section we are concerned with their use in reducing a matrix A to upper-triangular form, and in particular we wish to show how to update the decomposition of A when its columns are changed one by one. This will open the way to the implementation of efficient and stable algorithms for solving problems involving linear constraints.

Householder transformations are symmetric orthogonal matrices of the form $P_k = I - \beta_k u_k u_k^T$ where u_k is a vector and $\beta_k = 2/(u_k^T u_k)$. Their utility in this context is due to the fact that for any non-zero vector a it is possible to choose u_k in such a way that the transformed vector $P_k a$ is zero except for its first element. Householder [15] used this property to construct a sequence of transformations to reduce a matrix to upper-triangular form. In [29], Wilkinson describes the process and his error analysis shows it to be very stable.

Thus if $A = (a_1, \ldots, a_n)$ is an $m \times n$ matrix of rank r, then at the k-th stage of the triangularization ($k < r$) we have

$$A^{(k)} = P_{k-1} P_{k-2} \cdots P_0 A = \begin{pmatrix} R_k & S_k \\ 0 & T_k \end{pmatrix}$$

where R_k is an upper-triangular matrix of order k. The next step is to compute $A^{(k+1)} = P_k A^{(k)}$ where P_k is chosen to reduce the first column of T_k to zero except for the first component. This component becomes the last diagonal element of R_{k+1} and since its modulus is equal to the Euclidean length of the first column of T_k it should in general be maximized by a suitable interchange of the columns of $\begin{pmatrix} S_k \\ T_k \end{pmatrix}$. After r steps, T_r will be effectively zero (the length of each of its columns will be smaller than some tolerance) and the process stops.

Hence we conclude that if rank(A) = r then for some permutation matrix π the Householder decomposition (or "QR decomposition") of A is

$$QA\pi = P_{k-1}P_{k-2} \cdots P_0 A = \begin{pmatrix} \overbrace{R}^{r} & \overbrace{S}^{n-r} \\ 0 & 0 \end{pmatrix}$$

where $Q = P_{r-1}P_{r-2} \cdots P_0$ is an m × m orthogonal matrix and R is upper-triangular and nonsingular.

We are now concerned with the manner in which Q should be stored and the means by which Q, R, S may be updated if the columns of A are changed. We will suppose that a column a_p is deleted from A and that a column a_q is added. It will be clear what is to be done if only one or the other takes place.

Compact Method:

Since the Householder transformations P_k are defined by the vectors u_k the usual method is to store the u_k's in the area beneath R, with a few extra words of memory being used to store the β_k's and the diagonal elements of R. The product Qz for some vector z is then easily computed in the form $P_{r-1}P_{r-2} \cdots P_0 z$ where, for example, $P_0 z = (I - \beta_0 u_0 u_0^T)z = z - \beta_0(u_0^T z)u_0$. The updating is best accomplished as follows. The first p-1 columns of the new R are the same as before; the other columns p through n are simply overwritten by columns $a_{p+1}, \ldots, a_n, a_q$ and transformed by the product

$P_{p-1}P_{p-2} \cdots P_0$ to obtain a new $\begin{pmatrix} S_{p-1} \\ T_{p-1} \end{pmatrix}$; then T_{p-1} is triangularized as usual. This method allows Q to be kept in product form always, and there is no accumulation of errors. Of course, if $p = 1$ the complete decomposition must be re-done and since with $m \geq n$ the work is roughly proportional to $(m-n/3)n^2$ this can mean a lot of work. But if $p \cong n/2$ on the average, then only about 1/8 of the original work must be repeated each updating.

Explicit Method:

The method just given is probably best when $m \gg n$. Otherwise we propose that Q should be stored explicitly and that the updating be performed as follows:

(1) The initial Q can be computed by transforming the identity matrix thus:

$$P_{r-1}P_{r-2} \cdots P_0(A\pi \mid I_m) = \begin{pmatrix} R & S & \bigm| & Q \\ 0 & 0 & \bigm| & \end{pmatrix}.$$

(2) If a_q is added to A then compute $s_q = Qa_q$ and add it to the end of $\begin{pmatrix} S \\ 0 \end{pmatrix}$.

(3) Delete a_p where applicable $(p < r)$. This normally means just updating the permutation vector used to describe π.

(4) The initial situation

$QA\pi =$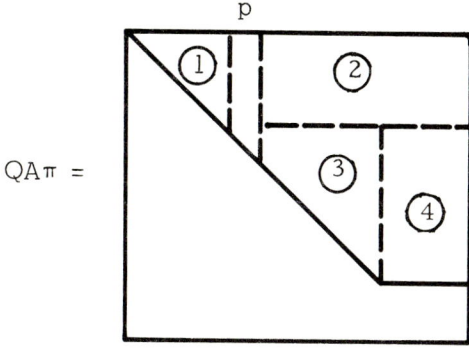

has thus been changed to

$Q\overline{A}\pi =$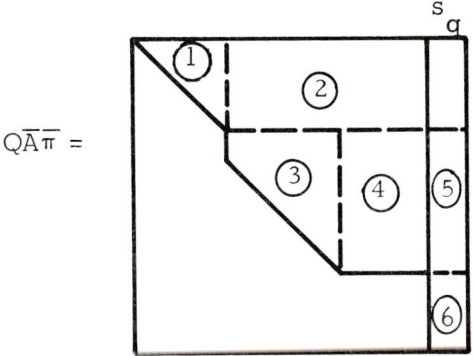

where the areas ①, ②, ③, ④ are the same as before. This is analogous to the Hessenberg form encountered in updating LU decompositions. We now employ a sequence of (r-p) plane rotations, as used by Givens and analyzed by Wilkinson [30], to reduce the subdiagonal of area ③ to zero. This changes areas ③, ④ and ⑤, and the corresponding rows of Q must also be transformed. Since the plane rotations are elementary orthogonal transformations, the latter step produces a new matrix Q^* which is also orthogonal, and the work necessary is approximately proportional to $2mn + n^2$.

(5) Finally, a single Householder transformation P_r is applied to produce $\overline{Q} = P_r Q^*$, where this transformation is the one which reduces area ⑥ to zeros except for the first element. The work involved is proportional to $2(m-n)m$.

Thus the transformation $\overline{Q}$ reduces $\overline{A\pi}$ to a new upper-triangular form, and the original transformations $P_0, \ldots, P_{r-1}$, the plane rotations, and the final Householder transformation may all be discarded since the required information is all stored in $\overline{Q}$. The total work involved is roughly proportional to $(2mn + n^2) + 2(m-n)m = 2m^2 + n^2$ and the stability of the orthogonal transformations is such that accumulation of rounding errors during repeated applications of the updating process should be very slight.

9. Projections

In optimization problems involving linear constraints it is often necessary to compute the projections of some vector either into or orthogonal to the space defined by a subset of the constraints (usually the current "basis"). In this section we show how Householder transformations may be used to compute such projections. As we have shown, it is possible to update the Householder decomposition of a matrix when the number of columns in the matrix is changed, and thus we will have an efficient and stable means of orthogonalizing vectors with respect to basis sets whose component vectors are changing one by one.

Let the basis set of vectors $a_1, a_2, \ldots, a_n$ form the columns of an $m \times n$ matrix A, and let S_r be the sub-space spanned by $\{a_i\}$. We shall assume that the first r vectors are linearly independent and that $\text{rank}(A) = r$. In general, $m \geq n \geq r$, although the following is true even if $m < n$.

Given an arbitrary vector z we wish to compute the projections

$$u = Pz, \quad v = (I - P)z$$

for some projection matrix P, such that

(a) $\quad z = u + v$

(b) $\quad u^T v = 0$

(c) $\quad u \in S_r$ (i.e., $\exists x$ such that $Ax = u$)

(d) $\quad v$ is orthogonal to S_r (i.e., $A^T v = 0$).

One method is to write P as AA^+ where A^+ is the $n \times m$ generalized inverse of A, and in [7] Fletcher shows how A^+ may be updated upon changes of basis. In contrast, the method based on Householder transformations does not deal with A^+ explicitly but instead keeps AA^+ in factorized form and simply updates the orthogonal matrix required to produce this form. Apart from being more stable and just as efficient, the method has the added advantage that there are always two orthonormal sets of vectors available, one spanning S_r and the other spanning its complement.

As already shown, we can construct an $m \times n$ orthogonal matrix Q such that

$$QA = \begin{pmatrix} \overbrace{R}^{r} & \overbrace{S}^{n-r} \\ 0 & 0 \end{pmatrix}$$

where R is an $r \times r$ upper-triangular matrix. Let

$$w = Qz = \begin{pmatrix} w_1 \\ w_2 \end{pmatrix} \begin{matrix} \}r \\ \}m-r \end{matrix} \qquad (9.1)$$

and define

$$u = Q^T \begin{pmatrix} w_1 \\ 0 \end{pmatrix}, \quad v = Q^T \begin{pmatrix} 0 \\ w_2 \end{pmatrix}. \qquad (9.2)$$

Then it is easily verified that u, v are the required projections of z, which is to say they satisfy the above four properties. Also, the x in (c) is readily shown to be

$$x = \begin{pmatrix} R^{-1}w_1 \\ 0 \end{pmatrix}.$$

In effect, we are representing the projection matrices in the form

$$P = Q^T \begin{pmatrix} I_r \\ 0 \end{pmatrix} (I_r \ 0) Q \qquad (9.3)$$

and

$$I - P = Q^T \begin{pmatrix} 0 \\ I_{m-r} \end{pmatrix} (0 \ I_{m-r}) Q \qquad (9.4)$$

and we are computing $u = Pz$, $v = (I - P)z$ by means of (9.1), (9.2). The first r columns of Q span S_r and the remaining $m-r$ span its complement. Since Q and R may be updated accurately and efficiently if they are computed using Householder transformations, we have as claimed the means of orthogonalizing vectors with respect to varying bases.

As an example of the use of the projection (9.4), consider the problem of finding the stationary values of $x^T A x$ subject to $x^T x = 1$ and $C^T x = 0$, where A is a real symmetric matrix of order n and C is an $n \times p$ matrix of rank r, with $r \leq p < n$. It is shown in [12] that if the usual Householder decomposition of C is

$$QC = \begin{pmatrix} \overbrace{R}^{r} & \overbrace{S}^{n-r} \\ 0 & 0 \end{pmatrix}$$

then the problem is equivalent to that of finding the eigenvalues and eigenvectors of the matrix $\hat{P}A$, where

$$\hat{P} = I - P = Q^T \begin{pmatrix} 0 & 0 \\ 0 & I_{n-r} \end{pmatrix} Q$$

is the projection matrix in (9.4). It can then be shown that if

$$QAQ^T = \begin{pmatrix} G_{11} & G_{12} \\ G_{12}^T & G_{22} \end{pmatrix}$$

where G_{11} is $r \times r$, then the eigenvalues of $\hat{P}A$ are the same as those of G_{22} and so the eigensystem has effectively been deflated by the number of independent linear constraints. Similar transformations can be applied if the quadratic constraint is $x^T B x = 1$ for some real positive definite matrix B.

10. Orthogonalization with Respect to Positive Definite Forms

Fletcher also shows in [7] how to update projection matrices when it is required to orthogonalize with respect to a given positive definite matrix D. We now show how to compute such projections using Householder transformations, and hence the comments made in the last section concerning changes of basis may also be applied here.

Given an arbitrary vector z it is required to find $u = Pz$, $v = (I - P)z$ for some P, such that

(a) $z = u + v$

(b) $u^T D v = 0$

(c) $\exists x$ such that $Ax = u$

(d) $(DA)^T v = 0$.

For simplicity we will assume that rank$(A) = n$. Then, rather than computing P explicitly as Fletcher does according to

$$P = A(A^T DA)^{-1} A^T D,$$

we obtain the Cholesky decomposition of D thus:

$$D = LL^T$$

where L is lower-triangular and non-singular if D is positive definite. We then compute $B = L^T A$ and obtain the decomposition

$$QB = \begin{pmatrix} R \\ 0 \end{pmatrix}.$$

Defining

$$w = QL^T z = \begin{pmatrix} w_1 \\ w_2 \end{pmatrix} \begin{matrix} \} n \\ \} m-n \end{matrix}$$

and

$$u = L^{-T} Q^T \begin{pmatrix} w_1 \\ 0 \end{pmatrix}, \quad v = L^{-T} Q^T \begin{pmatrix} 0 \\ w_2 \end{pmatrix}$$

it is easily verified that u, v are the required projections, and again the x in (c) is given by $x = R^{-1}w_1$. Since changing a column a_k of A is equivalent to changing the column $L^T a_k$ of B, the matrices Q and R may be updated almost as simply as before.

11. Linear Least Squares and Quadratic Programming

We first consider minimization of quadratic forms subject to linear equality constraints. The solution is given by a single system of equations and the algorithm we describe for solving this system will serve as a basic tool for solving problems with inequality constraints. It will also provide an example of how solutions to even strongly ill-conditioned problems may be obtained accurately if orthogonalization techniques are used.

Let A, G be given matrices of orders $m \times n$, $p \times n$ respectively and let b, h be given vectors of consistent dimension. The least squares problem to be considered here is

Problem LS: $\min \|b - Ax\|_2$

subject to $Gx = h$.

Similarly, let D be a given positive semi-definite matrix and c a given n-dimensional vector. The quadratic programming problem corresponding to the above is

Problem QP: $\min \frac{1}{2} x^T D x + c^T x$

subject to $Gx = h$.

Now we can obtain very accurately the following Cholesky decomposition of D:

$$D = A^T A =$$ [triangular matrix illustration]

where we deliberately use A again to represent the triangular factor. If D is semi-definite, a symmetric permutation of rows and columns will generally be required. If D is actually positive definite then A will be a non-singular triangular matrix.

With the above notation, it can be shown that the solutions of both problems satisfy the system

$$\begin{pmatrix} & & G \\ & I & A \\ G^T & A^T & \end{pmatrix} \begin{pmatrix} z \\ r \\ x \end{pmatrix} = \begin{pmatrix} h \\ b \\ c \end{pmatrix} \qquad (11.1)$$

where

$$c = 0, \quad r = b - Ax \qquad \text{for Problem LS},$$

$$b = 0, \quad r = -Ax \qquad \text{for Problem QP},$$

and z is the vector of Lagrange multipliers. In [2], [3] methods for solving such systems have been studied in depth. The method we give here is similar but more suited to our purpose. This method has been worked on independently by Leringe and Wedin [17]. The solution of (11.1) is not unique if the quantity rank $\begin{pmatrix} G \\ A \end{pmatrix}$ is less than n, but in such cases we shall be content with obtaining <u>one</u> solution rather than many. The important steps follow.

(1) Let Q_1 be the orthogonal matrix which reduces G^T to triangular form, and let Q_1 also be applied to A^T, thus:

$$Q_1(G^T \mid A^T) = \begin{pmatrix} R_1 & S \\ 0 & T \end{pmatrix}. \qquad (11.2)$$

As explained earlier, Q_1 can be constructed as a sequence of Householder transformations, and the columns of G^T should be permuted during the triangularization. This allows any redundant constraints in $Gx = h$ to be detected and discarded.

(2) Let Q_2 be the orthogonal matrix which reduces T^T to triangular form:

$$Q_2 T^T = \begin{pmatrix} R_2 \\ 0 \end{pmatrix}. \qquad (11.3)$$

Here we assume for simplicity that T is of full rank, which is equivalent to assuming that (11.1) has a unique solution, and again we suppress permutations from the notation.

(3) The combined effect of these decompositions is now best regarded as the application of an orthogonal similarity transformation to system (11.1), since the latter is clearly equivalent to

$$\begin{pmatrix} I & \\ & Q_2 \\ & & Q_1 \end{pmatrix} \begin{pmatrix} & G \\ I & A \\ G^T & A^T \end{pmatrix} \begin{pmatrix} I & \\ & Q_2^T \\ & & Q_1^T \end{pmatrix} \begin{pmatrix} z \\ Q_2 r \\ Q_1 x \end{pmatrix} = \begin{pmatrix} h \\ Q_2 b \\ Q_1 c \end{pmatrix}.$$

The resulting system consists of various triangular subsystems involving R_1, R_2, S, and can easily be solved.

(4) If desired, the solution thus obtained can be improved upon via the method of iterative refinement [9], since this just involves the solution of system (11.1) with

different right-hand sides, and the necessary decompositions are already available.

The algorithm just described has been tested on extremely ill-conditioned systems involving inverse Hilbert matrices of high order and with iterative refinement has given solutions which are accurate to full machine precision.

12. Positive Definite Programming

With the algorithm of the previous section available, we are now prepared to attack the following more general programming problems:

Problem LS: $\quad \min \|b - Ax\|_2$

subject to $G_1 x = h_1$,

$G_2 x \geq h_2$.

Problem QP: $\quad \min \frac{1}{2} x^T D x + c^T x$

subject to the same constraints.

Let G_1, G_2 be of orders $p_1 \times n$, $p_2 \times n$ respectively, and again suppose that D has the Cholesky decomposition $A^T A$. In this section we consider problems for which $\operatorname{rank} \begin{pmatrix} A \\ G_1 \end{pmatrix} = n$ (which is most likely to be true with least squares problems, though less likely in QP). In such cases the quadratic form is essentially invertible (but we exphasize that its inverse is not computed) and so x can be eliminated from the problem. With the notation of the preceding section the steps are as follows.

(1) Solve (11.1) with G_1, h_1 to get the solution $x = x_0$, then compute the vector $q = G_2 x_0 - h_2$.

(2) If $q \geq 0$ then x_0 is the solution. Otherwise, transform the inequality matrix using Q_1 from step (1), so that

$$Q_1(G_1^T \mid A^T \mid G_2^T) = \begin{pmatrix} R_1 & S & U \\ 0 & T & V \end{pmatrix} \begin{matrix} \} p_1 \\ \} n-p_1 \end{matrix}$$

(3) If $Q_2 T^T = \begin{pmatrix} R_2 \\ 0 \end{pmatrix}$ as before and if $M = R_2^{-T} V^T$ it can be shown that the active constraints are determined by the following linear complementarity problem (LCP):

$$w = q + M^T M z$$

$$w, z \geq 0, \quad z^T w = 0 .$$

(12.1)

w, z are respectively the slack variables and Lagrange multipliers associated with the inequality constraints.

(4) The active constraints (for which $w_i = 0$ in the solution of the LCP) are now added to the equalities $G_1 x = h_1$ and the final solution is obtained from (11.1).

We wish to focus attention on the method by which the LCP (12.1) is solved. Cottle and Dantzig's principal pivoting method [5] could be applied in a straightforward manner if $M^T M$ were computed explicitly, but for numerical reasons and because $M^T M$ ($p_2 \times p_2$) could be very large, we avoid this. Rather we take advantage of the fact that no more than $n-p_1$ inequalities can be active at any one time and work with a basis M_1 made up of k columns of M, where $1 \leq k \leq n-p_1$. The QR decomposition

$$QM_1 = \begin{pmatrix} R \\ 0 \end{pmatrix}$$

is maintained for each basis as columns of M are added to or deleted from M_1 and as we know, Q and R can be updated very quickly each change. Then just as in the LU method for linear programming, the new basic solution is obtained not by updating a simplex tableau but simply by solving the appropriate system of equations using the available decomposition.

As an example we show how complementary basic solutions may be obtained. Let the basis M_1 contain k columns of M and let M_2 be the remaining (non-basic) columns. The system to be solved is

$$\begin{pmatrix} 0 \\ w_B \end{pmatrix} = \begin{pmatrix} q_1 \\ q_2 \end{pmatrix} + \begin{pmatrix} M_1^T M_1 \\ M_2^T M_1 \end{pmatrix} z_B$$

with obvious notation. If we define $y = -M_1 z_B$ this is best written as

$$\begin{pmatrix} I & M_1 \\ M_1^T & \end{pmatrix} \begin{pmatrix} y \\ z_B \end{pmatrix} = \begin{pmatrix} 0 \\ q_1 \end{pmatrix} \qquad (12.2)$$

$$w_B = q_2 - M_2^T y \qquad (12.3)$$

and the solution of (12.2) is readily obtained from

$$u = R^{-T} q_1, \quad z_B = -R^{-1} u, \quad y = Q^T \begin{pmatrix} u \\ 0 \end{pmatrix} \begin{matrix} \}k \\ \}n-p_1-k \end{matrix}$$

The blocking variable when a non-basic variable is increased can be found from the solution of the same set of equations with the appropriate right-hand side. It is worth noting that the equations can be simplified if the basis is square (i.e., if there are as many constraints active as there are free variables). Since it seems very common for the basis to fill up during the iterations (even if the final solution does not have a full set of constraints) it is worth treating a full basis specially.

Almost-complementary solutions can be obtained in similar fashion (with somewhat more work required as the system is then not quite so symmetric). Thus an algorithm such as Cottle and Dantzig's can be implemented using these techniques, and convergence is thereby guaranteed.

Of special interest, however, is the following unpublished and apparently novel idea due to Yonathan Bard, with whose permission we report the results he has obtained. Almost-complementary bases are never allowed to occur; instead, if a basic variable is negative, then it is replaced by its complement <u>regardless of the effect on the other basic variables</u>. Bard has tried this method (carried to convergence) on hundreds of problems of the form $w = q + Mz$ and cycling has never occurred when the most negative element of q is chosen. In a series of tests on 100 random matrices of orders between 2 and 20, principal pivoting required a total of 537 pivots whereas the Cottle-Dantzig algorithm required 689.

The present authors' experience with fewer but larger problems confirms the above observation that convergence does actually occur and usually after a small number of iterations. Since the idea eliminates all work other than computation of complementary solutions it is particularly suited to the techniques of this section. At worst it should be used as a starting procedure to find a close-to-optimal basis quickly, and at best if the conjecture can be proven that it will <u>always</u> converge, then a lot of computer time could be saved in the future.

*Note added in proof at the end of the article.

13. Semi-Definite Programming

We now consider the more general problem in which the rank of the quadratic form combined with the equality constraints may be less than n. The method we propose is conceptually as simple as it is stable. It is analogous to the revised simplex method for linear programming in that the essential steps to be implemented are as follows:

(1) Find the current basic solution from a certain system of equations for which a decomposition is available.

(2) Determine according to a certain set of rules what modifications should be made to the system to obtain a new basis.

(3) If necessary, update the decomposition and return to step (1).

Thus, suppose that the current basis contains $G_B x = h_B$ as active constraints. As in (11.1) the corresponding basic solution is then given by

$$\begin{pmatrix} & & G_B \\ & I & A \\ G_B^T & A^T & \end{pmatrix} \begin{pmatrix} z_B \\ r \\ x \end{pmatrix} = \begin{pmatrix} h_B \\ b \\ c \end{pmatrix} \qquad (13.1)$$

and

$$w_B = \overline{h}_B - \overline{G}_B x . \qquad (13.2)$$

(Here, $\overline{G}_B x \geq \overline{h}_B$ are the currently inactive constraints, w_B the corresponding slack variables, and z_B the Lagrange multipliers or dual variables associated with the active constraints.) The elements of z_B corresponding to any equality constraints may be either positive or negative and need never be looked at. Ignoring these, the basic solution above is optimal if and only if

$$z_B \geq 0 \quad \text{and} \quad w_B \geq 0 .$$

A "QP algorithm" is now to be regarded as the "certain set of rules" mentioned in step (2) whereby z_B, w_B and possibly other information are used to determine which constraints should be added to or dropped from G_B. The efficiency of the method will depend on the speed with which this decision can be made and on the efficiency with which the decomposition of (13.1) can be updated.

Once again the most promising pivot-selection rule is that of Bard, as discussed in the previous section. The general idea in this context is as follows:

(a) Find $w_\alpha = \min w_i$, $z_\beta = \min z_i$ from those eligible elements of w_B, z_B.

(b) If $w_\alpha < 0$, constraint α could be added.

(c) If $z_\beta < 0$, constraint β could be dropped.

(d) If there are already n constraints active and $w_\alpha < 0$, constraint α could replace constraint β.

We do not consider here the question of convergence, but as already stated, this type of rule has been found to work.

The problem of updating the requisite decompositions is more relevant at present. We discuss this and other points briefly.

(1) The matrices Q_1, R_1 of Equation (11.2) can be updated efficiently using the methods of Section 8.

(2) Q_2, R_2 obtained from the matrix T in Equation (11.3) unfortunately cannot be updated, but the work needed to recompute them might often be very small, for the following reasons:

(a) In Problem LS, a preliminary triangularization of A (m × n) can be applied to obtain an equivalent problem for which $m \leq n$. The Cholesky factor of D in Problem QP already has this property.

(b) If there are many constraints active (up to n) then T has very few rows.

(c) If the rank of the system is low (relative to n) then T has very few columns.

(3) Hence the method is very efficient if close to n constraints are active each iteration, as should often be the case. It also has the property, along with Beale's algorithm [1], of being most efficient for problems of low rank.

(4) The procedure can be initiated with any specified set of constraints in the first basis, and an initial estimate of x is not required.

(5) Any number of constraints can be handled, in the same way that the revised simplex method can deal with any number of variables.

(6) If $D = 0$ the problem is a linear program and only bases containing n constraints need be considered. The method reduces to something like a self-dual simplex algorithm.

Finally we note that with semi-definite problems it is possible for some basic system (13.1) to be singular. If there are any solutions at all then there are many (this will always be the case with low rank least squares problems) but this does not matter, since z_B is still uniquely determined. However, a low rank quadratic program might be unbounded, and this is manifested by a singular system (13.1) proving to inconsistent. In general, this just means that there are not yet enough constraints in the basis, so that trouble can usually be avoided by initializing the procedure with a full set of constraints.

C. THE SVD AND NONLINEAR LEAST SQUARES

14. The Singular Value Decomposition

Let A be a real, $m \times n$ matrix (for notational convenience we assume that $m \geq n$). It is well known (cf. [16]) that

$$A = U \Sigma V^T \qquad (14.1)$$

where U, V are orthogonal matrices and

$$\Sigma = \begin{pmatrix} \begin{matrix} \sigma_1 & & 0 \\ & \ddots & \\ 0 & & \sigma_n \end{matrix} \\ \hline 0 \end{pmatrix} \begin{matrix} \\ \\ \end{matrix} (m-n) \times n$$

U consists of the orthonormalized eigenvectors of AA^T, and V consists of the orthonormalized eigenvectors of A^TA. The diagonal elements of Σ are the nonnegative square roots of the eigenvalues of A^TA; they are called <u>singular values</u> or <u>principal values</u> of A. We assume

$$\sigma_1 \geq \sigma_2 \geq \ldots \geq \sigma_n \geq 0.$$

Thus if rank(A) = r, $\sigma_{r+1} = \sigma_{r+2} = \ldots = \sigma_n = 0$. The decomposition (14.1) is called the <u>singular value decomposition</u> (SVD).

An $n \times m$ matrix X is said to be the <u>pseudo-inverse</u> of an $m \times n$ matrix A if X satisfies the following four properties:

(i) AXA = A, (ii) XAX = X, (iii) $(XA)^T = XA$, (iv) $(AX)^T = AX$.

We denote the pseudo-inverse by A^+. It can be shown that A^+ can always be determined and is unique (cf. [21]). It is easy to verify that $A^+ = V \Lambda U^T$ where Λ is the $n \times m$ matrix $\Lambda = \text{diag}[\sigma_1^{-1}, \sigma_2^{-1}, \ldots, \sigma_r^{-1}, 0, 0, \ldots, 0]$. There are many applications of the SVD in least squares problems (cf. [11]).

The SVD of an arbitrary matrix is calculated in the following way. First, a sequence of Householder transformations $\{P_k\}_{k=1}^n$, $\{Q_k\}_{k=1}^{n-1}$ is constructed so that

$$P_n P_{n-1} \ldots P_1 A Q_1 Q_2 \ldots Q_{n-1} \equiv P^T A Q = J$$

and J is an $m \times n$ bi-diagonal matrix of the form

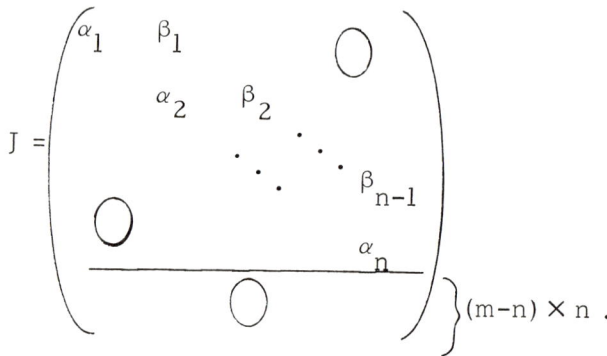

$$\}(m-n) \times n\,.$$

The singular values of J are the same as those of A.

Next the SVD of J is computed by an algorithm given in [11]. The algorithm is based on the highly effective QR algorithm of Francis [10] for computing eigenvalues. If the SVD of $J = X\Sigma Y^T$ then $A = PX\Sigma Y^T Q^T$ so that $U = PX$, $V = QY$.

15. Nonlinear Least Squares

Consider the nonlinear transformation $F(x) = y$ where $x \in E_n$ and $y \in E_m$ with $n \leq m$. We wish to consider the following problem:

$$\min \|b - F(x)\|_2$$

subject to

$$Gx = h, \qquad (15.1)$$

where G is a $p \times n$ matrix of rank p and $h \in E_p$. A very effective algorithm for solving such problems is a variant of the Levenberg-Marquardt algorithm [18, 19]; in this section we consider some of the details of the numerical calculation. Further extensions of the algorithm are given by Shanno [25] and Meyer [20].

Let us assume that we have an approximation $x^{(0)}$ which satisfies the relation $Gx^{(0)} = h$. Then at each stage of the iteration we determine $\delta^{(k)}$ so that

NUMERICAL TECHNIQUES IN PROGRAMMING

$$x^{(k+1)} = x^{(k)} + \delta^{(k)} \quad (15.2)$$

and

$$G\delta^{(k)} = 0 . \quad (15.3)$$

Again as in Section 11, we write $Q_1 G^T = \binom{R}{0}$ where Q_1 is the product of p Householder transformations and R is an upper triangular matrix. Let

$$Q_1 \delta^{(k)} = \begin{pmatrix} \xi^{(k)} \\ \eta^{(k)} \end{pmatrix} \begin{matrix} \} p \\ \} n-p \end{matrix} \quad (15.4)$$

Then from (15.3), we see that $\xi^{(k)} = 0$.

For notational convenience, let us drop the superscript k; we write $x^{(k)}$ as x_0 and $x^{(k+1)}$ as x_1.

In the Levenberg-Marquardt algorithm one determines the vector δ so that

$$\|r - J\delta\|_2^2 + \lambda \|\delta\|^2 = \min . \quad (15.5)$$

where

$$r = b - F(x_0) ,$$

J is the Jacobian evaluated at x_0, and λ is an arbitrary nonnegative parameter. From (15.4), we see that (15.5) is equivalent to determining η so that

$$\left. \begin{array}{c} \|r - JQ_1^T \begin{pmatrix} \xi \\ \eta \end{pmatrix}\|_2^2 + \lambda(\|\xi\|_2^2 + \|\eta\|_2^2) = \min . \\ \\ \text{subject to } \xi = 0 . \end{array} \right\} \quad (15.6)$$

163

Now let us write $JQ_1^T = [M, N]$ where N consists of the last $n-p$ columns of JQ_1^T. Then (15.6) is equivalent to finding η so that

$$\Phi(\eta) \equiv \|r - N\eta\|_2^2 + \lambda \|\eta\|_2^2 = \min.$$

Consider the SVD of N; namely

$$N = U \Sigma V^T.$$

Then

$$\Phi(\eta) = \|U^T r - \Sigma V^T \eta\|_2^2 + \lambda \|V^T \eta\|_2^2$$

$$= \|s - \Sigma \zeta\|_2^2 + \lambda \|\zeta\|_2^2 \tag{15.7}$$

where

$$s = U^T r, \quad \zeta = V^T \eta.$$

Writing out (15.7) explicitly, we have

$$\Phi(\zeta) = \sum_{j=1}^{\rho} (s_j - \sigma_j \zeta_j)^2 + \lambda \sum_{j=1}^{n-p} (\zeta_j)^2$$

where ρ is the rank of N. (Note ρ may change from iteration to iteration.) Then

$$\Phi(\hat{\zeta}) = \min$$

when

$$\hat{\zeta}_j = \frac{s_j \sigma_j}{\lambda + \sigma_j^2} \quad \text{for } j = 1, 2, \ldots, p,$$

$$= 0 \quad \text{for } j > p$$

and hence

$$\eta = \sum_{j=1}^{p} \frac{s_j \sigma_j}{\lambda + \sigma_j^2} v_j$$

where v_j is the j-th column of V. Thus

$$\delta = Q_1^T \begin{pmatrix} 0 \\ \eta \end{pmatrix}.$$

Note it is an easy matter to compute η (and hence δ) for various values of λ. The algorithm for computing the SVD can easily be organized so that s is computed directly ([11]).

There are several possible strategies for determining λ. One possibility is to choose $\hat{\lambda}$ so that

$$\|b - F(x_1(\hat{\lambda}))\|_2 \leq \|b - F(x_1(\lambda))\|_2.$$

This requires, of course, the evaluation of $F(x)$ at a great many points.

Another possibility is to choose δ such that

$$\left.\begin{array}{c} \|r - J\delta\|_2 = \min. \\ \text{subject to } \|\delta\|_2 \leq \alpha \end{array}\right\} \quad (15.8)$$

This is equivalent to determining λ such that

$$\|\eta\|_2^2 = \sum_{j=1}^{p} \left(\frac{s_j \sigma_j}{\lambda + \sigma_j^2} \right)^2 \le \alpha^2.$$

When $\lambda = 0$, we have the solution to the unconstrained problem and

$$\eta_0 = \sum_{j=1}^{p} \frac{s_j}{\sigma_j} v_j.$$

Let $\|\eta_0\|_2 = \beta$. If $\beta \le \alpha$, then we have the solution to (15.8). Otherwise, we must determine λ so that

$$\sum_{j=1}^{p} \left(\frac{s_j \sigma_j}{\lambda + \sigma_j^2} \right)^2 = \alpha^2. \qquad (15.9)$$

Let

$$u = \alpha^{-1} \begin{pmatrix} \sigma_1 s_1 \\ \sigma_2 s_2 \\ \vdots \\ \sigma_p s_p \end{pmatrix}, \quad \Omega = \mathrm{diag}(\sigma_1^2, \sigma_2^2, \ldots, \sigma_p^2);$$

we assume $s_j \ne 0$ for $j = 1, 2, \ldots, p$. By repeated use of the relationship

$$\det \begin{pmatrix} X & Y \\ Z & W \end{pmatrix} = \det(X) \det(W - ZX^{-1}Y) \quad \text{if } \det(X) \ne 0$$

we can show that (15.9) is equivalent to

$$\det((\Omega + \lambda I)^2 - uu^T) = 0 \qquad (15.10)$$

which has 2ρ roots; it can be shown that we need the largest real root, which we denote by λ^* ([8]). Let

$$\Gamma(\lambda) = \sum_{j=1}^{\rho} \left(\frac{s_j \sigma_j}{\lambda + \sigma_j^2} \right)^2 - \alpha^2$$

and assume that $\sigma_1^2 \geq \sigma_2^2 \geq \ldots \geq \sigma_\rho^2 > 0$. Note $\Gamma(0) = \beta^2 - \alpha^2 > 0$, and $\Gamma(\lambda) \to -\alpha^2$ as $\lambda \to \infty$, so that $0 \leq \lambda^* < \infty$ and it is the only root in that interval. We seek a more precise upper bound for λ^*. From (15.10) we see, using a Rayleigh quotient argument, that

$$\lambda^* \leq \max_{\|y\|_2=1} [-y^T \Omega y + \sqrt{(y^T \Omega y)^2 - y^T(\Omega^2 - uu^T)y}].$$

A short manipulation then shows that

$$0 < \lambda^* \leq \sqrt{\sigma_1^4 - \sigma_\rho^4 + u^T u} - \sigma_\rho^2. \qquad (15.11)$$

Thus, we wish to find a root of (15.10) which lies in the interval given by (15.11). Note that the determinantal equation (15.10) involves a diagonal matrix plus a matrix of rank one. In the next section we shall describe an algorithm for solving such problems.

16. Modified Eigensystems

As was pointed out in Section 15, it is sometimes desirable to determine some eigenvalues of a diagonal matrix which is modified by a matrix of rank one. Also, Powell [23] has recently proposed a minimization algorithm which requires the eigensystem of a matrix after a rank one modification. In this section, we give an algorithm for determining in $O(n^2)$ numerical operations some or all of the eigenvalues and eigenvectors of $D + \sigma uu^T$ where $D = \text{diag}(d_i)$ is a diagonal matrix of order n and $u \in E_n$.

Let $C = D + \sigma uu^T$; we denote the eigenvalues of C by $\lambda_1, \lambda_2, \ldots, \lambda_n$ and we assume $\lambda_i \geq \lambda_{i+1}$ and $d_i \geq d_{i+1}$. It can be shown (cf. [30]) that

(1) If $\sigma \geq 0$, $d_1 + \sigma u^T u \geq \lambda_1 d_1$, $d_{i-1} \geq \lambda_i \geq d_i$

$$(i = 2, \ldots, n),$$

(2) If $\sigma < 0$, $d_i \geq \lambda_i \geq d_{i-1}$ $(i = 1, 2, \ldots, n-1)$,

$$d_n \geq \lambda_n \geq d_n + \sigma u^T u.$$

Thus we have precise bounds on each of the eigenvalues of the modified matrix.

Let K be a bi-diagonal matrix of the form

$$K = \begin{pmatrix} 1 & r_1 & & & O \\ & 1 & \cdot & & \\ & & \cdot & \cdot & \\ & & & \cdot & r_{n-1} \\ O & & & & 1 \end{pmatrix}$$

and let $M = \text{diag}(\mu_i)$. Then

$$KMK^T = \begin{pmatrix} (\mu_1+\mu_2 r_1^2) & \mu_2 r_1 & & & & O \\ \mu_2 r_1 & \ddots & \ddots & & & \\ & \ddots & & & & \\ & & \mu_k r_{k-1} & (\mu_k+\mu_{k+1} r_k^2) & \mu_{k+1} r_k & \\ & & & \ddots & \ddots & \ddots \\ & & & & & & \mu_n r_{n-1} \\ O & & & & \mu_n r_{n-1} & \mu_n \end{pmatrix}$$

(16.1)

is a symmetric, tri-diagonal matrix.
  Consider the matrix equation

$$(D + \sigma u u^T) x = \lambda x . \tag{16.2}$$

Multiplying (16.2) on the left by K, we have

$$K(D + \sigma u u^T) K^T K^{-T} x = \lambda K K^T K^{-T} x$$

or

$$(KDK^T + \sigma K u u^T K^T) y = \lambda K K^T y \tag{16.3}$$

where $x = K^T y$. Let us assume that we have re-ordered the elements of u so that

$$u_1 = u_2 = \ldots = u_{p-1} = 0 \text{ and } 0 < |u_p| \leq |u_{p+1}| \leq \ldots \leq |u_n|.$$

Now it is possible to determine the elements of K so that

$$Ku = \begin{pmatrix} 0 \\ 0 \\ \vdots \\ 0 \\ u_n \end{pmatrix}. \qquad (16.4)$$

Specifically,

$$r_i = 0 \qquad (i = 1, 2, \ldots, p-1),$$

$$r_i = -u_i/u_{i+1} \qquad (i = p, p+1, \ldots, n),$$

and we note that $|r_i| \leq 1$. (This device of using a bi-diagonal matrix for annihilating $n-1$ elements of a vector has been used by Björck and Pereyra [4] for inverting Vandermonde matrices.) Therefore, if Ku satisfies (16.4), we see from (16.1) that $KDK^T + \sigma K u u^T K^T$ is a tri-diagonal matrix and similarly KK^T is a tri-diagonal matrix. Thus we have a problem of the form

$$Ay = \lambda By$$

where A and B are symmetric, tri-diagonal matrices and B is positive definite.

In [22], Peters and Wilkinson show how linear interpolation may be used effectively for computing the eigenvalues for such matrices when the eigenvalues are isolated. The algorithm makes use of the value of $\det(A - \lambda B)$. When A and B are tri-diagonal, it is very simple to evaluate $\det(A - \lambda B)$ for arbitrary λ. Once the eigenvalues are computed it is easy to compute the eigenvectors by inverse iteration.

In Section 15, we showed it was necessary to compute a parameter λ^* which satisfied the equation

$$\det((\Omega + \lambda I)^2 - uu^T) = 0 . \tag{16.5}$$

Again we can determine K so that Ku satisfies (16.4) and hence (16.5) is equivalent to

$$\det(K(\Omega + \lambda I)^2 K^T - Kuu^T K^T) = 0 . \tag{16.6}$$

The matrix $G(\lambda) = K(\Omega + \lambda I)^2 K^T - Kuu^T K^T$ is tri-diagonal so that it is easy to evaluate $G(\lambda)$ and det $G(\lambda)$. Since we have an upper and lower bound on λ^*, it is possible to use linear interpolation to find λ^*, even though $G(\lambda)$ is quadratic in λ. Numerical experiments have indicated it is best to compute $G(\lambda) = K(\Omega + \lambda I)^2 K^T - Kuu^T K^T$ for each approximate value of λ^* rather than computing $G(\lambda) = (K\Omega^2 K^T - Kuu^T K^T) + 2\lambda K\Omega K^T + \lambda^2 KK^T$.

The device of changing modified eigensystems to tri-diagonal matrices and then using linear interpolation for finding the roots can be extended to matrices of the form

$$C = \left(\begin{array}{c|c} D & u \\ \hline u^T & \sigma \end{array}\right)$$

Again we choose K so that Ku satisfies (16.4) and thus obtain the eigenvalue problem $Ay = \lambda By$ where

$$A = \left(\begin{array}{c|c} KDK^T & Ku \\ \hline u^T K^T & \sigma \end{array}\right), \quad B = \left(\begin{array}{c|c} KK^T & 0 \\ \hline 0 & 1 \end{array}\right)$$

so that A and B are both tri-diagonal and B is positive definite. Bounds for the eigenvalues of C can easily be established in terms of the eigenvalues of D and hence the linear interpolation algorithm may be used for determining the eigenvalues of C.

Note added in proof:

It has since been learned that Bard applied the principal-pivoting rule to LCP's of the somewhat special form in which

$$M = P^T P, \quad q = -P^T p$$

for some P, p. Problems of this form have been studied by Zoutendijk in [31, 32] where several pivot-selection rules are discussed. Finiteness is proven for one rule, but simpler methods (such as Bard's) are recommended in practice for efficiency.

The question of finiteness for the more general LCP remains open, and it is likely that somewhat more sophisticated rules (e.g., Cottle and Dantzig) will be required.

REFERENCES

1. Beale, E. M. L., "Numerical Methods," in <u>Nonlinear Programming</u>, J. Abadie (ed.). John Wiley, New York, 1967, pp. 133-205.

2. Björck, Å., "Iterative Refinement of Linear Least Squares Solutions II," <u>BIT</u> 8 (1968), pp. 8-30.

3. _____ and G. H. Golub, "Iterative Refinement of Linear Least Squares Solutions by Householder Transformations," <u>BIT</u> 7 (1967), pp. 322-37.

4. _____ and V. Pereyra, "Solution of Vandermonde Systems of Equations," Publicacion 70-02, Universidad Central de Venezuela, Caracas, Venezuela, 1970.

5. Cottle, R. W. and G. B. Dantzig, "Complementary Pivot Theory of Mathematical Programming," <u>Mathematics of the Decision Sciences, Part 1</u>, G. B. Dantzig and A. F. Veinott (eds.), <u>American Mathematical Society</u> (1968), pp. 115-136.

6. Dantzig, G. B., R. P. Harvey, R. D. McKnight, and S. S. Smith, "Sparse Matrix Techniques in Two Mathematical Programming Codes," <u>Proceedings of the Symposium on Sparse Matrices and Their Applications</u>, T. J. Watson Research Publication RA1, no. 11707, 1969.

7. Fletcher, R., "A Technique for Orthogonalization," <u>J. Inst. Maths. Applics.</u> 5 (1969), pp. 162-66.

8. Forsythe, G. E., and G. H. Golub, "On the stationary Values of a Second-Degree Polynomial on the Unit Sphere," <u>J. SIAM</u>, 13 (1965), pp. 1050-68.

9. _____ and C. B. Moler, <u>Computer Solution of Linear Algebraic Systems</u>, Prentice-Hall, Englewood Cliffs, New Jersey, 1967.

10. Francis, J., "The QR Transformation. A Unitary Analogue to the LR Transformation," <u>Comput. J.</u> 4 (1961-62), pp. 265-71.

11. Golub, G. H., and C. Reinsch, "Singular Value Decomposition and Least Squares Solutions," <u>Numer. Math.</u>, 14 (1970), pp. 403-20.

12. _____, and R. Underwood, "Stationary Values of the Ratio of Quadratic Forms Subject to Linear Constraints," Technical Report No. CS 142, Computer Science Department, Stanford University, 1969.

13. Hanson, R. J., "Computing Quadratic Programming Problems: Linear Inequality and Equality Constraints," Technical Memorandum No. 240, Jet Propulsion Laboratory, Pasadena, California, 1970.

14. _____, and C. L. Lawson, "Extensions and Applications of the Householder Algorithm for Solving Linear Least Squares Problems," Math. Comp., 23 (1969), pp. 787-812.

15. Householder, A. S., "Unitary Triangularization of a Nonsymmetric Matrix," J. Assoc. Comp. Mach., 5 (1958), pp. 339-42.

16. Lanczos, C., Linear Differential Operators. Van Nostrand, London, 1961. Chapter 3.

17. Leringe, Ö., and P. Wedin, "A Comparison Between Different Methods to Compute a Vector x Which Minimizes $\|Ax - b\|_2$ When $Gx = h$," Technical Report, Department of Computer Sciences, Lund University, Sweden.

18. Levenberg, K., "A Method for the Solution of Certain Non-Linear Problems in Least Squares," Quart. Appl. Math., 2 (1944), pp. 164-68.

19. Marquardt, D. W., "An Algorithm for Least-Squares Estimation of Non-linear Parameters," J. SIAM, 11 (1963), pp. 431-41.

20. Meyer, R. R., "Theoretical and Computational Aspects of Nonlinear Regression," P-1819, Shell Development Company, Emeryville, California.

21. Penrose, R., "A Generalized Inverse for Matrices," Proceedings of the Cambridge Philosophical Society, 51 (1955), pp. 406-13.

22. Peters, G., and J. H. Wilkinson, "Eigenvalues of $Ax = \lambda Bx$ with Band Symmetric A and B," *Comput. J.*, 12 (1969), pp. 398-404.

23. Powell, M. J. D., "Rank One Methods for Unconstrained Optimization," T.P. 372, Atomic Energy Research Establishment, Harwell, England (1969).

24. Rosen, J. B., "Gradient Projection Method for Nonlinear Programming. Part I. Linear Constraints," *J. SIAM*, 8 (1960), pp. 181-217.

25. Shanno, D. C., "Parameter Selection for Modified Newton Methods for Function Minimization," *J. SIAM, Numer. Anal.*, Ser. B, 7 (1970).

26. Stoer, J., "On the Numerical Solution of Constrained Least Squares Problems" (Private communication), 1970.

27. Tewarson, R. P., "The Gaussian Elimination and Sparse Systems," *Proceedings of the Symposium on Sparse Matrices and Their Applications*, T. J. Watson Research Publication RA1, no. 11707, 1969.

28. Wilkinson, J. H., "Error Analysis of Direct Methods of Matrix Inversion," *J. Assoc. Comp. Mach.*, 8 (1961), pp. 281-330.

29. ———, "Error Analysis of Transformations Based on the Use of Matrices of the Form $I - 2ww^H$," in *Error in Digital Computation*, Vol. ii, L. B. Rall (ed.), John Wiley and Sons, Inc., New York, 1965, pp. 77-101.

30. ———, *The Algebraic Eigenvalue Problem*, Clarendon Press, Oxford, 1965.

31. Zoutendijk, G., <u>Methods of Feasible Directions</u>, Elsevier Publishing Company, Amsterdam, 1960, pp. 80-90.

32. _____, "Nonlinear Programming, Computational Methods," in <u>Nonlinear and Integer Programming</u>, J. Abadie (ed.), North-Holland Publ. Co., 1970, pp. 37-86.

The work of the second author was supported in part by the U. S. Atomic Energy Commission. The work of the third author was supported in part by the U. S. Atomic Energy Commission and by the Department of Scientific and Industrial Research, Wellington, New Zealand.

A Superlinearly Convergent Method for Unconstrained Minimization

K. RITTER

ABSTRACT

A method is described for minimizing a continuously differentiable function $F(x)$ of n variables. It can be applied under the same general assumptions as the method of steepest descent. If $F(x)$ is twice continuously differentiable and the eigenvalues of the Hessian matrix $G(x)$ are always greater than a positive constant, then the algorithm generates a sequence of points which converges superlinearly to the unique minimizer of $F(x)$. No computation of second order derivatives is required.

1. Introduction

In [5] Goldstein and Price proposed an algorithm for constructing a sequence which converges superlinearly toward the unique minimizer of a function $F(x)$, provided $F(x)$ is twice continuously differentiable and the eigenvalues of the Hessian matrix of $F(x)$ are greater than some positive constant. Recently, Powell [7] showed that, under the same assumptions on $F(x)$, the sequence of points generated by the variable metric method [1], [3] converges superlinearly to the unique minimizer of $F(x)$. In both cases the convergence proof depends heavily on the assumptions stated above concerning the Hessian matrix. In most cases, however, it is difficult to verify this assumption. Therefore, it is desirable to have an algorithm which is superlinearly convergent in the above case and for which convergence (not necessarily superlinear) can also be established under the same general assumptions as in the method of steepest descent. Such an algorithm is described in this paper. In a subsequent paper the method will be extended to the case of a minimization problem with linear inequality constraints.

2. Formulation of the problem, definitions and notation

Let $x \in E^n$ and assume that $F(x)$ is a real valued function. If $F(x)$ is differentiable at a point x_i we denote its gradient at x_i by $\nabla F(x_i)$ or g_i. If $F(x)$ is twice differentiable at x_i we denote the Hessian matrix of $F(x)$ at x_i by $G(x_i)$ or G_i.

For $x \in E^n$ let $\|x\|$ denote the Euclidean norm of the column vector x. We say a sequence $\{x_j\} \subset E^n$ converges superlinearly to $z \in E^n$ if

$$\lim_{j \to 0} \frac{\|x_{j+1} - z\|}{\|x - z\|} = 0.$$

The purpose of this paper is to describe an algorithm which produces a sequence of points $\{x_j\}$ such that, under certain assumptions on $F(x)$,
1) $x_j \to z$, where z minimizes $F(x)$,
2) the convergence of $\{x_j\}$ to z is superlinear if, in a neighborhood of z, the smallest eigenvalue of $G(x)$ is greater than some positive constant,
3) the rate of the superlinear convergence can be established if, in a neighborhood of z, $G(x)$ satisfies a Hölder condition and its eigenvalues which are greater than some positive constant.

If M is a matrix, we denote its norm by $\|M\|$, i.e.,

$$\|M\| = \sup\{\|Mx\| \mid \|x\| = 1\}.$$

If M is symmetric, then [4]

$$\|M\| = \sup\{|x'Mx| \mid \|x\| = 1\}.$$

Assume that M is symmetric and that there are positive numbers μ and η such that

$$\mu \|x\|^2 \leq x'Mx \leq \eta \|x\|^2 \text{ for all } x \in E^n.$$

Then it follows from the above equation that

$$\|M\| \leq \eta$$

and

$$\frac{1}{\eta} \|x\|^2 \leq x'M^{-1}x \leq \frac{1}{\mu} \|x\|^2 \quad \text{for all} \quad x \in E^n.$$

In the following let $\{\beta_j\}$ and $\{\gamma_j\}$ denote sequence of real numbers such that

$$\beta_{j+1} \leq \beta_j, \quad \gamma_{j+1} \leq \gamma_j \quad \text{for all } j$$

and

$$\lim_{j \to \infty} \beta_j = 0, \quad \lim_{j \to \infty} \gamma_j = \gamma,$$

where $0 < \gamma$.

3. The algorithm

The algorithm generates either an infinite sequence of different points x_j or terminates after a finite number of cycles with an x_n such that $\nabla F(x_n) = 0$.

It is assumed that the algorithm starts with a point x_0 with the property that

$$S_0 = \{x \in E^n \mid F(x) \leq F(x_0)\}$$

is bounded and that $F(x)$ is continuously differentiable on some open convex set S containing S_0. Furthermore, it is assumed that $\nabla F(x_0) \neq 0$ and matrices

$$D_0^{-1} = (c_{10}, \ldots, c_{n0}), \quad c_{i0} \in E^n,$$

SUPERLINEARLY CONVERGENT MINIMIZATION

$$P'_0 = (p_{10}, \ldots, p_{n0}), \quad p_{in} \in E^m$$

are given such that $\|p_{i0}\| = 1$, $i = 1, \ldots, n$. Here and below c_{ij} and p_{ij} are column vectors and P'_j denotes the transpose of P_j.

We describe now a general cycle of the algorithm:

At the beginning of the jth cycle the following data is available: x_j, $\nabla F(x_j) = g_j \neq 0$,

$$D_j^{-1} = (c_{1j}, \ldots, c_{nj}), \quad P'_j = (p_{1j}, \ldots, p_{nj}),$$

$$\|p_{ij}\| = 1, \quad i = 1, \ldots, n.$$

Step I: Computation of direction of descent s_j

Let

$$\bar{s}_j = D_j^{-1} P_j g_j$$

and

$$s_j = \begin{cases} \bar{s}_j & \text{if } g'_j \bar{s}_j > \alpha_j \|g_j\|^2 \\ -\bar{s}_j & \text{if } -g'_j \bar{s}_j > \alpha_j \|g_j\|^2 \\ g_j & \text{if } |g'_j \bar{s}_j| \leq \alpha_j \|g_j\|^2 \end{cases}$$

where $\alpha_j = \min\{\alpha, \|g_j\|\}$ and $0 < \alpha < 1$.

Step II: Computation of step size σ_j

We consider two different methods for the computation of the step size σ_j

1) Let $0 < \delta < \frac{1}{2}$ and

$$h(x_j, \sigma) = \frac{F(x_j) - F(x_j - \sigma s_j)}{\sigma g_j' s_j}, \quad \sigma > 0$$

If $h(x_j, 1) \geq \delta$ put $\sigma_j = 1$; otherwise determine σ_j so that

$$0 < \sigma_j < 1 \text{ and } \delta \leq h(x_j, \sigma_j) \leq 1 - \delta$$

2) Let $\bar{\sigma}_j$ minimize $F(x_j - \sigma s_j)$ on the ray $\{x \in E^n \mid x = x_j - \sigma s_j, \sigma \geq 0\}$. Choose σ_j so that

$$|\sigma_j - \bar{\sigma}_j| \leq \beta_j \text{ and } F(x_j - \sigma_j s_j) < F(x_j)$$

Put

$$x_{j+1} = x_j - \sigma_j s_j,$$

where σ_j is determined by 1) or 2).
Compute g_{j+1}. If $g_{j+1} = 0$ stop, otherwise go to Step III.

<u>Step III</u>: Computation of P_{j+1} and D_{j+1}^{-1}
Let $c > 0$ be a given constant and

$$d_j = \frac{g_j - g_{j+1}}{\|\sigma_j s_j\|}.$$

If

$$|c_{nj}' d_j| \geq \gamma_j \|c_{nj}\| \text{ and } \|d_j\| \leq c$$

put

$$d_{1,j+1} = d_j,$$

otherwise compute

$$\bar{g}_{j+1} = \nabla F(x_j - \|\sigma_j s_j\| \frac{c_{nj}}{\|c_{nj}\|}) \text{ and } \bar{d}_j = \frac{g_j - \bar{g}_{j+1}}{\|\sigma_j s_j\|}$$

and set

$$d_{1,j+1} = \begin{cases} \bar{d}_j & \text{if } |c'_{nj}\bar{d}_j| \geq \gamma_j \|c_{nj}\| \text{ and } \|\bar{d}_j\| \leq c \\ \dfrac{c_{nj}}{\|c_{nj}\|} & \text{otherwise .} \end{cases}$$

Define

$$p_{1,j+1} = \begin{cases} \dfrac{s_j}{\|s_j\|} & \text{if } d_{1,j+1} = d_j \\ \dfrac{c_{nj}}{\|c_{nj}\|} & \text{if } d_{1,j+1} \neq d_j, \end{cases}$$

$$p_{i,j+1} = p_{i-1,j}, \quad i = 2, \ldots, n \text{ and}$$

$$P'_{j+1} = (p_{1,j+1}, \ldots, p_{n,j+1}).$$

Furthermore, let

$$c_{1,j+1} = \frac{c_{nj}}{c'_{nj} d_{1,j+1}},$$

$$c_{i,j+1} = c_{i-1,j} - \frac{c'_{i-1,j} d_{1,j+1}}{c'_{nj} d_{1,j+1}} c_{nj}, \quad i = 2, \ldots, n \quad \text{and}$$

$$D_{j+1}^{-1} = (c_{1,j+1}, \ldots, c_{n,j+1}).$$

Replace P_j, D_j^{-1}, g_j and x_j by P_{j+1}, D_{j+1}^{-1}, g_{j+1} and x_{j+1}, respectively, and go to Step I.

In the following the notation x_j, s_j, g_j, P_j and D_g^{-1} refers always to the vectors and matrices generated by the algorithm. Any statement on properties of the algorithm applies to both choices of σ_j unless the contrary is stated explicitly.

Lemma 1.
1) If $x_j \in S_0$ and s_j is determined according to Step I, then the choice of σ_j as prescribed by Step II is always possible.

2) Let $d_{ij} = d_{1,j+1-i}$, $i = 1, \ldots, n$; where $d_{1,j+1}$ are the vectors defined in Step III. Then

$$D'_j = (d_{1j}, \ldots, d_{nj})$$

for $j \geq n$.

3) The sequences

$$\{\|D_j^{-1}\|\} \quad \text{and} \quad \{\|s_j\|\}$$

are bounded.

Proof:
1) Since the set $S_0 = \{x \in E^n \mid F(x) \leq F(x_0)\}$ is closed and bounded and $x_j \in S_0$, it follows that there is $\bar{\sigma}_j$ which minimizes $F(x_j - \sigma s_j)$ on the ray $\{x \in E^n \mid x = x_j - \sigma s_j, \sigma \geq 0\}$. By the definition of s_j, $g_j \neq 0$ implies $g'_j s_j > 0$. Hence,

SUPERLINEARLY CONVERGENT MINIMIZATION

$$F(x_j - \bar{\sigma}_j s_j) < F(x_j)$$

and the choice of σ_j as prescribed by Step II.2 is possible.

To show that the choice of σ_j as prescribed by Step II.1 is possible, define

$$L_j(\sigma) = \{x \in E^n \mid x = x_j - \tau s_j, \ 0 \leq \tau \leq \sigma\}, \ \sigma \geq 0.$$

Since $g_j \neq 0$ implies $g_j' s_j > 0$, it follows that, for small $\sigma > 0$, $x_j - \sigma s_j \in S_0$. Hence, there is $\xi_j \in L_j(\sigma)$ such that

$$F(x_j) - F(x_j - \sigma s_j) = \sigma(\nabla F(\xi_j))' s_j =$$

$$= \sigma g_j' s_j + \sigma (\nabla F(\xi_j) - g_j)' s_j,$$

or by the definition of $h(x_j, \sigma)$,

$$h(x_j, \sigma) = 1 + \frac{(\nabla F(\xi_j) - g_j)' s_j}{g_j' s_j}$$

since $\nabla F(x)$ is continuous at x_j and $\|\xi_j - x_j\| \leq \sigma \|s_j\|$ it follows that

$$\|\nabla F(\xi_j) - g_j\| \to 0 \text{ as } \sigma \to 0.$$

Thus,

$$h(x_j, 0) = 1.$$

Suppose $h(x_j, 1) < \delta$. Then it follows from the continuity of $h(x_j, \sigma)$ that there is $0 < \sigma_j < 1$ such that $\delta \leq h(x_j, \sigma_j) \leq 1-\delta$.

2) Let $M'_0 = (d_{01}, \ldots, d_{0n})$ and $M'_0 D_0^{-1} = I$, where D_0^{-1} is the matrix associated with x_0. For $j = 0, 1, \ldots,$ let $d_{1,j+1}$ be the vector defined in Step III of the algorithm and define

$$M'_{j+1} = (d_{1,j+1}, \ldots, d_{n,j+1}) \quad j = 0, 1, \ldots,$$

where $d_{i,j+1}$ is the (i-1)th column of M'_j, $i = 2, \ldots, n$. Then

$$M_j D_j^{-1} = I, \quad j = 0, 1, 2, \ldots,$$

For $j = 0$ the assertion is a consequence of the definition of M_0.

Suppose that $M_j D_j^{-1} = I$ for some j, then it follows from the definition of M_{j+1} and D_{j+1}^{-1} that the element in the kth row and first column of $M_{j+1} D_{j+1}^{-1}$ is

$$d'_{k,j+1} c_{1,j+1} = \frac{d'_{k,j+1} c_{nj}}{c'_{nj} d_{1,j+1}} = \begin{cases} 1 & \text{if } k = 1 \\ 0 & \text{if } k = 2, \ldots, n \end{cases}$$

and the element in the kth row and ℓth column, $\ell = 2, \ldots, n$, of $M_{j+1} D_{j+1}^{-1}$ is

$$d'_{k,j+1} c_{\ell,j+1} = d'_{k,j+1} c_{\ell-1,j} - \frac{c'_{\ell-1,j} d_{1,j+1}}{c'_{nj} d_{1,j+1}} c'_{nj} d_{k,j+1} =$$

$$= \begin{cases} 0 & \text{if } \ell \neq k \\ 1 & \text{if } \ell = k = 2, \ldots, n \end{cases}.$$

Furthermore, for $j \geq n$,

$$M'_j = (d_{1,j}, \ldots, d_{nj}),$$

where $d_{ij} = d_{1, j+1-i}$, $i = 1, \ldots, n$.

3) Let j be arbitrary but fixed. By Step III of the algorithm, we have for $\ell = 1, \ldots, n$

$$\|c_{1, j+\ell}\| = \frac{\|c_{n, j+\ell-1}\|}{|c_{n, j+\ell-1} d_{1, j+\ell}|} \leq \frac{1}{\gamma}$$

and for $2 \leq i \leq \ell$

$$\|c_{i, j+\ell}\| \leq \|c_{i-1, j+\ell-1}\| + \frac{\|c_{i-1, j+\ell-1}\| \|d_{1, j+\ell}\|}{|c_{n, j+\ell-1} d_{1, j+\ell}|} \|c_{n, j+\ell-1}\|$$

$$\leq \|c_{i-1, j+\ell-1}\| (1 + \frac{c}{\gamma}).$$

Hence, it follows that

$$\|c_{i, j+n}\| \leq \frac{1}{\gamma}(1 + \frac{c}{\gamma})^{i-1} = \frac{(\gamma + c)^{i-1}}{\gamma^i}, \quad i = 1, \ldots, n,$$

which implies that

$$\|D^{-1}_{j+n}\| \leq \text{const.}$$

Thus

$$\|\bar{s}_j\| \leq \|D^{-1}_j\| \|P_j\| \|g_j\| \leq \text{const}$$

since $\|P_j\| \leq 1$ and $\|g_j\|$ is bounded on S_0.

The following theorem shows that the sequence $\{x_j\}$ generated by the algorithm converges to z minimizing $F(x)$ under the same general conditions as in the method of steepest descent. It is similar to results in [4].

Theorem 1. Assume $S_0 = \{x \in E^n \mid F(x) \le F(x_0)\}$ is bounded and let $\{x_j\}$ be the sequence of point generated by the algorithm. Then:

1) The sequence $F(x_j)$ is strictly monotone decreasing and the sequence $\|\nabla F(x_j)\|$ converges to zero.

2) Every cluster point of $\{x_j\}$ satisfies the equation $\nabla F(x) = 0$. If the roots of $\nabla F(x) = 0$ in S_0 are finite in number and if σ_j is chosen by Step II.1, then $\{x_j\}$ converges to one of them. If $\nabla F(x) = 0$ has a unique root in S_0, it minimizes $F(x)$.

3) If $F(x)$ is pseudo-convex, every cluster point of $\{x_j\}$ minimizes $F(x)$.

Proof:

1) If σ_j is chosen by Step II.2 it follows from the definition of x_{j+1} that $F(x_{j+1}) < F(x_j)$. If $g_j \ne 0$ and σ_j is determined by Step II.1, then $\sigma_j g_j' s_j > 0$ and, by the definition of $h(x_j, \sigma)$,

$$F(x_j) - F(x_j - \sigma_j s_j) \ge \delta \sigma_j g_j' s_j > 0.$$

Suppose there is $\alpha \ge \epsilon > 0$ and a subsequence $\{g_{j'}\}$ such that $\|g_{j'}\|^2 \ge \epsilon$. Then, by definition,

$$g_{j'}' s_{j'} \ge \alpha_{j'} \|g_{j'}\|^2 \ge \epsilon^3.$$

First assume σ_j is chosen according to Step II.1. We want to show that $\{\sigma_j\}$ is bounded away from zero. To this end suppose that a subsequence $\{\sigma_{j''}\}$ converges to zero and define

$$L(\sigma_j s_j) = \{x \in E^n | x = x_j - \tau(\sigma_j s_j), \ 0 \le \tau \le 1\}.$$

Then it follows from the uniform continuity of $\nabla F(x)$ on S_0 that with $\xi_{j''} \in L(\sigma_{j''} s_{j''})$

$$h(x_{j''}, \sigma_{j''}) = 1 + \frac{(\nabla F(\xi_{j''}) - g_{j''})' s_{j''}}{g_{j''} s_{j''}} \to 1$$

since $\{s_{j''}\}$ is bounded. This contradicts $h(x_j, \sigma_j) < 1-\delta$. Hence, there exists $\rho > 0$ such that $\sigma_{j'} > \rho$. Therefore,

$$F(x_{j'}) - F(x_{j'} - \sigma_{j'} s_{j'}) = h(x_{j'}, \sigma_{j'}) \sigma_{j'} g_{j'} s_{j'} \ge \delta \rho \epsilon^3 > 0, \tag{1}$$

in contradiction to the fact that $F(x)$ is bounded below.

Now if σ_j is chosen according to Step II. 2, it follows from the definition of $\bar{\sigma}_j$, $|\sigma_j - \bar{\sigma}_j| \le \beta_j$, $\beta_j \to 0$, and (1) that for j' sufficiently large

$$F(x_{j'}) - F(x_{j'} - \sigma_{j'} s_{j'}) \ge \bar{\epsilon} > 0$$

which again contradicts the fact that $F(x)$ is bounded below. Hence, we have $g_j \to 0$.

2) Let $\{x_j\}$ be a subsequence converging to a cluster point z. Since $\nabla F(x)$ is continuous on S and by 1) $g_j \to 0$,

$$g_j \to \nabla F(z) = 0 .$$

189

If the roots of $\nabla F(x)$ are finite in number, the number of cluster points of $\{x_j\}$ is finite too. Since $\|x_{j+1} - x_j\| \to 0$ it follows readily that $\{x_j\}$ converges to one of those cluster points. Since S_0 is compact and $F(x)$ is differentiable on S, $F(x)$ attains its minimum at a point in S_0 which is necessarily a root of $\nabla F(x) = 0$. Hence, if $\nabla F(x) = 0$ has a unique solution in S_0, it minimizes $F(x)$.

3) By definition [6] $F(x)$ is pseudo-convex if for every x_1 and x_2

$$(\nabla F(x_1))'(x_2 - x_1) \geq 0 \text{ implies } F(x_2) \geq F(x_1)$$

Now let z be a cluster point of $\{x_j\}$. Then $\nabla F(z) = 0$ and

$$F(x) \geq F(z) \text{ for every } x \in E^n.$$

4. Special convergence properties of the algorithm

Throughout this section we assume that there exist positive numbers $\mu \leq \eta$, a j_1 and an open convex set S_1 such that

$$\{x \in E^n \mid F(x) \leq F(x_{j_1})\} \subseteq S_1,$$

$F(x)$ is twice continuously differentiable on S_1 and

$$\mu \|x\|^2 \leq x'G(y)x \leq \eta \|x\|^2 \text{ for all } x \in E^n \text{ and all } y \in S_1.$$

We shall show that under these assumptions the sequence $\{x_j\}$ converges superlinearly toward a z minimizing $F(x)$, provided the constants γ and c are chosen in an appropriate way.

Lemma 2. Let $\mu > 0$ and

$$\mu \|x\|^2 \leq x'G(y)x \leq \eta \|x\|^2 \text{ for all } x \in E^n \text{ and all } y \in S_1.$$

SUPERLINEARLY CONVERGENT MINIMIZATION

Then there exists j_2 such that for $j \geq j_2$

1) $\|d_j\| \leq \eta$, $\|\bar{d}_j\| \leq \eta$
2) $d_{1,j+1} = d_j$ or $\bar{d}_j$, provided $\gamma < \mu$ and $c > \eta$.

Proof:
1) By definition,

$$\bar{d}_j = \frac{g_j - \bar{g}_{j+1}}{\|\sigma_j s_j\|}, \quad \bar{g}_{j+1} = \nabla F(x_j - \|\sigma_j s_j\| \frac{c_{nj}}{\|c_{nj}\|}).$$

Hence, by Taylor's theorem [2],

$$(*) \quad \bar{d}_j = (\int_0^1 G(x_j - t\|\sigma_j s_j\| \frac{c_{nj}}{\|c_{nj}\|}) dt) \frac{c_{nj}}{\|c_{nj}\|}.$$

Since, by Theorem 1, $\|\sigma_j s_j\| \to 0$, it follows that, for j sufficiently large,

$$x_j - \|\sigma_j s_j\| \frac{c_{nj}}{\|c_{nj}\|} \in S_1$$

and

$$\|\bar{d}_j\| \leq \sup_{0 \leq t \leq 1} \|G(x_j - t\|\sigma_j s_j\| \frac{c_{nj}}{\|c_{nj}\|})\| \leq \eta.$$

A similar argument shows that $\|d_j\| \leq \eta$ for j large enough.

2) It suffices to show that, for sufficiently large j,

$$|c'_{nj} \bar{d}_j| \geq \gamma_j \|c_{nj}\|.$$

By (*),

$$\frac{c'_{nj} \bar{d}_j}{\|c_{nj}\|} = \frac{c'_{nj} G(x_j) c_{nj}}{\|c_{nj}\|^2} + \frac{c'_{nj}}{\|c_{nj}\|} (\int_0^1 (G(x_j - t\|\sigma_j s_j\| \frac{c_{nj}}{\|c_{nj}\|}) - G(x_j)) dt) \frac{c_{nj}}{\|c_{nj}\|},$$

191

Thus, it follows that

$$\frac{|c'_{nj}\bar{d}_j|}{\|c_{nj}\|} \to \frac{c'_{nj}G(\gamma_j)c_{nj}}{\|c_{nj}\|^2} \geq \mu.$$

Since $\gamma_j \to \gamma$ this implies

$$|c'_{nj}\bar{d}_j| \geq \|c_{nj}\|$$

for j sufficiently large.

Theorem 2. Let $\mu > 0$ and

$$\mu\|x\|^2 \leq x'G(y)x \leq \eta\|x\|^2 \text{ for all } x \in E^n \text{ and all } y \in S_1,$$

and assume that $\gamma < \mu$ and $c > \eta$. Then:

1) $\|D_j^{-1}P_j - G_j^{-1}\| \to 0$ as $j \to \infty$

   $\|P_j^{-1}D_j - G_j\| \to 0$ as $j \to \infty$.

2) If there is $\theta > 0$ and $L > 0$ such that

$$\|G(x) - G(y)\| \leq L\|x - y\|^\theta \text{ for all } x, y \in S_1$$

then, for $j \geq j_1$,

$$\|D_j^{-1}P_j - G_j^{-1}\| \leq \text{const} \max_{i=1,\ldots,n} \{\|x_{j+1-i} - x_{j-i}\|^\theta\},$$

and

$$\|P_j^{-1}D_j - G_j\| \leq \text{const} \max_{i=1,\ldots,n} \{\|x_{j+1-i} - x_{j-i}\|^\theta\}.$$

Proof: Let $D_j^i = (d_{ij}, \ldots, d_{nj})$ be defined as in Lemma 1. By Lemma 2, we have for $j \geq j_2$,

$$d_{ij} = d_{1,j+1-i} = \hat{d}_{j-i}, \quad i = 1, \ldots, n,$$

where $\hat{d}_{j-i} = d_{j-i}$ or $\bar{d}_{j-i}$.
If we set

$$\hat{s}_{j-i} = \begin{cases} \sigma_{j-i} s_{j-i} & \text{if } \hat{d}_{j-i} = d_{j-i} \\ |\sigma_{j-i} s_{j-i}| \dfrac{c_{n,j-i}}{\|c_{n,j-i}\|} & \text{if } \hat{d}_{j-i} \neq d_{j-i}, \end{cases}$$

it follows from Taylor's theorem [2] and Step III of the algorithm that, for $i = 1, \ldots, n$ and $j \geq j_2$,

$$d_{ij} = \hat{d}_{j-i} = (\int_0^1 G(x_{j-i} - t\hat{s}_{j-i})dt) \frac{\hat{s}_{j-i}}{\|\hat{s}_{j-i}\|}$$

$$= G_{j-i} p_{ij} + (\int_0^1 (G(x_{j-i} - t\hat{s}_{j-i}) - G_{j-i})dt) p_{ij} \quad (1)$$

$$= G_{j-i} p_{ij} - e_{ij},$$

where $p_{ij} = p_{j-i} = \dfrac{\hat{s}_{j-i}}{\|\hat{s}_{j-i}\|}$ and

$$e_{ij} = -(\int_0^1 (G(x_{j-i} - t\hat{s}_{j-i}) - G_{j-i})dt) p_{ij}. \quad (2)$$

Furthermore, let

$$K_{ij} = G_j - G_{j-i}, \quad i = 1, \ldots, n,$$

then by (1),

$$G_j p_{ij} = G_{j-i} p_{ij} + K_{ij} p_{ij}$$
$$= d_{ij} + e_{ij} + K_{ij} p_{ij} \qquad (3)$$

for $i = 1, \ldots, n$ and $j \geq j_2$.

Since $G(x)$ is continuous on the compact set $\{x \in E^n \mid F(x) \leq F(x_{j_1})\}$ and $\|x_{j+1} - x_j\| \to 0$ as $j \to \infty$, $\|G_{j+1} - G_j\| \to 0$ as $j \to \infty$. Thus, for $i = 1, \ldots, n$,

$$\|K_{ij}\| = \|G_{ji} - G_{j-i}\| \to 0 \text{ as } j \to \infty, \qquad (4)$$

and, by (2),

$$\|e_{ij}\| \leq \sup_{0 \leq t \leq 1} (\|G(x_{j-i} - t\hat{s}_{j-i}) - G_{j-i}\|) \to 0 \qquad (5)$$

as $j \to \infty$.

For any $y \in E^n$ with $\|y\| = 1$, let $z' = y'D_j^{-1}$. Using (3) we obtain for $j \geq j_2$,

$$\|D_j^{-1} P_j - G_j^{-1}\| = \sup_{\|y\|=1} \|y'D_j^{-1}P_j - y'G_j^{-1}\|$$

$$= \sup_{\|z'D_j\|=1} \|z'P_j - z'D_j G_j^{-1}\|$$

$$= \sup_{\|z'D_j\|=1} \|\sum_{i=1}^{n} (p_{ij}(z)_i - G_j^{-1} d_{ij}(z)_i)\|$$

$$= \sup_{\|z'D_j\|=1} \|\sum_{i=1}^{n} G_j^{-1}(e_{ij} + K_{ij} p_{ij})(z)_i\|.$$

Since by Lemma 1, $\|D_j^{-1}\|$ is bounded, there is a constant c_1 such that for all j and all $y \in E_n$ with $\|y\|=1$

$$\|z\| = \|y'D_j^{-1}\| \le c_1 .$$

Therefore, for $j \ge j_2$,

$$\|D_j^{-1}P_j - G_j^{-1}\| \le c_1 \|G_j^{-1}\| \sum_{i=1}^{n} (\|e_{ij}\| + \|K_{ij}\|)$$

$$\le \frac{c_1}{\mu} \sum_{i=1}^{n} (\|e_{ij}\| + \|K_{ij}\|) , \qquad (6)$$

and it follows from (4) - (6) that for $j \to \infty$

$$\|D_j^{-1}P_j - G_j^{-1}\| \to 0 .$$

Observe that, for j sufficiently large,

$$\|(D_j^{-1}P_j - G_j^{-1})G_j\| < 1.$$

Thus we obtain from Lemma 2 in [5] that, for j sufficiently large, P_j^{-1} exists and

$$\|P_j^{-1}D_j - G_j\| \le \frac{\|D_j^{-1}P_j - G_j^{-1}\| \|G_j\|^2}{1 - \|D_j^{-1}P_j - G_j^{-1}\| \|G_j\|} \qquad (7)$$

which shows that

$$\|P_j^{-1}D_j - G_j\| \to 0 \text{ as } j \to \infty .$$

2) Now suppose $\|G(x) - G(y)\| \le L\|x - y\|^\theta$ for $x, y \in S_1$. By (4) - (6) we have for $j \ge j_2$,

$$\|D_j^{-1}P_j - G_j^{-1}\| \le \frac{c_1}{\mu} \sum_{i=1}^{n} (\|x_{j-1} - x_{j-i+1}\|^\theta + \|x_j - x_{j-i}\|^\theta)$$

$$\le \text{const} \max_{i=1,\ldots,n} \{\|x_{j-i} - x_{j-i+1}\|^\theta + \|x_j - x_{j-i}\|^\theta\} \quad (8)$$

$$\le \text{const} \max_{i=1,\ldots,n} \{\|x_{j+1-i} - x_{j-i}\|^\theta\}.$$

Because $\|G_j\| \le \eta$, it follows from (7) and (8) that, for j sufficiently large,

$$\|P_j^{-1}D_j - G_j\| \le \text{const} \|D_j^{-1}P_j - G_j^{-1}\|.$$

This completes the proof of the theorem.

Lemma 3. Let the assumptions of Theorem 2 be satisfied. Then

1) There exists j_3 such that for $j \ge j_3$

$$s_j = \bar{s}_j = D_j^{-1} P_j g_j .$$

2) If σ_j is chosen according to Step II.1), then there is j_4 with

$$\sigma_j = 1 \text{ for } j \ge j_4 .$$

If σ_j is chosen according to Step II. 2), then

$$\sigma_j \to 1 \text{ as } j \to \infty,$$

and if, in addition, $\|G(x) - G(y)\| \le L\|x - y\|^\theta$ for all $x, y \in S_1$, and $\beta_j \le \text{const} \max_{i=0,\ldots,n} \{\|g_{j-1}\|^\theta\}$, then

$$|1 - \sigma_j| \le \text{const} \max_{i=0,\ldots,n} \{\|g_{j-i}\|^\theta\} .$$

3) For any j,

$$\|\sigma_j s_j\| \le \text{const} \|g_j\| \quad \text{and}$$

$$\|g_j\| \le \text{const} \|\sigma_j s_j\| .$$

Proof:
1) Since $\nabla F(x)$ is bounded on S_0, it follows from Theorem 2 that

$$G_j^{-1} g_j \to D_j^{-1} P_j g_j = \bar{s}_j \quad \text{as } j \to \infty,$$

and therefore,

$$g_j' \bar{s}_j \to g_j' G_j^{-1} g_j' \quad \text{as } j \to \infty .$$

Since $g_j' G_j^{-1} g_j \ge \frac{1}{\eta} \|g_j\|^2$ and $\alpha_j \to 0$, this implies that, for j sufficiently large,

$$g_j' \bar{s}_j \ge \alpha_j \|g_j\|^2$$

and, hence, $s_j = \bar{s}_j = D_j^{-1} P_j g_j$.
2) By Taylor's theorem,

$$F(x_j - s_j) - F(x_j) = -g_j' s_j + \frac{1}{2} s_j' G(\xi_j) s_j,$$

where $\xi_j \in \{x \in E^n \mid x = x_j - ts_j, \; 0 \le t \le 1\}$. If we write $G(\xi_j) = G_j + (G(\xi_j) - G_j)$ and divide the above equation by $g_j' s_j$, it follows from the definition of $h(x_j, \sigma)$ that

$$h(x_j, 1) = 1 - \frac{1}{2} \frac{s_j' G_j s_j + s_j'(G(\xi_j) - G_j) s_j}{g_j' s_j}$$

or

$$\left| h(x_j, 1) - 1 + \frac{s_j' G_j s_j}{2 g_j' s_j} \right| = \left| \frac{s_j'(G(\xi_j) - G_j) s_j}{2 g_j' s_j} \right| \quad (1)$$

By the first part of the lemma and by Theorem 2,

$$s_j = D_j^{-1} P_j g_j \to G_j^{-1} g_j .$$

Thus

$$\frac{s_j' G_j s_j}{2 g_j' s_j} \to \frac{g_j' s_j}{2 g_j' s_j} = \frac{1}{2} \quad (2)$$

Furthermore,

$$\frac{|s_j'(G(\xi_j) - G_j) s_j|}{2 g_j' s_j} \le \frac{\|G(\xi_j) - G_j\| \, \|s_j\|^2}{2 g_j' s_j} \le$$

$$\frac{\|G(\xi_j) - G_j\| \, \|s_j\|^2}{2 \alpha_j \|g_j\|^2} \quad \text{for } j \ge j_3 \quad (3)$$

By the first part of the lemma and Theorem 2 this last expression converges to

$$\frac{\|G(\xi_j) - G_j\| \|G_j^{-1}g_j\|^2}{2\alpha_j \|g_j\|^2} \le \frac{\|G(\xi_j) - G_j\|}{2\alpha\mu^2}$$

Since $G(x)$ is uniformly continuous on $\{x \in E^n | F(x) \le f(x_{j_1})\}$ and $\|s_j\| \to 0$ as $j \to \infty$ the above expression goes to zero as $j \to \infty$. Hence, by (1) - (3),

$$h(x_j, 1) \to \frac{1}{2} > \delta .$$

Suppose $j \ge j_3$. Since $\bar{\sigma}_j > 0$ minimizes $F(x_j - \sigma s_j)$ on the ray $\{x \in E^n | x = x_j - \sigma s_j, \sigma \ge 0\}$ we have

$$(\nabla F(x_j - \bar{\sigma}_j s_j))' s_j = 0.$$

By Lagrange's formula [9] there exists $\xi_j \in \{x \in E^n | x = x_j - t(\sigma_j s_j), 0 \le t \le 1\}$ such that

$$0 = (\nabla F(x_j - \bar{\sigma}_j s_j))' s_j = -g_j' s_j + \bar{\sigma}_j s_j' G(\xi_j) s_j .$$

Hence,

$$\|\bar{\sigma}_j s_j\| = \frac{g_j' s_j \|s_j\|}{s_j' G(\xi_j) s_j} \le \frac{\|g_j\|}{\mu} . \quad (4)$$

From (4), Theorem 2 and the first part of the lemma we obtain, for j sufficiently large,

$$\bar{\sigma}_j = \frac{s_j' P_j^{-1} D_j s_j}{s_j' G(\xi_j) s_j} = 1 + \frac{s_j'(P_j^{-1} D_j - G(\xi_j)) s_j}{s_j' G(\xi_j) s_j} ,$$

or

$$|1 - \overline{\sigma}_j| \leq \frac{\|G(\xi_j) - G_j\| + \|G_j - P_j^{-1}D_j\|}{\mu} \to 0 \qquad (5)$$

as $j \to \infty$.

Since σ_j, as chosen by Step II, 2) converges to $\overline{\sigma}_j$ for $j \to \infty$, we have $\sigma_j \to 1$ as $j \to \infty$.

Now suppose that $\|G(x) - G(y)\| \leq L\|x - y\|^\theta$. Then it follows from (5) and Theorem 2 that

$$|1 - \overline{\sigma}_j| \leq \text{const}(\|x_{j+1} - x_j\|^\theta + \max_{i=1,\ldots,n} \{\|x_{j+1-i} - x_{j-i}\|^\theta\})$$

which, by the last part of the lemma gives

$$|1 - \overline{\sigma}_j| \leq \text{const} \max_{i=0,\ldots,n} \{\|g_{j-i}\|^\theta\}.$$

Since $|\sigma_j - \overline{\sigma}_j| \leq \text{const} \max_{i=0,\ldots,n} \{\|g_{j-i}\|^\theta\}$, this completes the proof of part 2 of the lemma.

3) Let $j \geq j_3$. Then $s_j = D_j^{-1}P_j g_j$ and

$$\|s_j\| \leq \|D_j^{-1}P_j\| \|g_j\|.$$

Since $\sigma_j \to 1$, $D_j^{-1}P_j \to G_j^{-1}$ and $\|G_j^{-1}\| \leq \frac{1}{\mu}$ for $j \geq j_1$, this implies

$$\|\sigma_j s_j\| \leq \text{const} \|g_j\| \text{ for all } j.$$

Furthermore, for j sufficiently large,

$$g_j = P_j^{-1}D_j s_j.$$

By Theorem 2, $P_j^{-1} D_j \to G_j$. Since $\|G_j\| \le \eta$ for $j \ge j_1$, we have

$$\|g_j\| \le \text{const } \|\sigma_j s_j\| \quad \text{for all } j.$$

The next theorem shows that the sequence of gradients $\{\nabla F(x_j)\}$ converges superlinearly to zero.

Theorem 3. Let

$$\mu \|x\|^2 \le x' G(y) x \le \eta \|x\|^2 \quad \text{for all } x \in E^n \text{ and all } y \in S_1,$$

and assume that $\gamma < \mu$ and $c > \eta$. Then

$$\frac{\|g_{j+1}\|}{\|g_j\|} \to 0 \quad \text{as } j \to \infty.$$

If, in addition, there is $\theta > 0$ and $L > 0$ such that

$$\|G(x) - G(y)\| \le L \|x - y\|^\theta \quad \text{for all } x, y \in S_1$$

and if $|\sigma_j - \bar{\sigma}_j| \le \text{const} \|g_{j-n}\|^\theta$ or σ_j is chosen according to Step II, 1), then

$$\frac{\|g_{j+1}\|}{\|g_j\|} \le \text{const} \|g_{j-n}\|^\theta.$$

Proof: For j sufficiently large, it follows from Taylor's theorem [2] and Lemma 3 that

$$g_{j+1} = g_j - (\int_0^1 G(x_j - t(\sigma_j s_j)) dt) \sigma_j s_j$$

$$= g_j - \sigma_j G_j D_j^{-1} P_j g_j - (\int_0^1 (G(x_j - t\sigma_j s_j) - G_j) dt) \sigma_j s_j.$$

Hence, writting $D_j^{-1} P_j = G_j^{-1} + (D_j^{-1} P_j - G_j^{-1})$ we have

$$\|g_{j+1}\| \leq |1 - \sigma_j| \|g_j\| + \sigma_j \|G_j\| \|G_j^{-1} - D_j^{-1} P_j\| \|g_j\| +$$

$$+ \sup_{0 \leq t \leq 1} \|G(x_j - t\sigma_j s_j) - G_j\| \|\sigma_j s_j\|,$$

or

$$\frac{\|g_{j+1}\|}{\|g_j\|} \leq |1 - \sigma_j| + \sigma_j \eta \|G_j^{-1} - D_j^{-1} P_j\| + \qquad (1)$$

$$+ \sup_{0 \leq t \leq 1} \|G(x_j - t\sigma_j s_j) - G_j\| \frac{\|\sigma_j s_j\|}{\|g_j\|},$$

which by Theorem 2 and Lemma 3 implies $\dfrac{\|g_{j+1}\|}{\|g_j\|} \to 0$

as $j \to \infty$.

Furthermore, this implies that for all j

$$\max_{i=0,\ldots,n} \|g_{j-i}\| \leq \text{const} \|g_{j-n}\|.$$

Hence, if $\|G(x) - G(y)\| \leq L \|x - y\|^\theta$ for all $x, y \in S_1$, it follows from Theorem 2 and Lemma 3 that for $j \geq j_1$

$$\|D_j^{-1} P_j - G_j^{-1}\| \leq \text{const} \|g_{j-n}\|^\theta$$

and

$$|1 - \sigma_j| \leq \text{const} \|g_{j-n}\|^\theta,$$

which by (1) implies

$$\frac{\|g_{j+1}\|}{\|g_j\|} \leq \text{const} \|g_{j-n}\|^\theta.$$

Theorem 4. Let the assumptions of Theorem 3 be satisfied. Then there is a unique $z \in E^n$ which minimizes $F(x)$ and

$$\frac{\|z - x_{j+1}\|}{\|z - x_j\|} \to 0 \text{ as } j \to \infty.$$

If, in addition, there is $\theta > 0$ and $L > 0$ with

$$\|G(x) - G(y)\| \leq L\|x - y\|^\theta \text{ for all } x, y \in S_1$$

and $|\sigma_j - \bar{\sigma}_j| \leq \text{const } \|x_{j-n+1} - x_{j-n}\|^\theta$ or σ_j is chosen according to Step II.1), then

$$\frac{\|z - x_{j+1}\|}{\|z - x_j\|} \leq \text{const } \|z - x_{j-n}\|^\theta.$$

Proof: The assumptions on $F(x)$ imply that $F(x)$ is strictly convex on S_1. Hence, there is a unique z minimizing $F(x)$, and, by Theorem 1, $x_j \to z$. By Taylor's theorem [2],

$$g_j = \nabla F(z) + (\int_0^1 G(x_j - t(x_j - z))dt)(x_j - z) \quad (1)$$

$$= 0 + G_j(x_j - z) + (\int_0^1 (G(x_j - t(x_j - z)) - G_j)dt)(x_j - z).$$

Hence, for j sufficiently large,

$$x_{j+1} - z = x_j - z - \sigma_j D_j^{-1} P_j g_j$$

$$= x_j - z - \sigma_j(x_j - z) - G_j^{-1}(\int_0^1 (G(x_j - t(x_j - z)) - G_j)dt)(x_j - z)$$

$$- (D_j^{-1} P_j - G_j^{-1})[G_j(x_j - z) + (\int_0^1 (G(x_j - t(x_j - z)) - G_j)dt)(x_j - x)],$$

or

$$\|x_{j+1} - z\| \le |1-\sigma_j| \ \|x_j - z\| + \|D_j^{-1}P_j - G_j^{-1}\| \ \|G_j\| \ \|x_j - z\| +$$

$$+ \|D_j^{-1}P_j - G_j^{-1}\| \sup_{0 \le t \le 1} \|G(x_j - t(x_j - z)) - G_j\| \ \|x_j - z\| +$$

$$+ \|G_j^{-1}\| \sup_{0 \le t \le 1} \|G(x_j - t(x_j - x)) - G_j\| \ \|x_j - z\| ,$$

from which we obtain

$$\frac{\|x_{j+1} - z\|}{\|x_j - z\|} \le |1 - \sigma_j| + \|D_j^{-1}P_j - G_j^{-1}\|(\eta +$$

$$+ \sup_{0 \le t \le 1} \|G(x_j - t(x_j - z)) - G_j\|) + \qquad (2)$$

$$+ \frac{1}{\mu} \sup_{0 \le t \le 1} \|G(x_j - t(x_j - z)) - G_j\| .$$

By Lemma 3, Theorem 2, and the uniform continuity of $G(x)$ on $\{x \in E^n \mid F(x) \le F(x_{j_1})\}$ the right hand side of this inequality converges to zero.

Suppose $\|G(x) - G(y)\| \le L\|x - y\|^\theta$ for all $x, y \in S_1$ and $|\sigma_j - \bar{\sigma}_j| \le \text{const} \|x_{j-n+1} - x_{j-n}\|^\theta$ or σ_j is chosen according to Step II.1). Then it follows from Lemma 3 that

$$|1 - \sigma_j| \le \text{const} \max_{i=0,\ldots,n} \{\|g_{j-i}\|^\theta\}$$

$$\le \text{const} \|g_{j-n}\|^\theta \qquad (3)$$

since, by Theorem 3, $\dfrac{\|g_j\|}{\|g_{j+1}\|} \to 0$.

Furthermore, by Theorem 2 and Lemma 3,

$$\|D_j^{-1} P_j - G_j^{-1}\| \le \text{const} \max_{i=1,\ldots,n} \{\|g_{j-i}\|^\theta\} \qquad (4)$$

$$\le \text{const} \|g_{j-n}\|^\theta .$$

Finally,

$$\sup_{0 \le t \le 1} \|G(x_j - t(x_j - z)) - G_j\| \le L \|x_j - z\|^\theta \qquad (5)$$

$$\le \text{const} \|x_{j-n} - z\|^\theta ,$$

because $\dfrac{\|z - x_{j+1}\|}{\|z - x_j\|} \to 0$.

Observe that by (1),

$$\|g_j\| \le \sup_{0 \le t \le 1} \|G(x_j - t(x_j - z))\| \, \|x_j - z\| \qquad (6)$$

$$\le \text{const} \|x_j - z\| \quad \text{for all } j .$$

Hence, it follows from (2) – (6) that

$$\dfrac{\|x_{j+1} - z\|}{\|x_j - z\|} \le \text{const} \|x_{j-n} - z\|^\theta .$$

REFERENCES

1. Davidon, W. C., Variable metric method for minimization, A. E. C. Research and Development Report, ANL-5990 (1959).

2. Dieudonnee, J., Foundations of modern analysis. New York, Academic Press, 1960.

3. Fletcher, R. and M. J. D. Powell, A rapidly convergent descent method for minimization. The Computer Journal 6, 163-168 (1963).

4. Goldstein, A. A., Constructive Real Analysis. New York, Harper and Row, 1967.

5. Goldstein, A. A. and J. F. Price, An Effective Algorithm for Minimization. Numerische Mathematik 10, 184-189 (1967).

6. Mangasarian, O. L., Pseudo-convex functions, J. SIAM Control 3, 281-290 (1965).

7. Powell, M. J. D., On the convergence of the variable metric algorithm, to appear.

8. Vainberg, M. M., Variational Methods for the Study of Nonlinear Operators. San Francisco, Holden-Day, 1964.

A Second Order Method for the Linearly Constrained Nonlinear Programming Problem

GARTH P. McCORMICK

ABSTRACT

An algorithm using second derivatives for solving the problem: minimize $f(x)$ subject to $Ax - b \geq 0$ is presented. Convergence to a Second-Order Kuhn Tucker Point is proved. If the strict second-order sufficiency conditions hold, the <u>rate</u> of convergence of the algorithm is shown to be superlinear or even quadratic with a Lipschitz condition on the second derivatives of $f(x)$.

GARTH P. McCORMICK

1. Introduction

The mathematical programming problem (LC) to be solved is:

$$\text{minimize } f(x) \qquad (1.1)$$

$$\text{subject to } Ax - b \geq 0 \qquad (1.2)$$

where $f(x)$ is a twice continuously differentiable function of x ($x \in E^n$). There are many published algorithms addressed to this problem. A partial listing of some of the best known are: a class of algorithms called Methods of Feasible Directions, [14], [11], the Convex-Simplex Method, [13], the Gradient Projection Method, [9], and its accelerated versions [5] and [10], the Reduced Gradient Method [3] and [12], and the Variable Reduction Method [8]. All of these methods explicitly use only values of the function $f(x)$ and its first derivatives to solve the programming problem (LC). This paper extends the Variable Reduction Method [8] by using second as well as first derivatives. Section 2 contains a statement of the algorithm. In Section 3 is a proof of convergence of the algorithm to a Second Order Kuhn-Tucker Point (one satisfying first and second-order necessary conditions for a local minimum of problem (LC)). The *rate* of convergence of the algorithm is analyzed in

SECOND ORDER METHOD

Section 4. The remainder of this section explains the notation to be used, and the first and second order necessary, and first and second order sufficiency conditions for a local minimum to problem (LC).

Notation:

E^n	Euclidean n-space
n	the number of variables
x	the vector of unknowns, $x \in E^n$
f(x)	the function to be minimized (1.1)
$x(k,\ell)$	The value of x at the beginning of the ℓth segment of the kth iteration
'	the symbol for transposition
A, b	the $m \times n$ matrix and $m \times 1$ vector defining the linear constraints (1.2)
$c(k,\ell)$	the $m \times 1$ vector of constraint values evaluated at $x(k, \ell)$
$g(k,\ell)$	the vector of first partial derivatives of f(x) evaluated at $x(k, \ell)$
$G(k,\ell)$	the $n \times n$ matrix of second partial derivatives of f(x) evaluated at $x(k, \ell)$
$S(k,1)$	an ordered set of indices corresponding to constraints exactly satisfied ("binding") at $x(k, 1)$
$S(k,\ell)$	for $\ell > 1$, the set of indices corresponding to constraints which have been "binding" at least at one point during iteration k
$r(k, \ell)$	the number of indices in $S(k, \ell)$
$[B(k,\ell), D(k,\ell)]$	an $r(k,\ell) \times n$ submatrix of A whose rows are determined by the indices in $S(k, \ell)$, where $B(k,\ell)$ is a square invertible matrix, and where columns correspond to the division into type 1 and type 2 variables mentioned below

$\begin{bmatrix} x_1(k,\ell) \\ x_2(k,\ell) \end{bmatrix}$	division of $x(k,\ell)$ into two parts, the top part considered as type 1 or "dependent" variables corresponding to the columns of $B(k,\ell)$, and the bottom part considered as type 2 or independent variables corresponding to the columns of $D(k,\ell)$
$g_1(k,\ell)$	the $r(k,\ell)$ by 1 vector of first partial derivatives of $f(x)$ corresponding to the division of variables explained above. Analogous notation is used for other quantities which depend on just the first or second group of variables
$H(k,\ell)$	an $[n-r(k,\ell)]$ by $[n-r(k,\ell)]$ matrix given by equation (2.8)
$h(k,\ell)$	an $n-r(k,\ell)$ by one vector given by (2.9)
$s(k,\ell)$	direction vector computed at $x(k,\ell)$
$p(k)$	the number of linear segments for iteration k
$\theta(k,\ell)$	the nonnegative scalar where

$$x(k,\ell+1) = x(k,\ell) + s(k,\ell)\theta(k,\ell),$$

$$\ell = 1, \ldots, p(k).$$

$x(k, p(k)+1)$	the terminal point for the k th iteration, also the starting point of the (k+1)th iteration, i.e.

$$x(k, p(k) + 1) = x(k + 1, 1).$$

$x^k(t)$	a continuous vector function giving the value of x for any increment during the kth iteration, where

$$x^k(0) \equiv x(k, 1),$$
$$x^k(T_k) \equiv x(k, p(k) + 1),$$
$$T_k \equiv \sum_{\ell=1}^{p(k)} \theta(k, \ell).$$

SECOND ORDER METHOD

We now formulate a regularity condition which will be used for all of the convergence theorems.

Linear Independence Assumption

Let $\bar{x}$ be any feasible point (point satisfying (1.2)). Let $\bar{\mathcal{S}} \equiv \{i \mid c_i(\bar{x}) = 0\}$. Then the set of gradients

$$\{a_i\}, \quad i \in \bar{\mathcal{S}} \quad \text{is \underline{linearly independent}.} \qquad (1.3)$$

We say $\bar{x}$ is a <u>Kuhn-Tucker Point</u> (KTP) if $\bar{x}$ satisfies the following well-known Kuhn-Tucker [6] conditions which are necessary for a linearly constrained programming problem for $\bar{x}$ to be a local minimum to problem (LC).

There exists $\bar{\lambda}$ such that

$$g(\bar{x}) - \sum_{i=1}^{m} a_i \bar{\lambda}_i = 0, \qquad (1.4)$$

$$\bar{\lambda}_i c_i(\bar{x}) = 0, \quad i = 1, \ldots, m, \qquad (1.5)$$

$$\bar{\lambda}_i \geq 0, \quad i = 1, \ldots, m, \qquad (1.6)$$

$$c_i(\bar{x}) \geq 0, \quad i = 1, \ldots, m. \qquad (1.7)$$

Second Order Conditions

The Kuhn-Tucker conditions (1.4)-(1.7) are "first-order" necessary conditions in that they only involve first derivatives of the objective function. Additional information is obtained when second derivative information is used. A detailed discussion of such conditions with references is contained in [4]. We state without proof two lemmas, one stating second-order necessary, the other stating second-order sufficient conditions concerning local minima of problem (LC).

Lemma 1. [Second Order Necessary Conditions]

If: i) $f(x)$ is twice continuously differentiable,
    ii) $\bar{x}$ is a local minimum,
    iii) the linear independence assumption (1.3) holds,

then: a) $\bar{x}$ is a KTP (i.e. there exists a $\bar{\lambda}$ such that $(\bar{x}, \bar{\lambda})$ satisfy (1.4)-(1.7), and

b) $$z'G(\bar{x})z \geq 0 \qquad (1.8)$$

for all z satisfying

$$z'a_i = 0, \quad \text{all } i \in \bar{B},$$

where

$$\bar{B} \equiv \{i \mid c_i(\bar{x}) = 0\}.$$

These are not the most stringent necessary conditions, but they are ones for which convergence can be proved. A point satisfying (1.4)-(1.8) will be called a <u>Second-Order Kuhn-Tucker Point</u> (SOKTP).

If a point x^* satisfies conditions which are similar to, but slightly stronger than those expressed in lemma 1, then that point is an isolated local minimum.

Lemma 2. [Second Order Sufficiency Conditions]

If: i) $f(x)$ is twice continuously differentiable,
    ii) associated with a point x^* is a λ^* such that

$$(x^*, \lambda^*) \text{ is a KTP,}$$

    iii) for that λ^*

$$z'G(x^*)z > 0 \qquad (1.9)$$

for all $z \neq 0$ such that

SECOND ORDER METHOD

$$z'a_i \geq 0, \quad \text{all } i \in (B^* - D^*),$$

$$z'a_i = 0, \quad \text{all } i \in D^*, \quad \text{where}$$

$$B^* \equiv \{i \mid c_i(x^*) = 0\},$$

$$D^* \equiv \{i \mid i \in B^*, \lambda_i^* > 0\},$$

then: a) x^* is an <u>isolated</u> local minimum of problem (LC).

2. The algorithm

2-1. Movement Along Piece-Wise Linear Segments

An iteration of the algorithm consists of movement along a continuous path made up of a finite number of linear segments. For this reason we denote the value of x at the beginning of the ℓth segment of the kth iteration to be $x(k, \ell)$. We let $p(k)$ denote the number of segments at the kth iteration. (This value is not known until the iteration is over). In all cases the direction vector $s(k, \ell)$ from which movement from $x(k, \ell)$ proceeds is given by (2.6). The steps of the algorithm are as follows.

Step (0, 1). Let $x(0, 1)$ be the given feasible starting point.

Step (k, 1), $k \geq 0$. Let $x(k, 1)$ denote the value of x at the start of the kth iteration. Let $S(k, 1)$ denote an ordered set of indices i for which $c_i(k, 1) = 0$, $i = 1, \ldots, m$. Let $r(k, 1)$ denote the number of indices in $S(k, 1)$. Let $[B(k, 1), D(k, 1)]$ denote the $r(k, 1)$ by n submatrix of A whose rows correspond to those constraints with indices in $S(k, 1)$. I.e., if i' is the ith index in $S(k, 1)$, then the ith row of the matrix $[B(k, 1), D(k, 1)]$ is the (i')th row of A. We assume the variables are ordered so that the $r(k, 1)$ by $r(k, 1)$ matrix $B(k, 1)$ has an inverse. The linear independence assumption made in (1.3) allows this to be done without loss of generality.

The vector of variables $x(k, 1)$ is split into two parts $x_1(k, 1)$ and $x_2(k, 1)$. The variables in $x(k, 1)$ correspond to the columns of $B(k, 1)$ and are called type one or "dependent" variables. The variables in $x_2(k, 1)$ correspond to the columns in $D(k, 1)$ and are called type 2 or "dependent" variables.

Given the values which define the direction vector in (2.6), a new point $x(k, 2)$ is obtained by moving along $s(k, 1)$ starting from $x(k, 1)$. We give this procedure for the general step (k, ℓ) since it is the same as that for step $(k, 1)$.

Generation of $x(k, \ell+1)$ from $x(k, \ell)$ and $s(k, \ell)$ proceeds as follows.

Solve the one dimensional programming problem:

$$\underset{\theta \geq 0}{\text{minimize}} \quad f[x(k, \ell) + s(k, \ell)\theta] . \qquad (2.1)$$

Let $\theta_1(k, \ell)$ be the value of θ closest to zero which is a local minimizing point of this problem.

Let $\theta_2(k, \ell)$ be the value of θ closest to zero which is a local minimum of

$$\text{maximize } \theta$$

$$\text{subject to} \quad c[x(k, \ell) + s(k, \ell)\theta] \geq 0 . \qquad (2.2)$$

Set $\theta(k, \ell) = \min[\theta_1(k, \ell), \theta_2(k, \ell)]$.

Since it was assumed that $x(0, 1)$ was a feasible point, then for all $\ell \geq 1$, and all $k \geq 0$ it follows that $\theta_2(k, \ell)$ (and hence $\theta(k, \ell)$) are both nonnegative.

Now if $\theta(k, \ell) = \theta_1(k, \ell)$, the iteration is over and $p(k) = \ell$, $x(k, p(k) + 1) = x(k, \ell) + s(k, \ell)\theta(k, \ell)$, and $x(k + 1, 1) = x(k, p(k) + 1)$. If

$$\theta(k, \ell) < \theta_1(k, \ell), \qquad (2.3)$$

then the iteration continues.

SECOND ORDER METHOD

Finding the value $\theta(k, \ell)$ assumes it is possible to find the local minimum closest to zero of a one-dimensional programming problem. That this can be done is an assumption of this paper. Further discussions of this problem can be found in [11] and [14] where the "step-size" function is examined.

Step $(k, \ell+1)$, $(\ell \geq 0)$.

Let $x(k, \ell+1)$ denote the value of x at the beginning of the $(\ell+1)$th segment at the kth iteration. Before generating the direction vector $s(k, \ell+1)$ it is necessary to construct the set $S(k, \ell+1)$ and the matrix $[B(k, \ell+1), D(k, \ell+1)]$ from $S(k, \ell)$ and $[B(k, \ell), D(k, \ell)]$.

All the indices corresponding to constraints which are binding at $x(k, \ell+1)$ but which are not already contained in the set $S(k, \ell)$ are added to it producing $S(k, \ell+1)$. The new matrix $[B(k, \ell+1), D(k, \ell+1)]$ is given as

$$[B(k,\ell+1), D(k,\ell+1)] \equiv \begin{bmatrix} B(k,\ell), & N_1(k,\ell), & N_2(k,\ell) \\ N_3(k,\ell), & N_4(k,\ell), & N_5(k,\ell) \end{bmatrix} \quad (2.4)$$

where

$$D(k,\ell) = [N_1(k,\ell), N_2(k,\ell)].$$

The rows $[N_3(k,\ell), N_4(k,\ell), N_5(k,\ell)]$ consist of the rows of A corresponding to the constraints just encountered (whose indices were not in the set $S(k, \ell)$). The columns $\begin{bmatrix} N_1(k,\ell) \\ N_4(k,\ell) \end{bmatrix}$ correspond to variables which are changed from type two to type one variables.

Now $r(k, \ell+1)$ is the number of indices in $S(k, \ell+1)$. It is also of course the number of rows in the matrix (2.4). If

215

$$\text{rank}[B(k, \ell+1), D(k, \ell+1)] < r(k, \ell+1), \qquad (2.5)$$

then the iteration is terminated and $p(k) = \ell$,

$$x(k+1, 1) \equiv x(k, \ell+1) = x(k, p(k) + 1).$$

Otherwise we can assume that the variables are so ordered $B(k, \ell+1)$ has an inverse, and continue step $(\ell+1)$ of iteration k.

This second cause for terminating an iteration occurs when the gradients corresponding to indices in the set $S(k, \ell+1)$ are linearly dependent. Note that the linear independence assumption (1.3) implies that not all of these constraints are binding.

2-2. Direction Vector Generation, all (k, ℓ), $\ell \geq 1$

In all cases for $\ell \geq 1$ the direction vector is given by

$$s(k,\ell) = \begin{bmatrix} -B(k,\ell)^{-1} D(k,\ell) \\ I \end{bmatrix} s_2(k,\ell) - \begin{bmatrix} B(k,\ell)^{-1} \\ 0 \end{bmatrix} u(k,\ell)$$

$$(2.6)$$

The various quantities involved in this expression will be defined below. The direction vector computations fall into two cases depending on the reason for termination of step $(\ell-1)$. (Recall that for a new segment to be used the reason for termination of movement along $s(k, \ell-1)$ must have been the encountering of a constraint boundary).

New Constraint Case (NCC)

Suppose that for iteration k the $(\ell-1)$th step terminated because the boundary of a constraint (s) whose index was not in $S(k, \ell-1)$ was encountered. (We also consider here the computation of the direction vector for

SECOND ORDER METHOD

$\ell = 1$). We assume that the set $S(k,\ell)$ and the matrices $B(k, \ell)$ and $D(k, \ell)$ have been computed as indicated in Section 2-1.

A matrix $H(k, \ell)$ is generated in the following way. Let

$$T(k,\ell) \equiv \begin{bmatrix} -B(k,\ell)^{-1}D(k,\ell) \\ I \end{bmatrix}. \qquad (2.7)$$

Now $\nabla^2 f(k, 1)$ is denoted by $G(k, 1)$. Let

$$H(k,\ell) \equiv T'(k,\ell)G(k,1)T(k,\ell) . \qquad (2.8)$$

Also, let

$$h(k,\ell) \equiv T'(k,\ell)g(k,\ell) . \qquad (2.9)$$

The computation of $s_2(k,\ell)$ is different according to four cases. Let $\delta(k, \ell)$ denote the smallest eigenvalue of $H(k, \ell)$.

NCC 1. The smallest eigenvalue of $H(k, \ell)$ is "very negative," i.e.

$$\delta(k,\ell) < -\epsilon_4 < 0$$

where ϵ_4 is a parameter of the algorithm. Let $e(k, \ell)$ denote an eigenvector associated with $\delta(k, \ell)$. Set

$$s_2(k,\ell) = e(k,\ell) \qquad (2.10)$$

where, without loss of generality it can be assumed that

$$-e'(k,\ell)h(k,\ell) \leq 0 \qquad (2.11)$$

Also for NCC1, set

$$u(k,\ell) = 0. \qquad (2.12)$$

NCC2. The smallest eigenvalue of $H(k,\ell)$ is negative but "nonzero," i.e.

$$-\epsilon_4 \leq \delta(k,\ell) < 0.$$

In this case let $e(k,\ell)$ denote an eigenvector associated with $\delta(k,\ell)$, and let

$$s_2(k,\ell) = e(k,\ell) + h(k,\ell) \qquad (2.13)$$

where

$$e'(k,\ell)h(k,\ell) \geq 0$$

can be assumed without loss of generality as in NCC1.
Let i' be the ith index in $\mathcal{S}(k,\ell)$ and let

$$\hat{\lambda}(k,\ell,i) \qquad (2.14)$$

denote the ith component of $B'(k,\ell)^{-1}g_1(k,\ell)$.
Let $u(k,\ell,i)$ denote the element of the $u(k,\ell)$ given by

$$u(k,\ell,i) = \begin{cases} 0 & \text{if } \hat{\lambda}(k,\ell,i) \geq 0 \text{ and } c_i(k,\ell) = 0 \\ c_{i'}(k,\ell) & \text{if } \hat{\lambda}(k,\ell,i) \geq 0 \text{ and } c_{i'}(k,\ell) > 0 \\ \hat{\lambda}(k,\ell,i) & \text{if } \hat{\lambda}(k,\ell,i) < 0 \end{cases}.$$

$$(2.15)$$

SECOND ORDER METHOD

NCC3. The smallest eigenvalue of $H(k, \ell)$ is "very positive", i.e.

$$\delta(k,\ell) > \epsilon_4.$$

Here,

$$s_2(k,\ell) = H(k,\ell)^{-1} h(k,\ell). \qquad (2.16)$$

Let $i' > i$, $\hat{\lambda}(k,\ell,i)$, and $u(k,\ell,i)$ be as defined above. For this case

$$u(k,\ell,i) = \begin{cases} 0 & \text{if } \hat{\lambda}(k,\ell,i) \geq 0, \ c_{i'}(k,\ell)=0 \\ 1 & \text{otherwise} \end{cases}. \qquad (2.17)$$

Note that the quantity in (2.14) is an estimate of the Kuhn-Tucker multiplier of (1.4) in that if $x(k,1)$ were a KTP, then (2.14) would be the $\overline{\lambda}_{i'}$ of the Kuhn-Tucker equation (1.4).

NCC4. The smallest eigenvalue of $H(k, \ell)$ is positive but "small", i.e.

$$0 \leq \delta(k, \ell) \leq \epsilon_4 \quad .$$

Here,

$$s_2(k, \ell) = h(k, \ell), \qquad (2.17.5)$$

and $u(k, \ell)$ is given as in NCC3, equation (2.17).

Old Constraint Case (OCC)

Suppose for iteration k the $(\ell-1)$th step terminated because only the boundary of a constraint(s) whose index was currently in $\mathcal{S}(k, \ell-1)$ was encountered. Here the

219

vector $s_2(k, \ell)$ remains unchanged, i.e.

$$s_2(k, \ell) = s_2(k, \ell-1). \qquad (2.18)$$

All components of $u(k, \ell)$ remain the same except those corresponding to the constraints the encountering of whose boundaries caused termination of step $(\ell-1)$. Let $\hat{i}$ be such an index. Then

$$u(k, \ell, \hat{i}) = 0. \qquad (2.19)$$

3. Convergence of the Algorithm

It is useful to define a continuous vector function $x^k(t)$ which gives the value of x for any finite step taken during the course of the kth iteration.

Given $t_1 \geq 0$, let $q \equiv q(k, t_1)$ be the smallest integer for which $\sum_{\ell=1}^{q} \theta(k, \ell) > t_1$. Let $\tau(k, t_1) = -\sum_{\ell=1}^{q-1} \theta(k, \ell) + t_1$. Then

$$x^k(t_1) = x(k,1) + \sum_{\ell=1}^{q-1} s(k,\ell)\theta(k,\ell) + s(k,q)\tau(k,t_1). \qquad (3.1)$$

Note that $x^k(0) = x(k, 1)$, and $x^k(T_k) = x^{k+1}(0) = x[k, p(k)+1]$ where $T_k = \sum_{\ell=1}^{p(k)} \theta(k, \ell)$.

The next lemma states, in effect, that if $\bar{x}$, any limit point of the sequence of points beginning each iteration is not a SOKTP, then consecutive limit points are "not close" to it.

Lemma 3. Suppose $\bar{x}$ is a limit point of $\{x(k,1)\}$. For simplicity of notation assume that $\{x(k,1)\}$ denotes also the subsequence comverging to $\bar{x}$.

SECOND ORDER METHOD

If: i) $f(x)$ is a twice continuously differentiable function,
    ii) the direction vectors $\{s(k,\ell)\}$ are as given by the algorithm,
    iii) the scalars $\{\theta(k,\ell)\}$ are as given by the algorithm,
    iv) the linear independence assumption (1.3) holds,
    v) $\liminf\limits_{k \to \infty} \max\limits_{0 \le t \le T_k} \| x^k(t) - x^k(0) \| = 0,$ (3.2)

(where $x^k(0)$ is also used to denote the subsequence of the original sequence converging to $\bar{x}$),

then: a) $\bar{x}$ is a SOKTP.

Proof: Since

$$\max_{0 \le t \le T_k} \| x^k(t) - x^k(0) \| \ge \| x^k(T_k) - x^k(0) \| =$$

$$= \| x[k, p(k)+1] - x(k, 1) \|,$$

(3.2) implies that

$$\liminf_{k \to \infty} \| x[k, p(k)+1] - x(k, 1) \| = 0.$$

Under assumption (v), for k large, all $x^k(t)$, $0 \le t \le T_k$ becomes close to $x^k(0)$. Let $\{x[k, p(k)+1]\}$ denote a converging subsequence of the successor sequence of $\{x(k, 1)\}$. There are two cases to consider. First, the iteration from $x(k, 1)$ to $x[k, p(k)+1]$ terminates infinitely often because the gradients of the constraints whose indices are in $S(k, p(k)+1)$ are linearly dependent. Assumption (v) implies that for k large, $S(k, p(k)+1)$ consists only of indices of constraints equal to zero at $\bar{x}$. Hence termination because of linear dependence cannot occur infinitely often since this would violate the linear independence assumption (1.3).

The second possibility is that the iteration terminates infinitely often because minimization of $f(x)$ from $x[k, p(k) + 1]$ in the direction $s[k, p(k) + 1]$ cannot continue. This implies that the directional derivative there is nonnegative, i.e.

$$g'[k, p(k) + 1] s[k, p(k) + 1] \geq 0. \tag{3.3}$$

It also implies that <u>if</u> equality holds in (3.3), the second rate of change of $f(x)$ along $s[k, p(k) + 1]$ from $x[k, p(k) + 1]$ is nonnegative, i.e.

$$s'[k, p(k) + 1] G[k, p(k) + 1] s[k, p(k) + 1] \geq 0. \tag{3.4}$$

Assumption (v) implies that $x^k(t) \to \bar{x}$, for all $0 \leq t \leq T_k$. Hence all elements of the matrix $B[k, p(k) + 1]$ correspond to gradients of constraints equal to zero at $\bar{x}$. Thus, since $x[k, p(k) + 1] \to \bar{x}$, (3.3) implies that (using the defining equation (2.6)),

$$-\bar{g}' \begin{bmatrix} -\bar{B}^{-1}\bar{D} \\ I \end{bmatrix} \bar{s}_i - \bar{g}' \begin{bmatrix} \bar{B}^{-1} \\ 0 \end{bmatrix} \bar{u} \geq 0 \tag{3.5}$$

where the bars indicate limits of matrices and vectors coming from (2.6). The existence of converging subsequence of the quantities involved follows from the twice continuous differentiability of the objective function, the definition of the vector $s_2(k, \ell)$ (equations (2.10), (2.13), (2.17), and (2.17.5)), the definition of $u(k, \ell)$ (equations (2.12), (2.16), (2.17), (2.17.3) and (2.19)), and the linear independence of any subset of gradients of constraints equal to zero at $\bar{x}$.

First, we show that NCC1 does not occur an infinite number of times (i.e. that $s_2(k, p(k) + 1)$ given by (2.10) and $u(k, p(k) + 1)$ by (2.12) does not happen more than a finite number of times). Assume the contrary. There are two possibilities. First, (3.4) occurs infinitely often

SECOND ORDER METHOD

when more constraints are considered at $x[k, p(k) + 1]$ then at $x[k, p(k)]$. I.e., there is at least one index in the set $S[k, p(k) + 1]$ which was not in the set $S(k, p(k))$. Now by the construction of $e[k, p(k) + 1]$, (see (2.11) and (2.12)),

$$g'[k, p(k) + 1]s[k, p(k) + 1] \leq 0. \qquad (3.6)$$

Since (3.3) also holds, it must also be true that (3.4) holds. Using (2.11), (2.12) and the definition (2.7) with (3.4) yields

$$e'[k, p(k) + 1]T'[k, p(k) + 1]G[k, p(k) + 1]T[k, p(k) + 1]e[k, p(k) + 1] \geq 0. \qquad (3.7)$$

Now $e[k, p(k) + 1]$ is an eigenvector of (2.8) where $\ell = p(k) + 1$. In fact it is an eigenvector having minimum eigenvalue. Because of the continuity of $G(x)$ (assumption (i)), and assumption (v), taking the limit in (3.7) as $k \to \infty$ yields the fact that the minimum eigenvalue of $\overline{T'GT}$ is greater than or equal to zero. Eventually then the criterion for NCC1 would not be met. Hence the first possibility leads to a contradiction.

The second possibility is that the same constraints are under consideration at $x[k, p(k) + 1]$, i.e. that $S[k, p(k)]$ and $S[k, p(k) + 1]$ are the same. Since $u[k, p(k)] = 0$ (see equation (2.12) it follows that $s[k, p(k) + 1] = s[k, p(k)]$. Using Taylor's theorem, (3.3) becomes

$$[g(k, p(k)) + G(\eta^k)s[k, p(k)]\theta(k, p(k))]'s(k, p(k)) \geq 0 \qquad (3.8)$$

where η^k is a convex combination of $x(k, p(k))$ and $x(k, p(k) + 1)$. Since the iteration did not terminate at

$x(k, p(k))$ we know that

$$g'(k, p(k))s(k, p(k)) \leq 0 \qquad (3.9)$$

Together with (3.8) this implies that

$$s'(k, p(k))G(\eta^k)s(k, p(k)) \geq 0. \qquad (3.10)$$

The same analysis following (3.7) applies and the second possibility also leads to a contradiction.

Under the assumptions of the lemma therefore, motion ceasing because of (3.3) an infinite number of time can come from computation of (2.6) from case NCC2, NCC3, NCC4, or OCC1.

It is clear that there are really only three cases NCC2, NCC3, and NCC4 to consider. Take case OCC1 simply sets a component(s) of $u(k, \ell)$ to zero for a constraint(s) whose multiplier estimate is nonnegative and whose constraint value is zero. Once set to zero, movement remains in that constraint boundary.

Suppose that $\bar{s}_2$ in (3.5) is a limit of $s_2(k, p(k)+1)$ vectors computed using (2.17). Then the first term in (3.5) has the form

$$\bar{\alpha}_1 = -\bar{g}'\bar{T}(\bar{H})^{-1}\bar{T}'\bar{g} \qquad (3.11)$$

Since the eigenvalues of $\bar{H}$ are bounded below by ϵ_4, it follows that

$$\bar{\alpha}_1 \leq 0. \qquad (3.12)$$

Because of (2.17) and (2.19) it follows that the second term of (3.5) is also non-positive, i.e.

$$\bar{\alpha}_2 \equiv -\bar{g}'\begin{bmatrix}\bar{B}^{-1}\\0\end{bmatrix}\bar{u} \leq 0. \qquad (3.13)$$

SECOND ORDER METHOD

If $\bar{s}$, in (3.5) is the limit of the application of (2.17.5) an infinite number of times, the first term of (3.5) has the form

$$\bar{\alpha}_1 = -\bar{h}'\bar{h} \leq 0 . \qquad (3.13.5)$$

The same $\bar{\alpha}_2$ as given in (3.13) applies for this case also.
Finally, if $\bar{s}_2$ in (3.5) is a limit of the application of (2.13) an infinite number of times, the first term of (3.5) has the form

$$\bar{\alpha}_1 = -\bar{h}'\bar{e} - \bar{h}'\bar{h} \qquad (3.14)$$

where

$$-\bar{h}'\bar{e} \leq 0 .$$

Hence

$$\bar{\alpha}_1 \leq 0 \qquad (3.15)$$

for this case. The $\bar{\alpha}_1$ for this case has the form (3.13) where $\bar{u}$ comes from limit of vectors generated by (2.16), and (2.19).

Using (3.13) for $\bar{\alpha}_2$ and either (3.14), (3.11), or (3.13.5) for $\bar{\alpha}_1$ (one of which most always apply), (3.5) then implies

$$\bar{\alpha}_1 = \bar{\alpha}_2 = 0. \qquad (3.16)$$

Hence, defining $\bar{\lambda} = (\bar{B}')^{-1}\bar{g}_1$, it follows that

$$\bar{\lambda}'\bar{u} = 0 \qquad (3.17)$$

For any $\bar{\lambda}_i \geq 0$, the limit of $\bar{u}_i$ (given by (2.16) is the value of the corresponding constraint at $\bar{x}$ which is zero.

The continuity of the derivatives of f imply that if any component of $\bar{\lambda}$ (say $\bar{\lambda}_i$) were negative, then for k large, $u(k, \bar{\ell}, i)$ would be strictly negative where $\bar{\ell}$ is the step when the last New Constraint Case occurred (see equation (2.16)). Because the constraints are linear, any constraint corresponding to a negative multiplier estimate would never be re-encountered after the last new constraint is picked up because the direction vector points <u>interior</u> to that constraint. Hence (2.19) would not be used. Hence (3.17) implies

$$\bar{\lambda} \geq 0 . \qquad (3.18)$$

The same analysis when $\bar{u}$ is given by (2.17) implies that (3.18) holds. [If $\bar{B}$ has rank n the analysis of the first term is bypassed.]

Now where $\bar{\alpha}_1$ is given by (3.11), $\bar{\alpha}_2 = 0$ implies

$$[-\bar{g}_1'\bar{B}^{-1}\bar{D} + \bar{g}_2'](\bar{H}^{-1})[-(\bar{B}')^{-1}\bar{D}'\bar{g}_1 + \bar{g}_2] = 0. \qquad (3.19)$$

But because $\bar{H}^{-1}$ is positive definite (3.19) implies

$$-\bar{g}_1'\bar{B}^{-1}\bar{D} + \bar{g}_2 = 0 .$$

From this, and the definition of $\bar{\lambda}$ it follows that

$$\begin{pmatrix} \bar{B}' \\ \bar{D}' \end{pmatrix} \bar{\lambda} = \bar{g} . \qquad (3.20)$$

Using the remark following (3.3), i.e. that only constraints equal to zero at $\bar{x}$ enter into the computation of $\bar{\lambda}$, it follows that complementary slackness holds – (1.5) is satisfied. Now (3.20) shows that (1.4) holds, and (3.18) implies (1.6) is satisfied. (That (1.7) remains satisfied if the initial point x_0 is feasible follows from (2.2). Thus we have shown that $\bar{x}$ is a KTP when $\bar{\alpha}_1$ is given by (3.11). Now for this case, the matrix $\bar{H}$, a limit matrix of (2.8)

SECOND ORDER METHOD

is positive definite (its minimum eigenvalue is greater than or equal to ϵ_2). Hence

$$\overline{H} = [-\overline{D}'(\overline{B}')^{-1}, I] \, \overline{G} \begin{bmatrix} -\overline{B}^{-1}\overline{D} \\ I \end{bmatrix} \tag{3.21}$$

is positive definite. Now any vector z such that

$$[\overline{B}, \overline{D}]z = 0 \tag{3.22}$$

has the form

$$z = \begin{bmatrix} -\overline{B}^{-1}\overline{D} \\ I \end{bmatrix} z_2 .$$

Thus multiplying $\overline{G}$ by any z satisfying (3.22) yields

$$z_2' \overline{H} z_2 ,$$

which, in this case is greater than zero. Hence (1.18) is satisfied and $\overline{x}$ is a SOKTP.
    The same conclusion follows from similar arguments if $\overline{\alpha}_1$ is given by (3.13.5).
    Suppose now that $\overline{\alpha}_1$ is given by (3.14). Then $\overline{\alpha}_1 = 0$ implies

$$-\overline{h}'\overline{e} - \|\overline{h}\|^2 = 0,$$

which, since both terms are nonpositive implies

$$\overline{h}'\overline{e} = 0$$

and

$$\|\overline{h}\|^2 = 0. \tag{3.23}$$

Now (3.23) is just

$$-\bar{g}_1' \bar{B}^{-1}\bar{D} + \bar{g}_2' = 0$$

and the same analysis used to show (3.20) applies. Hence $\bar{x}$ is a KTP. The analysis for the second order part is a little more complicated.

Consider now the case when (2.13) is used infinitely often at $x(k, p(k) + 1)$ to generate $s_2(k, p(k) + 1)$. Here the analysis is more complicated than that when (2.17) is used since $u(k, p(k) + 1)$ is not necessarily equal to zero. Again, there are two possibilities. First, consider that the vector $s(k, p(k) + 1)$ is different from $s(k, p(k) + 1)$. This could occur if there are more indices in $\mathcal{S}(k, p(k) + 1)$ than in $\mathcal{S}(k, p(k))$, or if the old constraint case OCC1 occurred at $x(k, p(k) + 1)$.

Define

$v(k, p(k) + 1)$

$$= -T(k, p(k)+1)[e(k, p(k)+1) + h(k, p(k)+1)]$$
$$- \begin{bmatrix} B(k, p(k)+1)^{-1} \\ 0 \end{bmatrix} [\gamma(k, p(k)+1)][B'(k, p(k)+1)^{-1}, 0] \quad (3.24)$$

$g(k, p(k) + 1)$

Now if termination in (3.3) resulted and the <u>new constraint case</u> applied, (3.6) holds, and the analysis following (3.6) can be applied. This is true because of (3.20) and (3.18), the terms $h(k, p(k)+1)$ and $u(k, p(k)+1)$ both go to zero. Hence (3.7) is true in the limiting case, implying the minimum eigenvalue of $\bar{H}$ is greater than or equal to zero. Hence $\bar{x}$ is a SOKTP.

SECOND ORDER METHOD

Now if termination in (3.3) resulted and the <u>old constraint case</u> applied, expand (3.3) using Taylor's theorem for the point where the last new constraint was encountered, (call it $x(k, k_1)$)

$$[g(k,k_1) + G(\xi^k)\{\sum_{\ell=k_1}^{p(k)} s(k,\ell)\theta(k,\ell)\}]s(k,p(k)+1) \geq 0. \quad (3.25)$$

Now $s(k, \ell)$, $\ell = k_1 + 1, \ldots, p(k) + 1$ differs from $s(k, k_1)$ only in that some of the components of $u(k, \ell)$ may be zero instead the value of the constraint at any $\lambda(k, \ell)$. (Recall $u(k, \ell)$ is given by (2.15) for this case.)

However, by assumption (v), they cannot be too far away from zero. Now because $g'(k, k_1)s(k, p(k)+1) \leq 0$, using this in (3.25) and dividing by $\sum_{\ell=k_1}^{p(k)} \theta(k,\ell)$ yields

$$[G(\xi^k)\{\sum_{\ell=k_1}^{p(k)} s(k,\ell)\hat{\mu}(k,\ell)\}]'s(k, p(k) + 1) \geq 0 \quad (3.26)$$

where $\{\hat{\mu}(k, \ell)\}$ are nonnegative and sum to one. The remarks after (3.24) point out that taking the limit as $k \to \infty$ yields the result that the minimum eigenvalue of $\overline{H}$ is greater than or equal to zero.

The third possibility is that $s(k, p(k) + 1)$ is the same as $s(k, p(k))$. Using the fact that $h(k, p(k) + 1) \to 0$ and $u(k, p(k) + 1) \to 0$ have already been shown, then the analysis at (3.8) can be used to complete the argument. Hence $\overline{x}$ is a SOKTP. Q.E.D.

The next lemma shows that the directions chosen by the algorithm are such that in a neighborhood of a point which is not a SOKTP, the outcome after one iteration of the algorithm will yield a value of f less than $f(\overline{x})$ if a finite distance is moved.

Lemma 4. Suppose $\overline{x}$ is a limit point of $\{x(k,1)\}$. Assume that $\{x(k,1)\}$ denotes also the subsequence converging to $\overline{x}$.

If: i)- iv) of Lemma 3 hold, and
        v)   for every $\epsilon > 0$ there is a t, $0 < t \leq \epsilon$,
such that for all k large

$$f[x^k(t)] \geq f(\bar{x}) \quad (\text{recall } x^k(0) \equiv x(k,1)), \quad (3.27)$$

then: a) $\bar{x}$ is a SOKTP.

Proof. We can assume ϵ is small enough so that only constraints binding at $\bar{x}$ are considered in what follows. With this restriction, it follows from the definition of $x^k(t)$ in (3.1) that $x^k(t)$ is defined even though t may be greater than T_k.

Expanding (3.37) using Taylor's theorem

$$f(\bar{x}) \leq f[x^k(t)] = f[x^k(0)] +$$

$$+ \{\sum_{\ell=1}^{q-1} s(k,\ell)\theta(k,\ell) + s(k,q)\tau(k,t)\}'g(\eta(k,t))$$

(although q depends upon k and t we drop this for notational convenience) where $\eta(k,t)$ is on the line segment joining $x^k(t)$ and $x^k(0)$.

Taking the limit as $k \to \infty$ yields

$$f(\bar{x}) \leq f(\bar{x}) + \{\sum_{\ell=1}^{\bar{q}-1} \bar{s}_\ell \bar{\theta}_\ell + \bar{s}_{\bar{q}} \bar{\tau}\}'g(\bar{\eta})$$

where $\bar{q} = \bar{q}(t)$, $\bar{\tau} = \bar{\tau}(t)$ and $\bar{\eta} = \bar{\eta}(t)$. That $\bar{q}$, a limit of $q(k,t)$ is finite follows because of the way in which the sets $S(k,\ell)$ are formed, and the fact that there are only a finite number of constraints to problem (LC).

Cancelling, dividing by t and taking the limit as $t \to 0$ yields

$$0 \leq \{\sum_{\ell=1}^{\tilde{q}} \tilde{s}_\ell \tilde{\mu}_\ell\}'g(\bar{x}) \quad (3.28)$$

SECOND ORDER METHOD

where $\sum_{\ell=1}^{\tilde{q}} \tilde{\mu}_\ell = 1$, $\tilde{\mu}_\ell \geq 0$, all ℓ and each $\tilde{s}_\ell$ is a direction vector of the form (2.6) computed from quantities evaluated at $\bar{x}$, and using only constraints equal to zero at $\bar{x}$. Note that the construction of (2.6) assures that

$$\tilde{s}_\ell' g(\bar{x}) \leq 0, \quad \ell = 1, \ldots, \tilde{q} \qquad (3.29)$$

in all four cases.

There are four cases to consider. First, suppose some $\tilde{\mu}_\ell > 0$ has an associated $\tilde{s}_\ell$ which was computed using (2.13), (2.16) and (2.19). Then we have from (3.28) and (3.29) that for some subset of constraints whose indices are in $\bar{\mathcal{S}}$,

$$-\bar{h}'(\bar{h} + \bar{e}) - \sum_{i \in \bar{\mathcal{S}}} \bar{\gamma}_i (\bar{\lambda}_i)^2 = 0 \qquad (3.30)$$

where $\bar{\gamma}_i = 0$ or 1. The remark preceding (3.18) applies here. If any $\bar{\lambda}_i < 0$, then the corresponding $\bar{\gamma}_i = 1$. Since $-\bar{h}'\bar{e} \leq 0$, then (3.30) implies $\bar{\gamma}_i = 0$ all $i \in \bar{\mathcal{S}}$ and $\bar{\lambda}_i > 0$. (If $\bar{\lambda}_i = 0$, $\bar{\gamma}_i$ maybe 1 but it obviously does not matter.) Hence, $\bar{h} = 0$, and $\bar{\lambda} \geq 0$, i.e. $\bar{x}$ is a KTP for NCC 2.

Second, assume that the $\tilde{s}_\ell$ associated with some $\tilde{\mu}_\ell > 0$ was computed using (2.16), (2.17), and (2.19). (NCC 3). Then

$$-\bar{h}'(\bar{H}^{-1})\bar{h} - \sum_{i \in \bar{\mathcal{S}}} \bar{\gamma}_i (\bar{\lambda}_i)^2 = 0 \qquad (3.31)$$

where the smallest eigenvalue of $\bar{H}$ is ϵ_4. Hence (3.31) easily implies that $\bar{x}$ is a SOKTP for this case.

Third, if (2.17.5), (2.17.3) and (2.19) are used, the conclusion that $\bar{x}$ is a SOKTP follows as above.

Next we show that no $\tilde{s}_\ell$ can be the limit of vectors of the form (2.6) chosen in accordance with (2.10), (2.12), and (2.19) and have an associated $\tilde{\mu}_\ell > 0$. (Note that (2.19) is now used in this case.)

Now for every k,

$$t = \sum_{\ell=1}^{q(k)} \theta(k, \ell) + \tau(k, t). \qquad (3.32)$$

For each iteration-step (k, ℓ) there is a set $\mathcal{S}(k, \ell)$. Starting at $x(k, 1)$ there are a finite number of possibilities of the order in which the sets $\mathcal{S}(k, 1)$, $\mathcal{S}(k, 2) \ldots$ can be formed. Assume therefore that all possible orderings which occur are infinite number of times $(k \to \infty)$ are grouped. That is, we have for one group

$$\begin{array}{c} \theta(k,1), \quad \ldots, \quad \theta(k_1, q_1), \; \tau(k_1, t) \\ \vdots \\ \theta(k_j,1), \quad \ldots, \quad \theta(k_j, q_1), \; \tau(k_j, t) \\ \vdots \end{array} \qquad (3.33)$$

and the sequence of indices in the sets $\mathcal{S}(k_j, 1), \ldots, \mathcal{S}(k_j, q_1)$ are all the same.

It is easy to show that if $H(k, \ell+1)$ has a negative eigenvalue $< -\epsilon_4$, so does $H(k, \ell)$. Hence for any array, if $s(k, \ell+1)$ comes from a computation using (2.10) and (2.12), so does $s(k, \ell)$.

Now for every array of the form (3.33) there is a smallest integer $\bar{\ell}$ such that

$$\limsup_{j \to \infty} \theta(k_j, \bar{\ell}-1) > 0.$$

There may be no such integer, but if there is there is a smallest one. If for any array the $(k_j, \bar{\ell}-1)$ sequence corresponds to direction vectors found by use of (2.10) and (2.12), select a subsequence for which $\theta(k_j, \bar{\ell}-1)$ converges to the lim sup. For notational convenience assume this is the original sequence. Now

SECOND ORDER METHOD

$$f(\bar{x}) \leq f[x^k(t)] \leq f[x(k, \bar{\ell})] = f[x(k, \bar{\ell}-1)$$

$$+ [s(k, \bar{\ell}-1)\theta(k, \bar{\ell}-1)]'g[x(k, \bar{\ell}-1)] \quad (3.34)$$

$$+ \frac{1}{2}\theta^2[k, \bar{\ell}-1][s(k, \bar{\ell}-1)]'\nabla^2 f[\xi(k, \bar{\ell})]s(k, \bar{\ell}-1) .$$

Since the iteration did not terminate at $x(k, \bar{\ell}-1)$ we know that $s'(k, \bar{\ell}-1)g[x(k, \bar{\ell}-1)] \leq 0$. Hence, continuing the inequality,

$$\leq f[x(k, \bar{\ell}-1)] + \frac{1}{2}\theta^2(k, \bar{\ell}-1)s'(k, \bar{\ell}-1)\nabla^2 f[\xi(k, \bar{\ell}-1)]s(k, \bar{\ell}-1).$$

Now since $\bar{\ell}$ was the smallest integer for which $\theta(h_j, \ell-1) \to 0$, and since only a finite number of steps occur from $(k, 1)$ to $(k, \bar{\ell}-1)$, it follows that $x(k, \bar{\ell}-1) \to \bar{x}$. Taking the limit in (3.34) then yields

$$f(\bar{x}) \leq f(\bar{x}) + \frac{1}{2}\theta^2(\bar{\ell}-1)\bar{s}'(\bar{\ell}-1)\nabla^2 f[\xi(\bar{\ell}-1)]\bar{s}(\bar{\ell}-1). \quad (3.35)$$

Cancelling, dividing by $\theta^2(\ell-1)/2$, and taking the limit as $t \to 0$ yields

$$\tilde{s}'G(\bar{x})\tilde{s} \geq 0 \quad (3.36)$$

where

$$\tilde{s} = \tilde{T}\tilde{e} , \quad (3.37)$$

$\tilde{e}$ an eigenvector of $\tilde{T}'G(\bar{x})\tilde{T}$ with smallest eigenvalue.
But then using the continuity of the second derivatives of f, (2.10) and (2.12) would not have been used more than a finite number of times. Hence, all θ's go to zero and thus no μ_ℓ can be nonzero for this case.

All cases have been considered except to show that use of (2.13), (2.15), and (2.19) for some $\tilde{s}_\ell$ associated with a nonzero μ_ℓ implies $\bar{x}$ is a SOKTP. Using (3.30) it has been shown that $\bar{x}$ is a KTP and only the second-order condition remains to be shown for this case.

The proof follows the same analysis as used to develop (3.34). The same conclusions (3.36) results but (3.37) is true only after using the fact the $\bar{x}$ a KTP has been proved previously for this case, and the construction of $s(k, \ell)$ using (2.13), (2.16), and (2.19) implies that $u(k, \bar{\ell}-1) \to 0$, $h(k, \bar{\ell}-1) \to 0$. Q.E.D.

Theorem 1. [Convergence of the Algorithm]

If:
   i) $f(x)$ is a twice continuously differentiable function,
  ii) the direction vectors are as given by the algorithm,
 iii) the scalars are as given by the algorithm,
 iv) the linear independence assumption (1.3) holds,

then: every limit point of $\{x(k, 1)\}$ is a SOKTP.

Proof: Suppose $\bar{x}$ is a limit point of $\{x(k, 1)\}$. Let $\{x(k, 1)\}$ denote also the subsequence converging to $\bar{x}$. Assume that $\bar{x}$ is <u>not</u> a SOKTP. Then from lemma 3 it follows that

$$\liminf_{k \to \infty} \max_{0 \le t \le T_k} \|x^k(t) - x^k(0)\| = \delta > 0. \quad (3.38)$$

Using the denial of (v) in lemma 4 we know that there is an $\epsilon > 0$ such that for any t with $0 < t \le \epsilon$ where k is large enough, (and $x(k, 1)$ feasible)

$$f[x^k(t)] < f(\bar{x}).$$

Let $t_1 > 0$ be smaller than one half the value for which δ is achieved in (3.38) and small enough so that lemma 4

SECOND ORDER METHOD

applies. Then eventually

$$f[x^k(T_k)] < f[x^k(t_1)] < f(\bar{x}).$$

But $f(\bar{x}) \leq f[x^k(T_k)]$ by the monotonicity of $\{f[x(k, 1)]\}$. This contradiction proves the theorem Q.E.D.

4. Rate of Convergence of the Algorithm

If there were no linear constraints on the problem, the algorithm would reduce to the Generalized Newton Method [4, p. 162] and [2] for minimizing an unconstrained function. This is very important in accelerating the <u>rate</u> at which the points generated by the algorithm converge to a local minimum.

The development of the proof of what the rate of convergence can be when certain additional assumptions are placed on the local minimizing point follows.

Lemma 5. If $\bar{x}$, a limit point of the sequence $\{x(k, 1)\}$ satisfies the second order sufficiency conditions (1.4)-(1.7), and (1.9), it is the <u>only</u> limit point of that sequence.

Proof: The proof follows by noting that (2.1) and (2.2) generate consecutive points in the sequence by finding the nearest local minimum along any direction chosen. Hence the fact that $\bar{x}$ is an <u>isolated</u> local minimum means the sequence of points during any iteration cannot get far away from $\bar{x}$ once a neighborhood of $\bar{x}$ is entered.

Another condition must be placed on the point $\bar{x}$ to ensure that the set of binding constraints remains the same when k is large. If this were not the case, the subspace in which the sequence of x's converging to $\bar{x}$ would never stay the same thus ruining attempts to accelerate the rate of convergence.

Strict Complementary Slackness

The term strict complementary slackness is said to apply to conditions (1.5) if

$$\overline{\lambda}_i > 0 \text{ whenever } c_i(\overline{x}) = 0. \tag{4.1}$$

Lemma 6. If at $\overline{x}$, a limit point of the sequence $\{x(k, 1)\}$, the linear independence assumption, the second order sufficiency conditions, and strict complementary slackness hold: then for k large enough, $p(k) = 1$, (i.e. each iteration consists of movement along only <u>one</u> segment), and $\mathcal{S}(k, 1)$ contains only the indices of the set of constraints which are binding at $\overline{x}$.

Proof. By Lemma 5, $\overline{x}$ is the unique limit point of $\{x(k, 1)\}$. Now because of the linear independence assumption for all k large the constraint gradients involved are linearly independent. Hence an iteration is terminated only because the directional derivative is nonnegative. This implies that

$$g'(k, p(k) + 1)s(k, p(k) + 1) \geq 0 \tag{4.2}$$

for all k large. Using the same arguments as those in lemma 3 it is possible to show that for every sequence of sets $\mathcal{S}(k, p(k) + 1)$ which occurs infinitely often there is a set of nonnegative $\{\hat{\lambda}_i\}$ such that (1.4) is satisfied. Furthermore, for those i where $i \not\in \mathcal{S}(k, p(k) + 1)$, $\hat{\lambda}_i = 0$. Since the $\hat{\lambda}_i$'s are unique (linear independence), and all strictly greater than zero for the binding constraints (strict complementarity slackness) it follows that for k large, $\mathcal{S}(k, p(k) + 1)$ must contain the indices of all constraints binding at $\overline{x}$.

The last possibility is that; of the constraints binding at the solution, not all are present at the start of an iteration, all have indices in $\mathcal{S}(k, p(k) +1)$ at the termination of an iteration, and for at least one constraint whose index is in $\mathcal{S}(k, p(k)+1)$, $c_i'(k, p(k) + 1) > 0$.

The reason this cannot happen is that the second term of (4.2) becomes

SECOND ORDER METHOD

$$-\sum_{i \in S(k, p(k)+1)} \hat{\lambda}(k, p(k)+1, i) \gamma_i(k, p(k)+1) \to 0$$

where $\gamma_i(k, p(k) + 1)$ is zero or one. This follows because (2.17.3) is used to generate $u(k, p(k) + 1)$ since under the assumption of this lemma either NCC3 or NCC4 applies. Now, given that all the $\bar{\lambda}_i$'s for binding constraints are strictly greater than zero, then $\gamma_i(k, p(k) + 1) = 0$ for k large. This means that for k large both $c_i\mathrm{'}(k, p(k)+1) = 0$ and $\hat{\lambda}(k, p(k)+1, i) > 0$ hold, hence at some point, all the constraints binding at $\bar{x}$ have zero value at the end of an iteration. Because the associated multiplier signs are strictly positive, linear independence, and the continuity of the derivatives of $f(x)$, they will remain at their zero value. Q.E.D.

The importance of the last two lemmas is that, after a finite number of iterations the direction vector generated is always of the form

$s(k, 1) =$

$$-\begin{bmatrix} -B(k,1)^{-1}D(k,1) \\ I \end{bmatrix} [H(k,1)]^{-1} [-D'(k,1)B'(k,1)^{-1}, I] g(k,1)$$

and the division of variables into dependent and independent remains the same. (See assumption (vi) of Theorem 2 for a slight rewording of this statement). Hence under our assumptions the algorithm amounts to the generalized Newton Method [4, p. 162 and [2]] in the reduced subspace. To see this we consider the total change of the algorithm function as a function of changes in the type 2 variables. This is most easily done by defining a new function $F(x_2)$ as

$$F(x_2) = f(x_1, x_2) \text{ where } x_1 = -B^{-1}Dx_2 + b_1.$$

Now we need to show that under the strict second order sufficiency conditions the Hessian of $F(x_2)$ is positive definite so that theorems on the rate of convergence can be proved.

Lemma 7.

If:
i) $f(x)$ is twice continuously differentiable,
ii) the linear independence assumption holds,
iii) the second order sufficiency condition holds at x^*,
iv) strict complementary slackness holds at x^*,

then: a) the matrix of second partial derivatives of $F(x_2)$ as indicated by the algorithm is positive definite in a neighborhood of x^*.

Proof: The gradient of F in the space of type 2 variables is given by the chain rule as

$$\nabla_{x_2} F = \frac{dx_1}{dx_2} \nabla_{x_1} f(x) + \nabla_{x_2} f(x)$$

$$= [-D'(B')^{-1}, I] g(x) .$$

In the method of "steepest descent" this would be the choice for the direction of the type 2 variables.

The second derivative matrix of F is

$$\nabla_{x_2 x_2} F = \begin{bmatrix} \frac{dx_1}{dx_2} , I \end{bmatrix} G(x) \begin{bmatrix} \frac{dx_1'}{dx_2} \\ I \end{bmatrix}$$

$$= [-D'(B')^{-1}, I] G(x) \begin{bmatrix} -B^{-1} D \\ I \end{bmatrix} . \quad (4.3)$$

We need to show that in a neighborhood of x^*, the triple matrix product given in (4.3) is positive definite. Pre and post multiplying by any z yields $z'G(x)z$ where $z' = z'[-D'(B')^{-1}, I]$. Now $(B,D)Z = 0$ which means Z is in the set orthogonal to all the binding constraint gradients.

SECOND ORDER METHOD

Our assumption (iv) implies all λ_i^* associated with binding constraints at x^* are strictly positive. Thus Z satisfies the requirement for $Z'G(x^*)Z > 0$ to hold. Continuity of $G(x)$ assures that is a neighborhood of x^* the strict inequality holds. Q.E.D.

The rule for computing H given in (2.8) is an attempt to approximate the matrix (4.3) at x^*. The direction vector generated by the algorithm is an attempt to minimize F in the space of type 2 variables using the generalized Newton method. We note that if $f(x)$ were a quadratic form, then the precise local minimum x^* would be found in one step once the binding constraints at x^* had been found.

As indicated, a proof of the rate of convergence of the SOVRM reduces to a proof of the rate of convergence of x_2^k to x_2^* using the generalized Newton method to minimize $F(x_2)$. Since proofs of this type are slight variations of the proof of the regular Newton method we will state, but not prove two theorems about the rate of convergence of the algorithm. The reader is referred to [2] for details on the Newton and generalized Newton rate of convergence.

Theorem 2 [Superlinear Rate of Convergence of the SOVRM].

If:
    i) $f(x)$ is twice continuously differentiable,
    ii)   the SOVRM is applied to problem (LC),
    iii) the linear independence assumption (1.3) holds,
    iv) the second order sufficiency conditions hold at x^*, a limit point of the algorithm,
    v)   strict complementarity slackness holds at x^*,
    vi) the smallest eigenvalue of $(T^*)'G^*T^*$ is greater than or equal to $2\epsilon_4$,

then:
    a) x^* is the unique limit point of the sequence of points generated by the algorithm,
    b) for all k large, each iteration consists of movement along only one segment,
    c) for all k large, eventually all movement is in the subspace of constraints binding at x^*,
    d) the division of variables x into type 1 and type 2 variables remains invariant for k large,

e) $\lim_{k \to \infty} \dfrac{\|x(k+1, 1) - x^*\|}{\|x(k, 1) - x^*\|} = 0,$ (4.4)

i.e. the algorithm has the <u>superlinear rate of convergence property</u>.

If we extend the conditions of f(x) (essentially to require that the third derivatives of f(x) be continuous) it follows that a quadratic rate of convergence is attainable.

Theorem 3 [Quadratic Rate of Convergence of the SOVRM].

If: i) The second derivatives of f(x) satisfy the Lipschitz conditions, i.e. there is an M such that for any x, y,

$$\|[G(x) - G(y)]z\| \le M \|x-y\| \, \|z\| \quad (4.5)$$

for all z,

ii) the SOVRM is applied to problem (LC),
iii) the linear independence assumption holds,
iv) the second order sufficiency conditions hold at x^*, a limit point of the algorithm,
v) strict complementary slackness holds at x^*,
vi) the smallest eigenvalue of $(T^*)'G^*T^*$ is greater than or equal to $2\epsilon_4$,

then: a)-d) of Theorem 2 hold, and
e) there is a N, independent of k so that for k large,

$$\|x(k+1, 1) - x^*\| \le N \|x(k, 1) - x^*\|^2. \quad (4.6)$$

We note that the requirement (vi), although awkward is necessary to ensure that (2.17) is used an infinite number of times to generate s_2. If $\epsilon_4 = 0$ were used, it is theoretically possible that convergence cannot be proved. Hence the requirement for a small threshold in the determinant of H

5. Discussion

The algorithm presented here is in some ways similar to those presented by others. In that at each iteration-step of the algorithm a direction is generated which both decreases the objective function and points into the feasible region or along its boundary, the method could qualify as a method of feasible directions [15]. The formulas by which movement of the boundary of constraints desired to remain satisfied are similar to those suggested by Wolfe in his Reduced Gradient Method [12]. An exact comparison of this is contained in [8].

Two important differences between the Reduced Gradient Method and the Variable Reduction Method are the rules for termination of an iteration, and the choice of the direction generation for the type 2, or independent variables. The former are very important for the prevention of zig-zagging, a phenomenon first noted by Zoutendijk [15]. In the Reduced Gradient Method the direction vector is the vector of "steepest descent." In the Variable Reduction Method, the vector is analogous to that prescribed by the Revised Optimum Newton Method [4]. The importance of this difference cannot be overstated since the main convergence and rate of convergence theorems of this paper cannot be obtained with a first-order steepest descent algorithm.

REFERENCES

1. Abadie, J., J. Carpentier, and C. Hensgen, "Generalization of the Wolfe Reduced Gradient Method to the Case of Nonlinear Constraints," paper presented at the Joint European Meeting of the Econometric Society/The Institute of Management Science, Warsaw, September, 1966.

2. Crockett, J. B., and H. Chernoff, "Gradient Methods of Maximization," Pacific J. Math., Vol. 5, No. 1, 1955.

3. Faure, P., and P. Huard, "Résolution des Programmes Mathématiques à Fonction Non-linéaire par la Méthode du Gradient Reduit," <u>Revue Francaise de Recherche Opérationelle</u>, Vol. 9, pp. 167-205, 1965.

4. Fiacco, A. V., and G. P. McCormick, <u>Nonlinear Programming: Sequential Unconstrained Minimization Techniques</u>, John Wiley and Sons, Inc., New York, 1968.

5. Goldfarb, D., "Extension of Davidon's Variable Metric Method to Maximization Under Linear Inequality and Equality Constraints," <u>SIAM J. Appl. Math.</u>, Vol. 17, No. 4, pp. 739-764, July, 1969.

6. Kuhn, H. W., and A. W. Tucker, "Non-linear Programming," in J. Neyman (Ed.), <u>Proceedings of the Second Berkeley Symposium on Mathematical Statistics and Probability</u>, University of California Press, Berkeley, pp. 481-493, 1951.

7. McCormick, G. P., "Anti-Zig-Zagging by Bending," <u>Management Science: Theory Series</u>, vol. 15, No. 5, January 1969.

8. McCormick, G., "The Variable Reduction Method for Nonlinear Programming," to appear in <u>Management Science</u>.

9. Rosen, J. B., "The Gradient Projection Method for Nonlinear Programming, Part I: Linear Constraints," <u>J. Soc. Ind. Appl. Math.</u>, 8(1): 181-217, 1960.

10. Shanno, D., "An Accelerated Gradient Projection Method for Linearly Constrained Nonlinear Estimation," <u>SIAM J. Appl. Math.</u>, Vol. 18, No. 2, March, 1970.

11. Topkis, D. M., and A. F. Veinott, Jr., "On the Convergence of Some Feasible Direction Algorithms for Nonlinear Programming," Technical Report No. 6, Stanford University, Dept. of Industrial Engineering, Stanford, California, 5 August 1966.

12. Wolfe, P., "Methods of Nonlinear Programming," Recent Advances in Mathematical Programming, in R. L. Graves and P. Wolfe (Eds.), McGraw-Hill Book Co., pp. 67-86, 1963.

13. Zangwill, W. I., "The Convex Simplex Method," Management Science, Series A, Vol. 14, No. 3, pp. 221-238, 1967.

14. Zangwill, W. I., Nonlinear Programming: A Unified Approach, Prentice-Hall, Englewood Cliffs, N. J., 1969.

15. Zoutendijk, G., Methods of Feasible Directions, Elsevier Publishing Company, Amsterdam and New York, 1960.

Convergent Step-Sizes for Gradient-Like Feasible Direction Algorithms for Constrained Optimization

JAMES W. DANIEL

ABSTRACT

We treat the problem of minimizing a function f over a set C by iterative methods of the form $x_{n+1} = x_n + t_n p_n$ where hopefully $\{x_n\}$ will converge to some solution x^*, the $\{p_n\}$ are directions often computed from $\{x_0, \ldots, x_n\}$ or x_n alone, and t_n is some suitable scalar step size. We shall examine, from a unified viewpoint, a wide variety of specific methods for selecting the step size t_n, some of which are very convenient computationally, and shall prove convergence, in some sense, for these methods. Included are modifications of essentially all the standard step size algorithms for unconstrained problems.

JAMES W. DANIEL

1. Introduction

We consider the problem of locating a point x^* in a set C such that $f(x^*)$ is the minimum value of the real valued function f as x ranges over the given set C; many algorithms have been proposed for solving this problem and, in very recent years, general analyses have been given of sufficient (and sometimes necessary) conditions for broad classes of methods to yield a solution [Topkis-Veinott (1967), Zangwill (1969)]. Many of the methods that have been analyzed or are widely used in practice compute iteratively a sequence $\{x_n\}_0^\infty$ of approximations to x^* where usually $x_{n+1} = x_n + t_n p_n$, p_n is some direction computed from x_n or $\{x_0, x_1, \ldots, x_n\}$, and t_n is some appropriate scalar step size. Unfortunately it is generally true that those rules for choosing t_n which have been analyzed are not exactly implementable computationally, so that the theory in many cases does not quite apply to the numerical methods, even ignoring rounding errors. In this paper we shall examine the problem of choosing the step size t_n and shall demonstrate that a number of computationally convenient choices yield convergent methods, assuming that suitable direction algorithms, that is, the methods of choosing p_n, are used. We do not give error estimates or include rounding errors in our analysis.

2. Gradient-like feasible direction algorithms

We shall restrict ourselves to methods which yield a sequence $\{x_n\}_0^\infty$ of points in the constraint set C; since $x_{n+1} = x_n + t_n p_n$ and we must determine t_n, it is reasonable to restrict ourselves to using directions which keep one inside C, at least temporarily. Since the scaling of the direction p_n is at our disposal, <u>we shall assume that</u> $p_n \equiv x_n' - x_n$ <u>where</u> $x_n + tp_n \in C$ <u>for</u> $0 \leq t \leq 1$ and in particular $x_n' \in C$; if C is convex, we need only assume $x_n' \in C$ of course. Such a direction is usually called "feasible." <u>We shall also assume that</u> p_n <u>is instantaneously a direction of non-increasing values of</u> f <u>at</u> x_n. Let us now provide the technical setting for our analysis.

Let X be a real Banach space with norm $\|\cdot\|$, C a subset of X, f a real valued function defined on X; we assume that f is Frechet differentiable at each $x \in X$ and we denote the gradient (derivative) by ∇f. Note that ∇f is a generally nonlinear mapping of X into the dual space X^* having norm $\|\cdot\|_*$; for any bounded linear functional $\ell \in X^*$ we denote its value, when applied to $y \in X$, by $<y, \ell>$ or by $<\ell, y>$ interchangeably.

<u>Remark.</u> If X is $\mathbb{R}^m$, finite dimensional Euclidean space, then ∇f exists and is the usual gradient if the first order partial derivatives of f exist and are continuous; in this case $X^* = X$ and one can take $<x, \ell> \equiv x^T \ell$ where T denotes the vector transpose.

Our hypothesis that p_n is a direction of non-increase now can be stated in the form: <u>we assume that</u> $<\nabla f(x_n), p_n> \leq 0$.

In the latter sections of this paper, the conclusions of our theorem will always take the form: $\lim_{n \to \infty} <\nabla f(x_n), p_n> = 0$. For the step size algorithms to be discussed to be useful when used in conjunction with some direction algorithm (such as gradient projection or the Frank-Wolfe method), the condition $<\nabla f(x_n), p_n> \to 0$ must be a useful one. If, for example, p_n is a function of x_n, that is $p_n = p(x_n)$, and if $<\nabla f(x), p(x)>$ is upper semicontinuous in some topology for which C is closed and every sequence in C has a subsequence $\{x_{n_i}\}$ converging to some point, say x', then $0 \geq\, <\nabla f(x'),\ p(x')> \geq \lim\sup_{i \to \infty} <\nabla f(x_{n_i}),\ p(x_{n_i})> = 0$; thus, if $<\nabla f(x),\ p(x)> = 0$ implies that x is an optimal point, we know that all limit points of $\{x_n\}_0^\infty$ are optimal. If the optimal point in C is unique, $\{x_n\}$ converges to it.

These properties of the direction sequence are essentially the properties assummed and exploited in [Topkis-Veinott (1967), Zangwill (1969), Zoutendijk (1960)]. For the purpose of example only, we will prove a theorem in Section 7 stating that for convex f, convex C, and special directions $\{p_n\}$, the condition $<\nabla f(x_n),\ p_n> \to 0$ implies that $f(x_n)$ converges to $f(x^*) = \min_{x \in C} f(x)$, and hence all limit points of $\{x_n\}_0^\infty$ minimize f if f is lower semicontinuous and C is closed in the relevant topology. More generally, however, the techniques of [Topkis-Veinott (1967), Zangwill (1969), Zoutendijk (1960)], insofar as the direction algorithms are concerned in $\mathbb{R}^m$, can be used to show that $<\nabla f(x_n),\ p_n> \to 0$ is an important result for many special direction methods such as Frank-Wolfe, gradient projection, Newton and other second order methods, cyclic coordinate descent, Arrow-Hurwicz-Uzawa, and various methods of Zoutendijk, with modifications in some cases; some similar results not restricted to $\mathbb{R}^m$ may be found in [Daniel (1970)] .

If one has any algorithm G for generating a sequence $\{x_n\}$ using directions determined from x_n only and such that $f(x_n)$ converges to its minimum over C, then clearly any other algorithm which generates a sequence $\{x_n\}$ with $f(x_{n+1}) \leq f(x_n)$ and such that infinitely many of the x_n are generated by the convergent algorithm G will also force $f(x_n)$ to converge to its minimum; this implies for example that various devices for accelerating convergence of a basic method can be used. We shall not repeat this obvious but often overlooked fact in the following discussions of basic methods, but the reader should keep it in mind.

Since we now assume that the condition $<\nabla f(x_n), p_n> \to 0$ is a useful one, we now turn our attention to step size algorithms which yield this conclusion. The excellent general analysis in [Zangwill (1969)] usually aasumes that t_n is chosen to minimize $f(x_n + tp_n)$, at least in the specific methods discussed; the excellent general analysis in [Topkis-Veinott (1967)] mentions some additional choices, more convenient computationally, but only gives details for choosing t_n via a quadratic approximation to f. We shall present and analyze several other methods.

3. General step size criteria

Our analysis throughout will make use of two concepts: a forcing function, and the reverse modulus of continuity.

Definition 3.1. A forcing function is a function d mapping $[0, \infty)$ into itself and such that t_n converges to zero whenever $d(t_n)$ converges to zero.

Definition 3.2. If ∇f is uniformly continuous on a set C, then the reverse modulus of continuity of ∇f, call it s, is defined as $s(t) = \inf\{\|x-y\|; x, y \in C, \|\nabla f(x) - \nabla f(x)\|_* \geq t\}$.

Remark. The reverse modulus of continuity is a monotonically non-decreasing forcing function, $\|x-y\| < s(\delta)$ implies $\|\nabla f(x) - \nabla f(y)\|_* < \delta$.

It has been observed [Cea (1969), Daniel (1970), Elkin (1968)] that forcing functions play a role in analyzing unconstrained minimization methods; in particular one usually proves convergence in these cases by showing that

$$f(x_n) - f(x_n + t_n p_n) \geq d(<-\nabla f(x_n), \frac{p_n}{\|p_n\|}>)$$ for a forcing function d, thus implying, if f is bounded below, that $<\nabla f(x_n), \frac{p_n}{\|p_n\|}> \to 0$, a very useful condition for unconstrained problems since typically $<-\nabla f(x_n), \frac{p_n}{\|p_n\|}> \geq c\|\nabla f(x_n)\|_*$ for some $c > 0$. For constrained minimization,

one might conceive of ignoring the constraints and computing a number t_n^u via an <u>unconstrained</u> step size method. If $x_n + t_n^u p_n \in C$, then we take $t_n = t_n^u$; otherwise we take $t_n = t_n^C$ for some easily computed t_n^C such that $x_n + t_n^C p_n \in C$, such as $t_n^C = 1$. That this is a useful technique for <u>convex</u> functions f appears to have been noticed first in special cases in [Cea (1969)].

Theorem 3.1. Let the convex functional f be bounded below on the bounded set C and, for some x_0 in C, let the set $\{x; f(x) \leq f(x_0)\}$ be bounded; let p_n define a feasible direction sequence and let $\|\nabla f(x)\|$ be uniformly bounded for $x \in C \cap \{x; f(x) \leq f(x_0)\}$. Let the numbers t_n^u be some steps satisfying $f(x_n) - f(x_n + t_n^u p_n) \geq d(<-\nabla f(x_n), \frac{p_n}{\|p_n\|}>)$ for a forcing function d. Let $z_{n+1} = x_n + t_n p_n$ where $t_n = t_n^u$ if $x_n + t_n^u p_n \in C$ and $t_n = t_n^C \geq \varepsilon > 0$ with $x_n + t p_n \in C$ for $0 \leq t \leq t_n$ otherwise. Choose $x_{n+1} \in C$ such that $f(x_n) - f(x_{n+1}) \geq \beta[f(x_n) - f(z_{n+1})]$ for fixed $\beta > 0$. Then $<\nabla f(x_n), p_n> \to 0$.

Proof: If $t_n = t_n^u$ then $f(x_n) - f(x_{n+1}) \geq \beta d(<-\nabla f(x_n), \frac{p_n}{\|p_n\|}>)$. If $t_n = t_n^C$ and $f(x_n + t_n^C p_n) \leq f(x_n + t_n^u p_n)$, then also $f(x_n) - f(x_{n+1}) \geq \beta d(<-\nabla f(x_n), \frac{p_n}{\|p_n\|}>)$; we consider the final case of $t_n = t_n^C$ and $f(x_n + t_n^C p_n) > f(x_n + t_n^u p_n)$. Since f is convex and $t_n^u > t_n^C$, we have $f(x_n + t_n^C p_n) =$
$f(x_n + \frac{t_n^C}{t_n^u} t_n^u p_n) \leq (1 - \frac{t_n^C}{t_n^u}) f(x_n) + \frac{t_n^C}{t_n^u} f(x_n + t_n^u p_n)$ and thus
$f(x_n) - f(x_n + t_n^C p_n) \geq \frac{t_n^C \|p_n\|}{t_n^u \|p_n\|} [f(x_n) - f(x_n + t_n^u p_n)]$. Since

$\{x;\ f(x) \le f(x_0)\}$ is bounded, there is a K such that $\|t_n^u p_n\| \le K$ and therefore $f(x_n) - f(x_n+1) \ge \beta t_n^c \|p_n\| \frac{1}{K} d(<-\nabla f(x_n),\ \frac{p_n}{\|p_n\|} >)$. Since $t_n^c \ge \varepsilon > 0$ and $\|\nabla f(x_n)\|_*$ is uniformly bounded and $<-\nabla f(x_n),\ p_n> = <-\nabla f(x_n),\ \frac{p_n}{\|p_n\|}> \|x_n' - x_n\| \le K <-\nabla f(x_n),\ \frac{p_n}{\|p_n\|}>$, from the three inequalities for $f(x_n) - f(x_{n+1})$ we deduce that $<\nabla f(x_n),\ p_n> \to 0$. Q.E.D.

Remarks. Since $p_n = x_n' - x_n$ for some $x_n' \in C$, $t_n^c = 1$ is always allowed so certainly $t_n^c \ge \varepsilon > 0$ is possible; in particular $t_n^c = \max\{t;\ x_n + tp_n \in C\}$ is possible. If C is itself bounded, by modifying and redefining f outside of C we can generally guarantee that $\{x;\ f(x) \le f(x_0)\}$ is bounded.

The hypothesis in Theorem 3.1 that f be convex is unduly restrictive; by considering various special choices of the step size, we can easily eliminate the convexity hypothesis. All of our analysis will make use of the following theorem on general step sizes; this theorem should be considered simply as a tool for our analysis. Throughout the rest of this paper, we shall let $W(x_0)$ denote the norm closed convex hull of $C \cap \{x;\ f(x) \le f(x_0)\}$ and $co(C)$ denote the norm closed convex hull of C.

Theorem 3.2. Let f be bounded below on $W(x_0)$, let ∇f be uniformly continuous and uniformly bounded on $co(C)$ with s the reverse modulus of continuity of ∇f on $co(C)$, let C be bounded, and let $p_n = p_n(x_n)$ define a feasible direction sequence for C. Let there exist functions $c_1(t)$ and $c_2(t)$ such that $c_1(t)$ and $t-c_2(t)$ are forcing functions. Let t_n^u be step sizes such that $c_1(<-\nabla f(x_n),\ \frac{p_n}{\|p_n\|}>) \le t_n^u \|p_n\| \le s[c_2(<-\nabla f(x_n),\ \frac{p_n}{\|p_n\|}>)]$, let t_n^c be step sizes such that $x_n + tp_n \in C$ for all t in $[0, t_n^C]$ and $t_n^C \|p_n\| \geq d_1(\|p_n\|)d_2(<-\nabla f(x_n), \frac{p_n}{\|p_n\|}>)$ for two forcing functions $d_1(t)$ and $d_2(t)$.

A) If we set $t_n = t_n^u$ if $x_n + tp_n \in C$ for all t in $[0, t_n^u]$ and $t_n = t_n^C$ otherwise, with $x_{n+1} = x_n + t_n p_n$, we conclude that $<\nabla f(x_n), p_n> \to 0$.

B) If t_n' is chosen as is t_n in part A from $t_n^{u'}$ and $t_n^{C'}$ and x_{n+1} is chosen in C such that $f(x_n) - f(x_{n+1}) \geq \beta[f(x_n) - f(x_n + t_n'p_n)]$ for a fixed $\beta > 0$, then $<\nabla f(x_n), p_n>$ converges to zero. The same conclusion follows if t_n' is chosen instead as in t_n in the first sentence of part B.

<u>Proof.</u> For notation we write $\gamma_n \equiv <-\nabla f(x_n), \frac{p_n}{\|p_n\|}>$.

First we consider part A. Since $t_n^u \|p_n\| \leq s[c_2(\gamma_n)]$, we have $|<\nabla f(x_n + tp_n) - \nabla f(x_n), \frac{p_n}{\|p_n\|}>| \leq c_2(\gamma_n)$ for $0 \leq t \leq t_n^u$, and hence $<-\nabla f(x_n + tp_n), \frac{p_n}{\|p_n\|}> \geq \gamma_n - c_2(\gamma_n)$ for $0 \leq t \leq t_n^u$. Since $f(x_n) - f(x_n + t_n^u p_n) = <-\nabla f(x_n + tp_n), t_n^u p_n>$ for <u>some</u> $t \in (0, t_n^u)$ and $t_n^u \|p_n\| \geq c_1(\gamma_n)$, we conclud that $f(x_n) - f(x_n + t_n^u p_n) \geq t_n^u \|p_n\| [\gamma_n - c_2(\gamma_n)] \geq c_1(\gamma_n)[\gamma_n - c_2(\gamma_n)]$. If $t_n = t_n^u$, we then have $f(x_n) - f(x_{n+1}) \geq c_1(\gamma_n)[\gamma_n - c_2(\gamma_n)]$. If $t_n = t_n^C$ then $t_n^C \|p_n\| < t_n^u \|p_n\|$ and arguing as for t_n^u we get $f(x_n) - f(x_n + t_n^C p_n) \geq t_n^C \|p_n\| [\gamma_n - c_2(\gamma_n)] \geq d_2(\gamma_n)d_1(\|p_n\|)[\gamma_n - c_2(\gamma_n)]$. Thus "$\gamma_n \to 0$ or $\|p_n\| \to 0$"; since $\|p_n\| = \|x_n' - x_n\|$ and

$\|\nabla f(x_n)\|$ are bounded, this gives $<\nabla f(x_n), p_n> \to 0$.
Part B follows easily from the estimates of part A. Q.E.D.

Remark. Since $x_n + 1 \cdot p_n \in C$, $t_n^c \equiv 1$ is possible.

4. Step sizes based on minimization

Perhaps the most natural method of choosing t_n is so as to minimize $f(x_n + tp_n)$ on the feasible region of t-values. Obviously, locating a near minimum is nearly as good; one way to describe this near minimization is via minimization of the nearby function $f(x_n + tp_n) - \alpha_n t <\nabla f(x_n), p_n>$ for some $\alpha_n \in [0,1)$, since making the derivative of this zero is equivalent to reducing the derivative of $f(x_n + tp_n)$ to a factor α_n of its value at $t = 0$.

Theorem 4.1. Let f be bounded below on $W(x_0)$, let ∇f be uniformly continuous and uniformly bounded on co(C) where C is bounded and norm closed, and let p_n define a feasible direction sequence for C. For numbers $\alpha_n \in [0, \alpha]$ with $\alpha < 1$, let $f_n(t) \equiv f(x_n + tp_n) - \alpha_n t <\nabla f(x_n), p_n>$ and let t_n be chosen as any number satisfying

i) $x_n + tp_n \in C$ for all $t \in [0, t_n]$ and
ii) $f_n(t) \geq f_n(t_n)$ for $0 \leq t \leq t_n$, and either
iii) $0 = \frac{d}{dt} f_n(t)\big|_{t=t_n} = <\nabla f(x_n + t_n p_n), p_n> - \alpha_n <\nabla f(x_n), p_n>$ or
iv) $t_n = \sup\{t;\ x_n + \tau p_n \in C \text{ for all } \tau \in [0, t]\}$.

Let $x_{n+1} \in C$ be any point such that $f(x_n) - f(x_{n+1}) \geq \beta [f(x_n) - f(x_n + t_n p_n)]$ for a fixed $\beta > 0$. Then $<\nabla f(x_n), p_n> \to 0$.

Proof. Let s be the reverse modulus of continuity of ∇f on co(C). By part A of Theorem 3.2 with $c_1(t) \equiv s(ct)$ and $c_2(t) \equiv ct$ for any fixed $c \in (0, 1-\alpha)$, $d_1(t) \equiv t$ and $d_2(t) \equiv 1$, the algorithm with t_n' determined from

$$t_n^{u'} \|p_n\| \equiv s[c<-\nabla f(x_n), \frac{p_n}{\|p_n\|} >],\ t_n^{c'} \equiv 1$$

gives the desired convergence. Assume iii) holds; we claim that $t_n \|p_n\| \geq s(c\gamma_n)$, where $\gamma_n \equiv <-\nabla f(x_n), \frac{p_n}{\|p_n\|}>$. If not, we have $|<[\nabla f(x_n + t_n p_n) - \alpha_n \nabla f(x_n)] - [\nabla f(x_n) - \alpha_n \nabla f(x_n)], \frac{p_n}{\|p_n\|}>| \leq c\gamma_n$, that is, $(1-\alpha)\gamma_n \leq (1-\alpha_n)\gamma_n \leq c\gamma_n$, a contradiction to $c \in (0, 1-\alpha)$. Thus $t_n \|p_n\| \geq s(c\gamma_n) \equiv t_n^{u'} \|p_n\|$ which implies $x_n + t_n^{u'} \|p_n\| \in C$ and $t_n' = t_n^{u'}$. If, on the other hand, iv) holds, clearly $t_n \geq t_n'$; hence in either case, $t_n \geq t_n'$. Therefore, by ii), $\frac{1}{\beta}[f(x_n) - f(x_{n+1})] \geq f(x_n) - f(x_n + t_n p_n) \geq f(x_n) - f(x_n + t_n' p_n) + \alpha_n(t_n - t_n')<-\nabla f(x_n), p_n> \geq f(x_n) - f(x_n + t_n' p_n)$, which implies the rest of the theorem by part B of Theorem 3.2. Q. E. D.

This theorem proves convergence for several well-known choices of step size, as we see in the following theorem which follows directly from the preceding.

      Theorem 4.2. Let f, C, and α_n be as in Theorem 4.1. Let t_n be chosen in any of the following ways:

    i)   as any point minimizing $f_n(t) \equiv f(x_n + tp_n) - \alpha_n t<\nabla f(x_n), p_n>$ over the set $T_n = \{t; x_n + \tau p_n \in C \text{ for all } \tau \text{ in } [0, t]\}$;

    ii)   as the smallest positive number providing a local minimum for $f_n(t)$ over T_n;

    iii) as the first positive root r of $\frac{d}{dt} f_n(t) = 0$ in T_n if such exists and as $\sup \{t; t \in T_n\}$ otherwise;

    iv)   as $t_n = \sup\{\tau; \frac{d}{dt} f_n(t) \leq 0 \text{ for } 0 \leq t \leq \tau \text{ and } \tau \in T_n\}$.

Let $x_{n+1} \in C$ be a point such that $f(x_n) - f(x_{n+1}) \geq \beta[f(x_n) - f(x_n + t_n p_n)]$ for fixed $\beta > 0$. Then $<\nabla f(x_n), p_n> \to 0$.

      Remark. For $\alpha_n \equiv \alpha \equiv 0$, these methods are certainly well-known and are often called:

    i)   optimal step size
    ii)   first local optimum step size
    iii) Curry step size [Curry (1944)], and
    iv)   Altman's step size [Altman (1966a, 1966b)].

Yet another way of indicating that one need not perform exact minimization is given in the following theorem, in essence saying that we may introduce a relaxation factor [Elkin (1968)] or proportion of the distance we need proceed towards the minimum.

Theorem 4.3. Let f, C, α, α_n, and f_n be as in Theorem 4.1, and let t_n be chosen as any number satisfying i) and ii) and either iii) or iv) of Theorem 4.1 as well as

v) $\frac{d}{dt} f_n(t) \leq 0$ for $0 \leq t \leq t_n$. Let $\hat{t}_n = \lambda_n t_n$ where $d(<-\nabla f(x_n), \frac{p_n}{\|p_n\|}>) \leq \lambda_n \leq 1$ for some forcing function $d(t)$, and let $x_{n+1} \in C$ be such that, for some fixed $\beta > 0$, $f(x_n) - f(x_{n+1}) \geq \beta[f(x_n) - f(x_n + \hat{t}_n p_n)]$. Then $<\nabla f(x_n), p_n> \to 0$.

Proof. Let $\gamma_n = <-\nabla f(x_n), \frac{p_n}{\|p_n\|}>$, and for any $c \in (0, 1-\alpha)$ let $t_n^{u''}\|p_n\| \equiv d(\gamma_n) s(c\gamma_n)$, $t_n^{c''} \equiv \lambda_n$, $c_1(t) \equiv d(t)s(ct)$, $c_2(t) \equiv ct$, $d_1(t) \equiv t$, $d_2(t) \equiv d(t)$. The method determined by $t_n^{u''}$ via $t_n^{u''}$ and $t_n^{c''}$ is clearly convergent by part A of Theorem 3.1. Recalling the definition of $t_n^{'}$ in the proof of Theorem 4.1 and the fact that $t_n \geq t_n^{'}$, we have $t_n \geq \hat{t}_n = \lambda_n t_n \geq \lambda_n t_n^{'} = t_n^{''}$. Therefore, by v), we see that $f_n(\hat{t}_n) \leq f_n(t_n^{''})$, and the result follows from part B of Theorem 3.1 just as did Theorem 4.1. Q.E.D.

Remark. We do not state the obvious result when Theorem 4.3 is combined with Theorem 4.2; the reader should note that the relaxation factor analysis is not necessarily applicable when the optimal step size is used since one then need not have v), namely, $\frac{d}{dt} f_n(t) \leq 0$ for $0 \leq t \leq t_n$.

We wish to emphasize that the algorithms above are not intended to be implemented by using certain choices of α_n; the introduction of the α_n was merely to reveal the fact that one need only move well towards the minimum, say by sufficiently reducing $\frac{d}{dt}f_n(t)$. This is important from the numerical viewpoint since it says that we need not compute exactly; any of a wide range of values of the step size will do.

If ∇f is Lipschitz continuous, that is, if $\|\nabla f(x) - \nabla f(y)\| \leq L\|x-y\|$, then $s(t) = \frac{t}{L}$ and the preceding theorems can be used to yield a range of values for t_n which lead to convergence. As has often been found in other analyses [Altman (1966a, 1966b), Cea (1969), Elkin (1968), Goldstein (1964a, 1964b, 1965, 1966), Levitin-Poljak (1966)], this range can be doubled by more precise analysis.

Theorem 4.4. Let f be bounded below on C, ∇f satisfy $\|\nabla f(x) - \nabla f(y)\| \leq L\|x-y\|$ for x, y in C, and $p_n = p_n(x_n)$ define a feasible direction sequence. Pick δ_1, δ_2, δ_3 all greater than zero and let numbers γ_n lie in

$$[\min(\delta_1, \frac{\delta_2 \|p_n\|^2}{-<\nabla f(x_n), p_n>}), \frac{2}{L} - \delta_3]$$

for all n. For each n let $x_{n+1} = x_n + t_n p_n$ where t_n is defined via

$$t_n = \min(1, \frac{\gamma_n <-\nabla f(x_n), p_n>}{\|p_n\|^2}).$$

Then $f(x_n)$ decreases to a limit. If $\|p_n\|$ is uniformly bounded, for example if C is bounded, then

$$\lim_{n \to \infty} <\nabla f(x_n), p_n> = 0.$$

If $\|p_n\| \to 0$ implies $<\nabla f(x_n), \frac{p_n}{\|p_n\|}> \to 0$, then $\lim_{n \to \infty} <\nabla f(x_n), \frac{p_n}{\|p_n\|}> = 0$.

Proof:

$$f(x_{n+1}) - f(x_n) \leq \langle \nabla f(x_n), x_{n+1} - x_n \rangle +$$

$$+ \int_0^1 \langle \nabla f(x_n + \lambda t_n p_n) - \nabla f(x_n), t_n p_n \rangle d\lambda$$

$$\leq \langle \nabla f(x_n), x_{n+1} - x_n \rangle + \frac{L}{2} t_n^2 \|p_n\|^2$$

$$\leq -t_n \langle -\nabla f(x_n), p_n \rangle + \frac{L}{2} t_n^2 \|p_n\|^2 .$$

If $1 \leq \gamma_n \dfrac{\langle -\nabla f(x_n), p_n \rangle}{\|p_n\|^2}$ then $t_n = 1$, x_{n+1} is in C, and

$$f(x_{n+1}) - f(x_n) \leq \langle -\nabla f(x_n), p_n \rangle [-1 + \frac{L}{2} \frac{\|p_n\|^2}{\langle -\nabla f(x_n), p_n \rangle}]$$

$$\leq \langle -\nabla f(x_n), p_n \rangle [-1 + \frac{L \gamma_n}{2}] \leq \frac{-\delta_3 L}{2} \langle -\nabla f(x_n), p_n \rangle \leq 0.$$

If however $1 > t_n = \gamma_n \dfrac{\langle -\nabla f(x_n), p_n \rangle}{\|p_n\|^2}$, then x_{n+1} is in C and

$$f(x_{n+1}) - f(x_n) \leq \gamma_n \frac{\langle -\nabla f(x_n), p_n \rangle^2}{\|p_n\|^2} +$$

$$+ \frac{L}{2} \|p_n\|^2 \gamma_n^2 \frac{\langle -\nabla f(x_n), p_n \rangle^2}{\|p_n\|^4}$$

so

$$f(x_{n+1}) - f(x_n) \le \frac{<-\nabla f(x_n), p_n>^2}{\|p_n\|^2} \left[\frac{\gamma_n^2 L}{2} - \gamma_n\right] \le \text{either}$$

$$\frac{-\delta_1 \delta_3 L}{2} \frac{<-\nabla f(x_n), p_n>^2}{\|p_n\|^2} \quad \text{or} \quad \frac{-\delta_2 \delta_3 L}{2} <-\nabla f(x_n), p_n>.$$

In either case $f(x_{n+1}) - f(x_n) \le 0$ and $f(x_n)$ decreases to a limit. If $\|p_n\| = \|x_n' - x_n\|$ is bounded, then from the three inequalities bounding the decrease in f we obtain $\delta > 0$ such that

$$f(x_n) - f(x_{n+1}) \ge \delta <-\nabla f(x_n), p_n>^r,$$

for $r = 1$ or $r = 2$, which implies $\lim_{n \to \infty} <-\nabla f(x_n), p_n> = 0$.

Since $<\nabla f(x_n), \frac{p_n}{\|p_n\|}> = \frac{<\nabla f(x_n), p_n>}{\|p_n\|}$, the final conclusion also follows. Q.E.D.

The above theorem is a natural extension of the well-known results for unconstrained problems; a similar but more special result appears in [Levitin-Poljak (1966)]. Other simple range theorems in terms of the Lipschitz constant can be easily derived; however we prefer to proceed to a consideration of some different techniques for providing a range of values for t_n, following the ideas of [Elkin (1968), Goldstein (1964a, 1964b, 1965, 1966)] in the unconstrained case.

5. Step sizes based on a range function

We determine admissible values of t_n in terms of the <u>range function</u>

$$g(x, t, p) \equiv \frac{f(x) - f(x + tp)}{-t < \nabla f(x), p>}.$$

Given a feasible direction sequence defined by p_n, a real number $\delta \in (0, \frac{1}{2}]$ and a forcing function $d(t) \leq \delta t$, we move from x_n to x_{n+1} as follows. If, for $t_n = 1$ and $x'_n = x_n + p_n$ we find

$$g(x_n, t_n, p_n) \geq \frac{d(<-\nabla f(x_n), p_n>)}{<-\nabla f(x_n), p_n>} \qquad (5.1)$$

we set $x_{n+1} = x'_n$; otherwise find t_n in $(0, 1)$ satisfying Equation 5.1 and also

$$|g(x_n, t_n, p_n) - 1| \geq \frac{d(<-\nabla f(x_n), p_n>)}{<-\nabla f(x_n), p_n>} \qquad (5.2)$$

and set $x_{n+1} = x_n + t_n p_n \in C$ since $x_n + p_n \in C$. We observe that the algorithm is well defined. Since $g(x_n, 0, p_n) = 1$ and $1 - \frac{d(t)}{t} \geq \frac{d(t)}{t}$ for all t, if we have $g(x_n, 1, p_n) < \frac{d(z)}{z}$ where $z = <-\nabla f(x_n), p_n>$, then by continuity of $g(x_n, t, p_n)$ in t and the fact that $x_n + tp_n$ is in C for t in $[0, 1]$ since p_n is a feasible direction there exists t_n in $(0, 1)$ with $\frac{d(z)}{z} \leq g(x_n, t_n, p_n) \leq 1 - \frac{d(z)}{z}$ which certainly satisfies Equations 5.1 and 5.2.

Theorem 5.1. Let f be bounded below on C, ∇f be uniformly continuous on $co(C)$, and p_n determine a feasible direction sequence such that $<-\nabla f(x_n), \frac{p_n}{\|p_n\|}>$ tends to zero if $\frac{d(<-\nabla f(x_n), p_n>)}{\|p_n\|}$ does so, where $d(t)$ is a forcing function with $d(t) \leq \delta t$ for δ in $(0, \frac{1}{2}]$. Let $\{p_n\}$ be bounded, for example C be bounded. Let the algorithm described above be applied. Then $\lim_{n \to \infty} <\nabla f(x_n), p_n> = 0$.

Proof. From Equation 5.1 we have

$$f(x_n) - f(x_{n+1}) \geq t_n d(<-\nabla f(x_n), p_n>) . \qquad (5.3)$$

If $t_n = 1$ does not satisfy Equation 5.1 then $t_n \in (0, 1)$; for these n we write

$$f(x_{n+1}) - f(x_n) = <\nabla f(x_n + \lambda_n t_n p_n), p_n> \text{ for some } \lambda_n \in (0, 1).$$

Thus, from Equation 5.2,

$$\frac{d(<-\nabla f(x_n), p_n>)}{<-\nabla f(x_n), p_n>} \leq |g(x_n, t_n, p_n) - 1|$$

$$\leq \left| \frac{<\nabla f(x_n + \lambda_n t_n p_n) - \nabla f(x_n), p_n>}{<\nabla f(x_n), p_n>} \right|$$

$$\leq \frac{\|\nabla f(x_n + \lambda_n t_n p_n) - \nabla f(x_n)\|_* \|p_n\|}{<-\nabla f(x_n), p_n>} ,$$

that is,

$$\|\nabla f(x_n + \lambda_n t_n p_n) - \nabla f(x_n)\|_* \geq \frac{d(<-\nabla f(x_n), p_n>)}{\|p_n\|} .$$

Therefore, if s is the reverse modulus of continuity of ∇f on $co(C)$,

$$t_n \|p_n\| = \|x_{n+1} - x_n\| \geq \|\lambda_n t_n p_n\|$$

$$\geq s(\|\nabla f(x_n + \lambda_n t_n p_n) - \nabla f(x_n)\|_*)$$

$$\geq s\left[\frac{d(<-\nabla f(x_n), p_n>)}{\|p_n\|}\right].$$

Thus, from Equation 5.3, we conclude

$$f(x_n) - f(x_{n+1}) \geq \frac{d(<-\nabla f(x_n), p_n>)}{\|p_n\|} s\left[\frac{d(<-\nabla f(x_n), p_n>)}{\|p_n\|}\right].$$

If, on the other hand, we have $t_n = 1$, then Equation 5.3 yields $f(x_n) - f(x_{n+1}) \geq d(<-\nabla f(x_n), p_n>)$. Since $\{\|p_n\|\}$ is bounded and $f(x_n) - f(x_{n+1})$ must tend to zero, the last two inequalities for $f(x_n) - f(x_{n+1})$ imply that $<\nabla f(x_n), p_n>$ tends to zero. Q.E.D.

Remark. Computationally one might commonly take $d(t) = \delta t$ for δ in $(0, \frac{1}{2}]$; in this case clearly $<-\nabla f(x_n), \frac{p_n}{\|p_n\|}>$ tends to zero whenever $\frac{d(<-\nabla f(x_n), p_n>)}{\|p_n\|}$ does so, as required in the theorem.

The step size algorithm above has appeared to be very useful for unconstrained optimization [Goldstein (1964a, 1964b, 1965, 1966)] and certainly should also be so for the constrained case. However the algorithm is not completely computational in that it may be numerically difficult to locate a t_n in $(0, 1)$ satisfying Equations 5.1 and 5.2. For the unconstrained case a very simple computational scheme is known which circumvents this difficulty [Armijo (1966), Elkin (1968)]; we show that the scheme is applicable for the constrained case as well.

Theorem 5.2. Let f, C, p_n, and d be as in Theorem 5.1 and let $\alpha \in (0, 1)$. Let t_n be chosen as the first number from the sequence $\alpha^0, \alpha^1, \alpha^2, \ldots$ such that Equation 5.1 is satisfied; then $<\nabla f(x_n), p_n>$ tends to zero.

Proof. Let α^j be the first such power, which exists by the discussion preceding Theorem 5.1. If $j = 0$ we have $f(x_n) - f(x_{n+1}) \geq d(<-\nabla f(x_n), p_n>)$ which says that, for these n, $<\nabla f(x_n), p_n>$ tends to zero; we must examine the more difficult case of $j > 0$. Let $\bar{x}_n \equiv x_n + \alpha^{j-1} p_n$. Then we have

$$f(x_n) - f(\bar{x}_n) < \alpha^{j-1} d(<-\nabla f(x_n), p_n>)$$

$$f(x_n) - f(x_{n+1}) \geq \alpha^j d(<-\nabla f(x_n), p_n>)$$

Therefore, for some $\lambda_n \in (0, 1)$,

$$(1-\alpha)\alpha^{j-1} d(<-\nabla f(x_n), p_n>) > f(x_{n+1}) - f(\bar{x}_n)$$

$$= <\nabla f(\lambda_n \bar{x}_n + (1-\lambda_n) x_{n+1}), x_{n+1} - \bar{x}_n>,$$

which yields

$$<\nabla f(\lambda_n \bar{x}_n + (1-\lambda_n) x_{n+1}), p_n> > - d(<-\nabla f(x_n), p_n>)$$

$$\geq -\delta <-\nabla f(x_n), p_n>.$$

Hence

$$\|\nabla f(\lambda_n \bar{x}_n + (1-\lambda_n) x_{n+1}) - \nabla f(x_n)\|_* \|p_n\|$$

$$\geq <\nabla f(\lambda_n \bar{x}_n + (1-\lambda_n) x_{n+1}) - \nabla f(x_n), p_n>$$

$$> (1-\delta) <-\nabla f(x_n), p_n>.$$

We then have

$$\|x_{n+1} - x_n\| \geq \alpha \|\lambda_n \bar{x}_n + (1-\lambda_n)x_{n+1} - x_n\| \geq$$

$$\geq \alpha s[\,(1-\delta) < -\nabla f(x_n), \frac{p_n}{\|p_n\|} >]$$

where $\delta(t)$ is the reverse modulus of continuity of ∇f on $co(C)$. From this and Equation 5.1 we deduce

$$f(x_n) - f(x_{n+1}) \geq \frac{\alpha}{\|p_n\|} s[\,(1-\delta) < -\nabla f(x_n), \frac{p_n}{\|p_n\|}]\, d(<-\nabla f(x_n), p_n >)$$

from which, along with the inequality found for $j = 0$, and the hypothesis on p_n, we conclude that $<\nabla f(x_n), p_n>$ tends to zero. Q. E. D.

6. Step sizes based on a search procedure

Many step size algorithms used in practice attempt to use minimization along $x_n + tp_n$, as theoretically analyzed in Section 4; however, since computers must generally deal with discrete data, one usually can at best minimize over $x_n + tp_n$ for some finite set of t values. To present a theoretical analysis of the effect of this, we restrict ourselves to <u>strictly unimodal</u> functions f, that is, to functions which have a unique local minimizing point along each straight line. It is easy to see that this condition implies that f is <u>strongly quasi-convex</u>, that is, that $f(\lambda x_1 + (1-\lambda)x_2) < \max\{f(x_1), f(x_2)\}$ for all $\lambda \in (0, 1)$ whenever $x_1 \neq x_2$. It is simple to isolate a minimizing point on a line for such a function since, given three t-values $t_1 < t_2 < t_3$ such that $f(x + t_2 p) < f(x + t_1 p)$ and $f(x + t_2 p) < f(x + t_3 p)$, we can conclude that $f(x + tp)$ has a local minimum for a t-value in (t_1, t_3).

Theorem 6.1. Let f be strongly quasi-convex and bounded below on $W(x_0)$, let ∇f be uniformly continuous and uniformly bounded on $co(C)$, let C be bounded and norm closed, and let p_n define a feasible direction sequence for C. Suppose that for each n there are values $t_{n,1} < t_{n,2} < \ldots < t_{n,k_n+1}$ such that

    i) $t_{n,k_n+1} \leq A_n \equiv \sup\{t; x_n+\tau p_n \in C$ for all $\tau \in [0, t]\}$, and

    ii) $\dfrac{t_{n,k_n-1}}{t_{n,k_n+1}} \geq \lambda > 0$ for some fixed λ, and

    iii) $f(x_n) > f(x_n + t_{n,1}p_n) > \ldots > f(x_n + t_{n,k_n}p_n)$, and either

    iv) $f(x_n + t_{n,k_n}p_n) \leq f(x_n + t_{n,k_n+1}p_n)$ or

    v) $t_{n,k_n+1} = A_n$.

Then, setting t_n equal to any number in $[t_{n,k_n-1}, t_{n,k_n}]$ and $x_{n+1} = x_n + t_n p_n$ yields a sequence $\{x_n\}$ such that $\langle \nabla f(x_n), p_n \rangle$ tends to zero.

    Proof. The point $\bar{t}_n$ providing the first local minimum of $f(x_n + tp_n)$ over $T_n \equiv [0, A_n]$ must satisfy $t_{n,k_n-1} \leq \bar{t}_n \leq t_{n,k_n+1}$ under the above hypotheses. Therefore $t_{n,k_n-1} = \lambda_n \bar{t}_n$ where $\lambda_n \equiv \dfrac{t_{n,k_n-1}}{\bar{t}_n} \geq \dfrac{t_{n,k_n-1}}{t_{n,k_n+1}} \geq \lambda$ and $\lambda_n \leq 1$. Thus, by Theorem 4.3 with $\alpha_n \equiv \alpha = 0$ and $d(t) \equiv \lambda$, we conclude that setting $t_n = t_{n,k_n-1}$ forces $\langle \nabla f(x_n), p_n \rangle$ to zero. Since $f(x_n+t_n p_n) < f(x_n+t_{n,k_n-1}p_n)$ for all t_n in $[t_{n,k_n-1}, t_{n,k_n}]$ choosing such a t_n gives a still larger decrease in f and hence forces $\langle \nabla f(x_n), p_n \rangle$ to zero as in part B of Theorem 3.2. Q.E.D.

    Corollary 6.1. Under the hypotheses of Theorem 6.1, if in addition $t_{n,i+1} - t_{n,i} = h_n$ for all i and n, then

$k_n \geq 2$ is sufficient to guarantee that choosing $t_n \in [(k_n-1)h_n, k_n h_n]$ will force $<\nabla f(x_n), p_n>$ to zero.

Proof. We have $\dfrac{t_{n,k_n-1}}{t_{n,k_n+1}} = \dfrac{k_n-1}{k_n+1} \geq \dfrac{1}{3} \equiv \lambda$. Q.E.D.

These results indicate that one has a convergent method merely by isolating the minimizing point accurately; clearly if one proceeds further, as in the common practice, by using some interpolation scheme to locate the minimizing point in the interval more accurately, the convergence is not disturbed so long as the function values are only decreased. Thus we have proved convergence for many useful computational schemes for step sizes based on searches. We shall now develop a special search routine which is very simple and appears of great use. Some preliminaries are necessary.

Theorem 6.2. Let f be strongly quasi-convex and bounded below on $W(x_0)$, let ∇f be uniformly continuous and uniformly bounded on $co(C)$, let C be bounded and norm closed, and let p_n define a feasible direction sequence for C. Suppose that for each n there exists a positive number η_n such that

i) $x_n + tp_n \in C$ for all $t \in [0, 2\eta_n]$, and

ii) $f(x_n + \eta_n p_n) \leq f(x_n + \frac{1}{2} \eta_n p_n) < f(x_n)$, and either

iii) $f(x_n + 2\eta_n p_n) \geq f(x_n + \eta_n p_n)$ or

iv) $2\eta_n = \sup\{t; x_n + \tau p_n \in C$ for all $\tau \in [0, t]\}$.

Then, setting $t_n = \eta_n$ and $x_{n+1} = x_n + t_n p_n$ forces $<\nabla f(x_n), p_n>$ to converge to zero.

Proof. If $f(x_n + \frac{3}{2} \eta_n p_n) < f(x_n + \eta_n p_n)$, we have an example of Corollary 6.1 with $k_n = 3$ and $h_n = \frac{1}{2}\eta_n$. On the other hand, if $f(x_n + \frac{3}{2} \eta_n p_n) \geq f(x_n + \eta_n p_n)$, we have an example of Corollary 6.1 with $k_n = 2$ and $h_n = \frac{1}{2}\eta_n$. Q.E.D.

We shall combine the results into an algorithm in a moment; since a simplification is possible if f is in fact convex, we first derive one more result.

JAMES W. DANIEL

Theorem 6.3. [Cea (1969)]. Let f, C, and p_n be as in Theorem 6.2, and let f be convex on C. Suppose that for each n there is an $\eta_n > 0$ such that
 i) $x_n + tp_n \in C$ for $t \in [0, 2\eta_n]$, and
 ii) $f(x_n + \eta_n p_n) \leq f(x_n + 2\eta_n p_n) \leq f(x_n)$.
Then $t_n = \eta_n$ and $x_{n+1} = x_n + t_n p_n$ yields a sequence such that $<\nabla f(x_n), p_n>$ converges to zero.

Proof. The point $\bar{t}_n$ minimizing $f(x_n + tp_n)$ for $0 \leq t \leq A_n = \sup\{t; x_n + \tau p_n \in C$ for all $\tau \in [0, t]\}$ must satisfy $0 \leq \bar{t}_n \leq 2\eta_n$; we observe that setting $t_n = \bar{t}_n$ would force $<\nabla f(x_n), p_n>$ to zero by Theorem 4.2. Since f is convex, for $0 \leq t \leq \eta$, we have

$$f(x_n + tp_n) \geq 2f(x_n + \eta_n p_n) - f(x_n + 2\eta_n p_n) +$$
$$+ \frac{f(x_n + 2\eta_n p_n) - f(x_n + \eta_n p_n)}{\eta_n} t$$

$$\geq 2f(x_n + \eta_n p_n) - f(x_n + 2\eta_n p_n)$$

$$\geq 2f(x_n + \eta_n p_n) - f(x_n).$$

Arguing similarly for $\eta_n \leq t \leq 2\eta$, we deduce

$$f(x_n + tp_n) \geq 2f(x_n + \eta_n p_n) - f(x_n) .$$

Using these inequalities for $t \equiv \bar{t}_n$ yields

$$f(x_n + \bar{t}_n p_n) \geq 2f(x_n + \eta_n p_n) - f(x_n)$$

and hence

$$f(x_n) - f(x_n + \eta_n p_n) \geq \frac{1}{2}[f(x_n) - f(x_n + \bar{t}_n p_n)]$$

which proves the theorem by use of Theorem 4.2. Q.E.D.

CONVERGENT STEP-SIZE ALGORITHMS

We can now describe a search routine which is computationally simple; this is an adaptation to constrained problems of a method for unconstrained problems presented in [Cea (1969)]. For convenience we write in a pseudo-ALGOL language.

We assume that we are given x_n, p_n and a number h $(= h_n)$ such that $x_n + tp_n \in C$ for $0 \le t \le h$; for example h = 1 is satisfactory.

Search Routine

start: if $f(x_n + hp_n) < f(x_n)$ then go to first;

reduce: $h \leftarrow \frac{h}{2}$;

           if $f(x_n + hp_n) \ge f(x_n)$ then go to reduce;

           if $f(x_n + \frac{h}{2} p_n) \ge f(x_n + hp_n)$ then

               EXIT FROM ROUTINE NOW WITH t_n = h;

           if f IS CONVEX then EXIT FROM ROUTINE NOW WITH $t_n = \frac{h}{2}$;

           $h \leftarrow \frac{h}{2}$;

loop: while $f(x_n + \frac{h}{2} p_n) < f(x_n + hp_n)$ do $h \leftarrow \frac{h}{2}$;

           EXIT FROM ROUTINE NOW WITH t_n = h;

first: if $x_n + tp_n$ IS IN C FOR $0 \le t \le 2h$ then go to inside;

           comment If C is convex one need only ask if $x_n + 2hp_n$ is in C;

           $h \leftarrow \frac{h}{2}$;

           go to start;

inside: <u>if</u> $f(x_n + 2hp_n) \geq f(x_n + hp_n)$ <u>then go to</u> oldway;
$t \leftarrow 2h$;

change: <u>while</u> $f(x_n + (t+h)p_n) < f(x_n + tp_n)$ <u>and</u>

$x_n + \tau p_n$ IS IN C FOR $0 \leq \tau \leq t+h$

<u>do</u> $t \leftarrow t + h$;

<u>comment</u> If C is convex one need only ask if

$x_n + (t+h)p_n$ is in C;

EXIT FROM ROUTINE NOW WITH $t_n = t$;

oldway: <u>if</u> f IS CONVEX <u>and</u> $f(x_n + 2hp_n) \leq f(x_n)$

<u>then</u> EXIT FROM ROUTINE NOW WITH $t_n = h$;

<u>go to</u> loop;

<u>end</u> Search Routine;

It is now straightforward to prove the following theorem using the preceding results.

    <u>Theorem 6.4.</u> Let f, C, and p_n be as in Theorem 6.2. Let the Search Routine above be used to determine t_n, starting with $h = h_n$ such that $x_n + tp_n \in C$ for $0 \leq t \leq h$, and let $x_{n+1} = x_n + t_n p_n$. Then $\langle \nabla f(x_n), p_n \rangle$ converges to zero.

7. Example of directions: variable metric gradient projections

As we mentioned in Section 2, the conclusions of most of our theorems, namely that $\langle \nabla f(x_n), p_n \rangle$ tends to zero, can be shown to be a useful condition for essentially all of the common direction algorithms; fundamentally, this is the kind of result to be found in [Topkis-Veinott (1967), Zangwill (1969), Zoutendijk (1960)]. For primarily illustrative purposes and because the result, while straightforward, does not seem to have appeared, we show the condition to be useful for the directions generated by a variable metric version of the well known gradient projection or projected gradient method.

The steepest descent method for unconstrained problems, in which $p_n = -\nabla f(x_n)$, has been a popular method for many years, for some applications undeservedly. For constrained problems that direction need not point into the constraint set C so it is not directly applicable. Perhaps the most successful way of handling this has been to "project" the direction onto C; more precisely one proceeds in the direction $p_n = x'_n - x_n$ where x'_n is the orthogonal projection onto C of $x_n - \alpha_n \nabla f(x_n)$ for some scalar $\alpha_n > 0$. This is the well known **gradient projection** method [Rosen (1960-61)]. In view of the numerical evidence that certain so-called **variable metric** methods are much better than steepest descent for unconstrained problems [Fletcher-Powell (1963)] and the growing interest in such methods for constrained problems [Goldfarb (1966, 1969a, 1969b), Goldfarb-Lapidus (1968)] we consider an analogous **variable metric projected gradient** method. We suppose that $\{A_n\}$ is a uniformly bounded, uniformly positive definite family of self-adjoint linear operators on the space X, that is, that there are $m > 0$, $M < \infty$ such that $m <x, x> \leq <A_n x, x> \leq M <x, x>$ for all x in X. For each n, let x'_n be the projection, with respect to the variable metric $<., A_n .>$, of $x_n - \alpha_n A_n^{-1} \nabla f(x_n)$ onto C; that is, x'_n minimizes $<x-(x_n - \alpha_n A_n^{-1} \nabla f(x_n)), A_n[x-(x_n-\alpha_n A_n^{-1}\nabla f(x_n))]>$ over x in C. If C is norm closed and convex a unique such x'_n exists. By the usual necessary condition, the variational definition of x'_n means that for all x in a convex C we must have

$$<x-x'_n, A_n(x'_n - w_n)> \geq 0 \qquad (7.1)$$

where $w_n \equiv x_n - \alpha_n A_n^{-1} \nabla f(x_n)$. If we set $x \equiv x_n$ in this inequality, we obtain

$$0 \geq \langle x_n - x'_n, A_n(2_n - x'_n)\rangle = \langle x_n - x'_n, A_n(w_n - x_n)\rangle +$$

$$+ \langle x_n - x'_n, A_n(x_n - x'_n)\rangle$$

and since $w_n - x_n = -\alpha_n A_n^{-1} \nabla f(x_n)$ we obtain

$$\langle x_n - x'_n, -\alpha_n \nabla f(x_n)\rangle \leq -\langle x_n - x'_n, A_n(x_n - x'_n)\rangle$$

or

$$\alpha_n \langle -\nabla f(x_n), p_n \rangle \geq \langle p_n, A_n p_n \rangle. \qquad (7.2)$$

Therefore the direction sequence is feasible. We now show that the condition $\lim_{n \to \infty} \langle \nabla f(x_n), p_n \rangle = 0$ is useful.

Theorem 7.1. Let f be convex, bounded below on the norm closed, bounded, convex set C, and attain its minimum over C at x^*. Let x_n be a sequence in C such that the projected gradient directions p_n defined above satisfy $\lim_{n \to \infty} \langle \nabla f(x_n), p_n \rangle = 0$ and $\alpha_n \geq \varepsilon > 0$. Then $\{x_n\}$ is a minimizing sequence, that is, $f(x_n) \to f(x^*)$.

Proof. We write

$$0 \leq f(x_n) - f(x^*) \leq \langle \nabla f(x_n), x_n - x^* \rangle$$

$$\leq \langle \nabla f(x_n), x_n - x'_n \rangle + \langle \nabla f(x_n), x'_n - x^* \rangle$$

$$\leq \langle -\nabla f(x_n), p_n \rangle + \frac{1}{\alpha_n} \langle x_n - \alpha_n A_n^{-1} \nabla f(x_n)$$

$$- x'_n, A_n(x^* - x'_n)\rangle + \frac{1}{\alpha_n} \langle x_n - x'_n, A_n(x'_n - x^*)\rangle$$

$$\leq \langle -\nabla f(x_n), p_n \rangle + \frac{1}{\alpha_n} \langle x_n - x'_n, A_n(x'_n - x^*)\rangle$$

by Equation 7.1. Therefore

$$0 \leq f(x_n) - f(x^*) \leq <-\nabla f(x_n), p_n> + \frac{1}{\alpha_n} M \|p_n\| \|x'_n - x^*\|$$

$$\leq <-\nabla f(x_n), p_n> + \frac{1}{\alpha_n} M \|x'_n - x^*\| [\frac{\alpha_n}{m} <-\nabla f(x_n), p_n>]^{\frac{1}{2}}$$

using Equation 7.2 and the positive definiteness of A_n. Thus

$$0 \leq f(x_n) - f(x^*) \leq <-\nabla f(x_n), p_n> + \frac{M \|x'_n - x^*\|}{\varepsilon^{\frac{1}{2}} m^{\frac{1}{2}}} <-\nabla f(x_n), p_n>^{\frac{1}{2}}$$

which tends to zero. Q.E.D.

Remark. If C is a polyhedron in $\mathbb{R}^\ell$, if $f(x) = <h-x, A(h-x)>$, if $A_n \equiv A$, and if $\alpha_n = 1$, then $x'_n = h$; thus, as in the unconstrained case, one might reasonably use an estimate of the Hessian matrix of f to define the new metric.

We note that our projected gradient method for $A_n = I$, $X = \mathbb{R}^\ell$, and C a polyhedral set, is not quite the same as the gradient projection method originally described in [Rosen (1960-61)] since that requires that x'_n be the projection onto one of the faces to which x_n belongs or, in some implementations [Cross (1968)], onto a small neighborhood of x_n in C. The computational versions of gradient projection in use apply a special technique near edges of C which turns out to be essentially equivalent to bounding α_n away from zero but keeping it small enough so that the projection is always very near x_n. Thus it is clear that a simple convergence proof for Rosen's original computational gradient projection method can be fashioned in this way from our results above; this has been done [Kreuser (1969)]. If one however does not take α_n small, one needs a good, efficient method for projection, in an arbitrary quadratic metric, onto a full polyhedral set. Such an algorithm has been brought to our attention [Golub-

Saunders (1969)] and raises the possibility of using larger α_n which may well be more powerful than the original gradient projection approach, at least far away from the solution.

REFERENCES

1. Altman, M. (1966a), "Generalized gradient methods of minimizing a functional," Bull. Acad. Polon. Sci., vol. 14, 313-318.

2. Altman, M. (1966b), "A generalized gradient method for the conditional minimum of a functional," Bull. Acad. Polon. Sci., vol. 14, 445-451.

3. Armijo, L. (1966), "Minimization of functions having Lipschitz continuous first partial derivatives," Pacific J. Math., vol. 16, 1-3.

4. Cea, J. (1969), "Methode numerique d'optimisation," lecture notes from l'Ecole d' Ete Analyse Numerique, France.

5. Cross, K. E. (1968), "A gradient projection method for constrained optimization," Union Carbide Nuclear Division Report K-1746.

6. Curry, H. B. (1944), "The method of steepest descent for nonlinear minimization problems," Quart. Appl. Math., vol. 2, 258-263.

7. Daniel, J. W. (1970), <u>Theory and methods for the approximate minimization of functionals</u>, Prentice-Hall.

8. Elkin, R. M. (1968), "Convergence theorems for Gauss-Seidel and other minimization algorithms," Computer Sci. Report #68-59, U. of Maryland, College Park.

9. Fletcher, R., Powell, M. (1963), "A rapidly convergent descent method for minimization," Computer J., vol. 6, 163-168.

10. Goldfarb, D. (1966), "A conjugate gradient method for nonlinear programming," Dissertation, Princeton Univ.

11. Goldfarb, D. (1969a), "Extension of Davidon's va variable metric method to maximization under linear inequality and equality constraints," SIAM J. Appl. Math., vol. 17, 739-764.

12. Goldfarb, D. (1969b), "Sufficient conditions for the convergence of a variable metric algorithm " in Optimization, R. Fletcher (ed.), Academic Press, London.

13. Goldfarb, D., Lapidus, L. (1968), "Conjugate gradient method for nonlinear programming problems with linear constraints," I. and E. C. Fundamentals, vol. 7, 142-151.

14. Goldstein, A. A., (1964a), "Convex programming in Hilbert space," Bull. Amer. Math. Soc., vol 70, 709-710.

15. Goldstein, A. A. (1964b), "Minimizing functionals on Hilbert space," 159-166 in Computing methods in optimization problems, ed. by Balakrishnan and Neustadt, Academic Press, New York.

16. Goldstein, A. A. (1965), "On steepest descent," J. SIAM Control, vol. 3, 147-151.

17. Goldstein, A. A. (1966), "Minimizing functionals on normed linear spaces," J. SIAM Control, vol. 4, 81-89.

18. Golub, G. H., Saunders, M. A. (1969), "Linear least squares and quadratic programming," Comp. Sci. Tech Rep. CS 134, Stanford University, Stanford, Calif.

19. Kreuser, J. (1969), private communication.

20. Levitin, E. S., Poljak, B. T. (1966), "Constrained minimization methods," (Russian), Zh. vych. Mat. mat. Fiz., vol. 6, 787-823. Also translated in USSR Comput. Math. Math. Phys., vol. 6, 1-50.

21. Rosen, J. B. (1960-61), "The gradient projection method for nonlinear programming. Part I: linear constraints," J. SIAM, vol. 8, 181-217. "_____. Part II: nonlinear constraints," J. SIAM, vol. 9, 514-532.

22. Topkis, D. M., Veinott, A. F., Jr. (1967), "On the convergence of some feasible direction algorithms for nonlinear programming," SIAM J. Control, vol. 5, 268-279.

23. Zangwill, W. I. (1969), <u>Nonlinear programming: a unified approach</u>, Prentice-Hall.

24. Zoutendijk, G (1960), <u>Methods of feasible directions</u>, Elsevier, Amsterdam.

Acknowledgement

The general outline of this material and occasional complete passages are taken from the author's forthcoming book <u>The Approximate Minimization of Functionals</u> to be published in the Prentice-Hall Series in Automatic Computation; the author thanks the publishers for their permission to include this material.

On the Implementation of Conceptual Algorithms

E. POLAK

ABSTRACT

For our purposes, we define an algorithm to be conceptual, if each of its iterations, when specified constructively, requires an infinite number of function evaluations and arithmetical operations. We shall discuss general and highly efficient procedures for implementing conceptual algorithms, i.e, for transforming them into an algorithm with iterations requiring only a finite number of function evaluations and arithmetical operations. We shall discuss both adaptive and open loop methods for truncating infinite calculations.

1. Introduction

   For our purposes, we define an algorithm to be conceptual if each of its iterations is made up of an arbitrary number of arithmetical operations and function evaluations. We define an algorithm to be implementable if each of its iterations is made up of a finite number of arithmetical operations and function evaluations. We shall assume that, in the case of implementable algorithms, a function evaluation is an operation performable in finite time on a digital computer.

   An alternative way of looking at this matter is to consider every algorithm as being made up of "outer" and of "inner" iterations. An implementable algorithm is one in which only a finite number of inner iterations are required for each outer iteration. An algorithm which specified an infinite number of inner iterations for each outer iteration is a conceptual algorithm.

   In practice, the above definition is too restrictive if taken literally, since very few functions, indeed, can even be evaluated exactly on a digital computed in finite time. Hence, we shall adopt the following convention. We shall assume that we can compute acceptable approximations to the values of such functions as $\sin x$ or e^x in finite time, but that if $f(x)$ is the smallest positive root of the equation $\sin(x + \alpha) + e^{-\alpha} = 0$, then we shall assume that the computation of $f(x)$ requires an infinite amount of time, since any evaluation of an acceptable approximation to $f(x)$, by an algorithm such as the Newton-Raphson method, requires a very large number of evaluations of $e^{-\alpha}$ and of $\sin(x + \alpha)$.

It is common in the literature to present algorithms in conceptual form, for this simplifies matters a great deal. When it comes to applying such algorithms, the user has a choice of either "doing one's best" in complying with the "non implementable" operations in the algorithm or of somehow modifying the algorithm to make it implementable. In the author's opinion, "doing one's best" is the simplest, but not the best idea. It is computationally far more efficient to first modify a conceptual algorithm. The obvious question is: but how?

A few methods for doing this have been developed by the author [1], [2], [3], and a few in collaboration with Dr. G. Meyer [4], a former graduate student. In this paper, we shall present some of the most crucial features of the author's own work.

2. Conceptual algorithms

Most conceptual algorithms of nonlinear programming are of the form of the following model, which solves the problem below.

2.1. <u>Abstract Problem</u>: Given a closed subset T of a Banach space B (with norm $\| \cdot \|_B$), construct points in T with property P.

In keeping with the terminology of previously published work, we shall say that a point $z \epsilon T$ is <u>desirable</u> if it has the property P. We shall assume that we have a test for determining whether any given $z \epsilon T$ is desirable.

2.2. <u>Algorithm Model</u>. (A: $T \to 2^T$)
    <u>Step 0</u>: Compute a $z_0 \epsilon T$, set $i = 0$.
    <u>Step 1</u>: Compute a $z_{i+1} \epsilon A(z_i)$.
    <u>Step 2</u>: If z_{i+1} is desirable, stop; else, set $i = i+1$ and go to Step 1.

The convergence properties of the above model are illuminated to a great extent by the following result.

2.3. __Theorem (Polak [1], [2])[†]__: Suppose that (i) there exists a function c: $T \to \mathbb{R}^1$ such that either $c(\cdot)$ is continuous at all $z \in T$ which are not desirable, or else $c(z)$ is bounded from below on T; and suppose that (ii) for every $z \in T$ which is not desirable, there exists an $\varepsilon(z) > 0$ and a $\delta(z) < 0$ such that

2.4 $\quad c(z'') - c(z') \leq \delta(z) < 0 \text{ for all } z' \in B(z, \varepsilon) =$

$= \{z' \in T \mid \|z' - z\| \leq \varepsilon(z)\}$, and for all $z'' \in A(z')$.

Then either the sequence $\{z_i\}$, constructed by algorithm (2.2), in solving the problem (2.1), is finite and its last element is desirable, or else it is infinite and every accumulation point of $\{z_i\}$ is desirable.

__Proof__: The case of finite sequences is trivial. Hence let us suppose that $\{z_i\}$ is infinite and that $z_i \to \hat{z}$, as $i \to \infty$, for $i \in K \subset [0, 1, 2, \ldots]$, where $\hat{z}$ is not desirable. Then there exists an $\hat{\varepsilon} > 0$ and $\hat{\delta} < 0$ such that for all $z' \in B(\hat{z}, \hat{\varepsilon})$,

2.5 $\quad\quad\quad c(z'') - c(z') \leq \hat{\delta} < 0 \text{ for all } z'' \in A(z')$.

Since $z_i \to \hat{z}$ as $i \to \infty$, for $i \in K$, there exists an integer $k \geq 0$ such that

2.6 $\quad\quad\quad c(z_{i+1}) - c(z_i) \leq \hat{\delta} \text{ for all } i \in K, i \geq k,$

[†]This theorem applies to a larger number of algorithms than a related and well known convergence theorem due to W. I. Zangwill (see W. I. Zangwill, Nonlinear Programming: A Unified Approach," Prentice Hall, 1969, pp. 91.) Whenever the assumptions of Zangwill's theorem are satisfied, the assumptions of theorem (2.5) are also satisfied, but the converse is not true.

and hence, if i, $i+j$ are now consecutive integers in K, with $i \geq k$, we have

2.7 $\quad c(z_{i+j}) - c(z_i) = [c(z_{i+j}) - c(z_{i+j-1})] + \ldots +$

$$[c(z_{i+1}) - c(z_i)] .$$

Now, since $\{z_i\}$ is infinite, none of the z_i are desirable, and hence (2.4) implies that

2.8 $\quad c(z_{i+1}) - c(z_i) < 0 \text{ for } i = 0, 1, 2, \ldots$

We therefore conclude that if i, $i+j$ are consecutive indices in K, with $i \geq k$ then

2.9 $\quad c(z_{i+j}) - c(z_i) < \hat{\delta} < 0 .$

However for $i \in K$, the monotonically decreasing sequence $\{c(z_i)\}$ must converge because of assumption (i). Since this is contradicted by (2.9), we conclude that $\hat{z}$ must be desirable.

To illustrate both the applicability of the model (2.2) and the way in which one uses theorem (2.3), let us consider two famous conceptual algorithms. The first one is a method of centers due to Huard [5] and the second one is the Frank and Wolfe algorithm [6]. In the next two sections we shall show how one can transform these conceptual algorithms into an implementable form.

Consider the problem

2.10 $\quad \min\{f^0(z) \mid f^i(z) \leq 0, \ i = 1, 2, \ldots, m\}$

where $f^i : \mathbb{R}^n \to \mathbb{R}^1$ are strictly convex, continuously differentiable functions, and suppose that for some $z_0 \in T = \{z \mid f^i(z) = 0, \ i = 1, 2, m\}$ the set

2.11 $\quad C(z_0) = \{z \mid f^0(z) - f^0(z_0) \leq 0;$

$\qquad f^i(z) \leq 0, \ i = 1, 2, \ldots, m\}$

is compact and has in interior.
  The following algorithm solves problem (2.10).

2.12. **Modified Method of Centers** (Huard [5]).
  <u>Step 0:</u> Compute a point z_0 satisfying $f^i(z_0) \leq 0$ for $i = 1, 2, \ldots, m$, such that $C(z_0)$ is compact, and set $i = 0$.
  <u>Step 1:</u> Set $z = z_i$.
  <u>Step 2:</u> Compute $(h^0(z), h(z))(h^0(z) \in \mathbb{R}^1, h(z) \in \mathbb{R}^n)$ as a solution of the linear programming problem

2.13 $\quad \min\{h^0 \mid -h^0 + <\nabla f^0(z), h> \leq 0; \ -h^0 + f^i(z)$

$\qquad + <\nabla f^i(z), h> \leq 0, \ i = 1, 2, \ldots, m;$

$\qquad |h^j| \leq 1, \ j = 1, 2, \ldots, m\}.$

  <u>Step 3:</u> If $h^0(z) = 0$, set $z_{i+1} = z_i$ and stop; else, go to Step 4.
  <u>Step 4:</u> Compute $\mu(z, h(z))$ to be the smallest positive scalar such that

2.14 $\quad d(z + \mu(z, h(z))h(z), z) = \min\{d(z + \mu h(z), z) \mid \mu \geq 0\},$

where $d: \mathbb{R}^n \times \mathbb{R}^n \to \mathbb{R}^1$

is defined by

2.15 $\quad d(z', z'') = \max\{f^0(z') - f^0(z''); \ f^i(z'),$

$\qquad\qquad i = 1, 2, \ldots, m\}.$

IMPLEMENTATION OF CONCEPTUAL ALGORITHMS

Step 5: Set $z_{i+1} = z + \mu(z, h(z))h(z)$, set $i = i+1$, and go to Step 1.

Remark: Since the computation indicated in (2.14) cannot be implemented (in the sense indicated earlier), we see that algorithm (2.12), is a conceptual algorithm.

We begin by noting that $h^0(z)$, as determined in Step 2 of (2.12), satisfies $h^0(z) = 0$ if and only if z is optimal for problem (2.10), under the convexity assumptions stated. (See [1], [2]). Hence the stop condition in Step 3 of (2.12) is the same as the stop condition in Step 2 of (2.2).

Next, we define

2.16 $$T = \{z \mid f^i(z) \leq 0, \ i = 1, 2, \ldots, m\},$$

we define $\hat{z} \in T$ to be desirable if it is an optimal solution of (2.10), and we define A: $T \to 2^T$ as follows:

2.17
$$A(z) = \{z\} \text{ if } z \text{ is optimal for (2.10)}$$
$$A(z) = \{z' \mid z' = z + \mu(z, h(z))h(z)\} \text{ otherwise,}$$

where $h(z)$ is part of an optimal solution of (2.13) and $\mu(z, h(z))$ is then determined by (2.14). Since $h(z)$ is not necessarily unique, we see that $A(z)$ may consist of more than one point. With T and $A(\cdot)$ defined as above, the identification of algorithm (2.12) with the model (2.2) is completed.

Finally, if we set

2.18 $$c(z) = f^0(z) \text{ for all } z \in T,$$

we can show that the assumptions of theorem (2.3) are satisfied (for a proof see [1]).

Next, suppose that $t \in \mathbb{R}^n$ and that T is a compact, convex subset of $\mathbb{R}^n$, $t \notin T$, and consider the problem

2.19 $$\min\{\|t - z\| \mid z \in T\}$$

A solution to this problem can be obtained by the method below.

2.20. **The Frank and Wolfe Algorithm [6].**
<u>Step 0</u>: Compute a $z_0 \in T$, and set $i = 0$.
<u>Step 1</u>: Compute a $\omega_i \in T$ as a solution of the problem.

2.21 $$\max\{<t - z_i, \omega> \mid \omega \in T\}$$

<u>Step 2</u>: Compute $z_{i+1} \in [z_i, \omega_i] \stackrel{\Delta}{=} \{z \mid z = \lambda z_i + (1-\lambda)\omega_i, \lambda \in [0, 1]\}$, such that

2.22 $$\|t - z_{i+1}\| = \min\{\|t - z\| \mid z \in [z_i, \omega_i]\}$$

<u>Step 3</u>: If $\|t - z_{i+1}\| = \|t - z_i\|$, stop; else set $i = i+1$ and go to Step 1.

Remark: We shall consider the operation (2.21) to be not implementable. Now, it is not difficult to see that the stop condition in Step 3 can be satisfied, if and only if, z_i is optimal for (2.19), hence it agrees with the stop condition in Step 2 of (2.2), provided we define $z \in T$ to be desirable if it is optimal for (2.19). Next, we define

2.23 $A(z) = \{z\}$ if z is desirable

$$A(z) = \bigcup_{\omega' \in W(z)} \{z' \mid z' \in [z, \omega'], \|t - z'\| = \min\{\|t - z''\| \mid z'' \in [z, \omega']\}\},$$

otherwise,

where

2.24 $\quad W(z) = \{\omega' \in T \mid <t-z, \omega'> = \max\{<t-z, \omega''> \mid \omega'' \in T\}$.

This completes the identification of algorithms (2.20) with the model (2.2). If we now set

2.25 $\quad\quad\quad c(z) = \|t-z\|$,

then we can show that the assumptions of theorem (2.3) are satisfied (see theorem (5.3.26) in [3]).

We can now proceed to examine methods for making algorithms such as (2.12) and (2.20) implementable.

3. Adaptive Procedures for Implementation

As we have seen from the two examples described, our difficulty with conceptual models of the form of algorithm (2.2) stem from the fact that we are unable to find points in $A(z_i)$ by means of a finite subprocedure. The following model gets around this problem by introducing an approximation to the set $A(z_i)$. Also, since according to theorem (2.3) we need to have a function $c: T \to R^1$ with certain properties, if we are to be sure that algorithm (2.2) is convergent in the sense of theorem (2.3), and since from the examples given we saw that such a $c(\cdot)$ is trivially constructed, we can include it into the model below, which solves problem (2.1)

3.1. <u>Algorithm Model:</u> (A: $R^+ \times T \to 2^T$, c: $T \to R^1$,
$\varepsilon_0 > 0$, $\varepsilon' \in (0, \varepsilon_0)$, $\alpha > 0$, $\beta \in (0, 1)$).
<u>Step 0:</u> Compute a $z_0 \in T$, set $i = 0$.
<u>Step 1:</u> Set $\varepsilon = \varepsilon_0$.
<u>Step 2:</u> Compute a $y \in A(\varepsilon, z_i)$.

Step 3: If $c(y) - c(z_i) \leq \alpha\varepsilon$, set $z_{i+1} = y$, set $i = i+1$ and go to Step 1; else, go to Step 4.

Step 4: If $\varepsilon > \varepsilon'$, set $\varepsilon = \beta\varepsilon$ and go to Step 2; else, go to Step 5.

Step 5: Check whether z_i is desirable.

Step 6: If z_i is desirable, set $z_{i+1} = z_i$ and stop; else, set $\varepsilon = \beta\varepsilon$ and go to Step 2.

When the test for desirability is quite difficult to perform, we may prefer the following approach.

3.2. Algorithm Model: (A: $\mathbb{R}^+ \times T \to 2^T$, c: $T \to \mathbb{R}^1$, $\varepsilon_0 > 0$, $\alpha > 0$, $\beta \in (0, 1)$).

Step 0: Compute a $z \in T$, set $i = 0$.

Step 1: Set $\varepsilon = \varepsilon_0$.

Step 2: Compute a $y \in A(\varepsilon, z_i)$.

Step 3: If $c(y) - c(z_i) \leq -\alpha\varepsilon$, set $z_{i+1} = y$, set $i = i+1$ and go to Step 1; else, set $\varepsilon = \beta\varepsilon$, set $z_{i+1} = z_i$, set $i = i+1$ and go to Step 2.

Finally, if there is reason to believe that returning from Step 3 to Step 1, in the algorithms (3.1) or (3.2), causes a large loss of computing time in the process of reducing ε to an acceptably low value, we may return from Step 3 to Step 2, without loss of the convergence properties to be stated in the theorem below.

3.3. Theorem: Suppose that (i) $c(\cdot)$ is either continuous at all undesirable points in T, or else $c(z)$ is bounded from below for $z \in T$; and that (ii) for every $z \in T$ which is not desirable, there exist an $\varepsilon(z) > 0$, a $\delta(z) < 0$ and a $\gamma(z) > 0$ such that

3.4 $$c(z'') - c(z') \leq \delta(z) < 0$$

for all $z' \in T$, $\|z' - z\|_B \leq \varepsilon(z)$, for all $z'' \in A(\gamma, z')$, for all $\gamma \in [0, \gamma(z)]$.

If $\{z_i\}$ is a sequence constructed by algorithm (3.1) or (3.2) or the time varying version of these algorithms obtained by changing the command in Step 3 from "go to Step 1" to "go to Step 2", then either $\{z_i\}$ is finite and its last element is desirable, or else it is infinite and every accumulation point of $\{z_i\}$ is desirable.

Proof: We shall only give a proof for the case of algorithm (3.2) (as stated), since the other cases can be established in a similar manner.

First, suppose that for some i, $z_i = z_{i+1} = z_{i+2} = \ldots$, i.e., that the algorithm jams up. Then it must be constructing a sequence of vectors $y_j \in A(\beta^j \varepsilon_0, z_i)$, $j = 0, 1, 2, \ldots$ such that $c(y_j) - c(z_i) > -\alpha\beta^j \varepsilon_0$, $j = 0, 1, 2, \ldots$. We therefore conclude assumption (ii) that z_i is desirable. Also note that in this case $z_j \to z_i$ as $j \to \infty$.

Hence, suppose that there is no i such that $z_i = z_{i+1} = z_{i+2} = \ldots$, and suppose that $\hat{z}$ is an accumulation point of $\{z_i\}$, i.e. $z_i \to \hat{z}$ as $i \to \infty$ for $i \in K \subset \{0, 1, 2, \ldots\}$, where the z_i are distinct points. Suppose that $\hat{z}$ is not desirable. Then, by (ii) there exist $\hat{\varepsilon} > 0$, $\hat{\delta} < 0$, and $\hat{\gamma} > 0$ for which (3.4) is satisfied. Since $z_i \to \hat{z}$ for $i \in K$ there must exist an integer $k \geq 0$ such that $\|z_i - \hat{z}\|_B \leq \hat{\varepsilon}$ for all $i \in K$, $i \geq k$ and $\max\{\beta^k \varepsilon_0, \alpha\beta^k \varepsilon_0\} \leq \min\{\hat{\gamma}, -\hat{\delta}\}$. Let z_i, z_{i+j} be consecutive points of the sequence $\{z_i\}_{i \in K}$, and let $\ell < j$ be such that $z_i = z_{i+1} = \ldots = z_{i+\ell} \neq z_{i+\ell+1}$. Clearly $\ell < k+1$ because $\ell \geq k+1$ would imply $\varepsilon = \beta^{\ell-1}\varepsilon_0 \leq \beta^k\varepsilon_0 \leq \hat{\gamma}$ at $z_{i+\ell-1}$ and therefore $z_{i+\ell} \neq z_{i+\ell-1}$ by (3.4) and the choice of k, $\hat{\varepsilon}$, and $\hat{\gamma}$. Since $\ell < k+1$ we can write

$$3.5 \quad c(z_{i+j}) - c(z_i) = [c(z_{i+j}) - c(z_{i+j-1})] + \ldots +$$
$$+ [c(z_{i+1}) - c(z_i)] \leq -\alpha\beta^\ell \varepsilon_0 \leq -\alpha\beta^k \varepsilon_0 \leq \hat{\delta}$$

which shows that the sequence $\{c(z_i)\}$ for $i \in K$ is not Cauchy and hence not convergent. But the monotonically decreasing sequence $\{c(z_i)\}$ must converge because of assumption (i) and hence we have a contradiction; i.e. the accumulation point $\hat{z}$ must be desirable.

We shall now show how we can use these models to make implementable the method of centers discussed in the preceding section. Since the functions $f^i(\cdot)$, $i = 0, 1, 2, \ldots, m$ in problem (2.10) are convex, we can use a Fibonacci search to find a $\mu(z, h(z))$ satisfying (2.14). In the algorithm below, this search is truncated so as to give an algorithm of the form (3.1), which solves problem (2.10) under the assumptions stated.

3.6. **Algorithm**: (Implementation of (2.12)).
   Step 0: Compute a point z_0 satisfying $f^i(z_0) \leq 0$ for $i = 1, 2, \ldots, m$, such that $C(z_0)$ is compact; select an $\varepsilon_0 > 0$, and set $i = 0$.
   Step 1: Set $\varepsilon = \varepsilon_0$.
   Step 2: Set $z = z_i$.
   Step 3: Compute $(h^0(z), h(z))$ by solving (2.13).
   Step 4: If $h^0(z) = 0$, set $z_{i+1} = z_i$ and stop; else go to Step 5.
   Step 5: Use the Fibonacci search procedure to find $0 \leq \mu' < \mu'' < \infty$ such that $\mu'' - \mu' \leq \varepsilon$ and such that $\mu(z, h(z))$ (as defined in (2.14)) satisfies $\mu(z, h(z)) \in [\mu', \mu'']$.

Comment: the search for μ', μ'' is a finite process.
   Step 6: If $f^0(z + \mu'h(z), z) - f^0(z) \leq -\varepsilon$, set $z_{i+1} = z + \mu'h(z)$, set $i = i+1$ and go to Step 1; else, set $\varepsilon = \varepsilon/2$ and go to Step 5.

It is not too difficult to show that the algorithm (3.6) satisfies the assumptions of theorem (3.3)[+], with $c(\cdot) = f^0(\cdot)$, $T = C(z_0)$, $z \in T$ defined to be desirable if $h^0(z) = 0$ (i.e. if it is optimal for (2.10)), and $A(\cdot, \cdot)$ defined by the instructions in (3.6).

[+] Note that $f^0(\cdot)$ is uniformly continuous on $C(z_0)$ defined by (2.11) and that for every $z' \in A(\varepsilon, z)$ there is a $z'' \in A(z)$ defined by (2.17) such that $\|z' - z''\| \leq \varepsilon$.

4. Open Loop Procedures for Implementation

The idea of open loop truncation is due to G. Meyer [4]. However, the specific results to be presented below are due to the author. As the reader will observe, the models given below bear a strong resemblance to the ones introduced in the preceding section, with one notable exception: the quality of approximation to the set A(z) in algorithm (2.2) is not governed by an ε test. Rather, it is continuously improved as the iterations progress. The advantage of such an "open loop" approach lies in the fact that it reduces the need for comparisons which may require lengthy preliminary calculations. The open loop approach is particularly effective when used in a situation where the same problem has to be resolved over and over again, with relatively minor changes in its parameters.

As in the preceding section, we make use of a map $c: T \to \mathbb{R}^1$ and of a map $A(\cdot, \cdot)$. However, we now define A on $N \times T$, where N is the set of all positive integers ($0 \in N$). In addition, we shall make use of truncation functions. We say that $\ell: N \to N$ is a <u>truncation function</u> if for every $k \in N$ there exists a $k' \in N$ such that $\ell(i) > k$ for all $i \geq k'$.

The following algorithm solves problem (2.1).

4.1. <u>Algorithm Model</u>: (A: $N \times T \to 2^T$, c: $T \to \mathbb{R}^1$, $\ell_1: N \to N$, $\ell_2: N \to N$).

<u>Step 0</u>: Compute $z_0 \in T$, set i = 0, set j = 0.
<u>Step 1</u>: Set $z = z_i$, set $j = \ell_1(i)$.
<u>Step 2</u>: Compute a $y \in A(j, z)$.
<u>Step 3</u>: If $c(y) < c(z)$, set $z_{i+1} = y$, set i = i+1 and go to Step 1; else, set $j = \ell_2(j)$ and go to Step 2.

When the utility of the test $c(y) < c(z)$ in Step 3 of algorithm (4.1) is more than cancelled out by the effort in computing $c(y)$, we may prefer to use a direct test for desirability and to modify algorithm (4.1) as follows.

4.2. **Algorithm Model:** (A: $N \times T \to 2^T$, $\ell: N \to N$)
    Step 0: Compute $z_0 \in T$, set $i = 0$.
    Step 1: Set $j = \ell(i)$.
    Step 2: Compute a $y \in A(j, z)$: if y is desirable, stop; else go to Step 3.
    Step 3: Set $z_{i+1} = y$, set $i = i+1$ and go to Step 1.

4.3. **Theorem:** Consider algorithm (4.1). Suppose that (i) $c(\cdot)$ is either continuous at all nondesirable $z \in T$, or else $c(z)$ is bounded from below for $z \in T$, and that (ii) for every $z \in T$ which is not desirable, there exists an $\varepsilon(z) > 0$, a $\delta(z) < 0$ and an integer $k(z) \geq 0$ such that

4.4 $$c(z'') - c(z') \leq \delta(z) < 0$$

for all $z' \in T$, $\|z' - z\|_B \leq \varepsilon(z)$, for all $z'' \in A(j, z')$, for all $j \geq k(z)$. Finally, suppose that $\ell_1(\cdot)$, $\ell_2(\cdot)$ are truncation functions, and that $\ell_2(j) > j$ for all $j \in N$.

    If $\{z_i\}$ is an infinite sequence constructed by algorithm (4.1), then every accumulation point of $\{z_i\}$ is desirable.

    Proof: First suppose that there is an integer i such that the algorithm jams up at z_i cycling between steps 2 and 3. Then, because of (ii), z_i must be desirable. Hence, we need not worry about this case.

    Next, suppose that $\{z_i\}$ is infinite and that $z_i \to \hat{z}$ as $i \to \infty$ for $i \in K \subset \{0, 1, 2, \ldots\}$. Furthermore, suppose that $\hat{z}$ is not desirable. Then there exist an $\hat{\varepsilon} > 0$, a $\hat{\delta} < 0$ and a $\hat{k} \in N$ such that

4.5 $$c(z'') - c(z') \leq \hat{\delta}$$

for all $z' \in T$, $\|z' - \hat{z}\|_B \leq \hat{\varepsilon}$, for all $z'' \in A(j, z')$, $j \geq \hat{k}$. Since $z_i \to \hat{z}$ for $i \in K$, and $\ell_1(\cdot)$, $\ell_2(\cdot)$ are truncation functions, there exists an integer $k \geq 0$ such that for all $i \in K$, $i \geq k$, $\|z_i - \hat{z}\|_B \leq \hat{\varepsilon}$ and $j(i) \geq \hat{k}$. Hence, if i, $i+j$ are consecutive elements in K, with $i \geq k$, then we must have

4.6
$$c(z_{i+j}) - c(z_i) < \hat{\delta}.$$

Since (4.6) contradicts the convergence of the sequence $\{c(z_i)\}$, $i \in K$, we are done

4.7. Theorem: Consider algorithm (4.2). If there exists a function $c: T \to \mathbb{R}^1$ such that assumptions (i) and (ii) of theorem (4.3) are satisfied and $\ell(\cdot)$ is a truncation function, then every accumulation point of a sequence $\{z_i\}$ constructed by algorithm (4.2), and satisfying $c(z_{i+1}) < c(z_i)$ for $i = 0, 1, 2, \ldots$ must be desirable.

To illustrate the use of the above model, we implement the Frank and Wolfe method by modifying it to correspond to the form (4.2). Suppose that the set T in (2.19) has an interior and is defined by convex, continuously differentiable inequalities, as follows:

4.8
$$T = \{z \in \mathbb{R}^n \mid f^i(z) \leq 0, i = 1, 2, \ldots, m\}.$$

Let

4.9
$$f^0(\omega, z) = \langle \omega, z - t \rangle \text{ for all } z, \omega \in \mathbb{R}^n.$$

Then we can solve the subproblem (2.21) in Step 1 of algorithm (2.20) by a method of feasible directions, which we truncate as shown below.

4.10. Algorithm for Computing $\omega(z, y, j)$, **given** $y, z \in T$,
  given $j \in N$, $\beta \in (0, 1)$.
  Step 0: Set $k = 0$, set $x = y$.
  Step 1: Compute $(h^0(x), h(x))(h^0 \in \mathbb{R}^1, h \in \mathbb{R}^n)$ by solving the linear programming problem
  $\min\{h^0 \mid -h^0 + \langle (z - t), h \rangle \leq 0;$
  $-h^0 + f^p(x) + \langle \nabla f^p(x), h \rangle \leq 0$, $p = 1, 2, \ldots, m$; $|h^j| \leq 1$, $j = 1, 2, \ldots, n\}$,
  (with f^p as in (4.8)).
  Step 2: Find smallest positive integer q such that $f^p(x + \beta^q(h(x))) \leq 0$ for $p = 1, 2, \ldots, m$.

Step 3: If $k = j - 1$, set $\omega(z, y, j) = x + \beta^q h(x)$ and stop; else set $x = x + \beta^q h(x)$, set $k = k + 1$ and go to Step 1.

The Frank and Wolfe method now yields the following algorithm which corresponds to the model (4.2).

4.11. <u>Algorithm</u>: (Implementation of (2.20)).
  <u>Step 0</u>: Compute a $z_0 \in T$; select a truncation function $\ell(\cdot)$, and set $i = 0$.
  <u>Step 1</u>: Set $j = \ell(i)$.
  <u>Step 2</u>: Compute $\omega_i = \omega(z_i, z_i, j)$ by means of (4.10).

Comment: Note that this process is finite.
  <u>Step 3</u>: Compute $z_{i+1} \in [z_i, \omega_i]$ such that

4.12 $\qquad \|t - z_{i+1}\| = \min\{\|t - z\| \mid z \in [z_i, \omega_i]\}$

Comment: Note that (4.12) can be solved trivially and hence the computation of z_{i+1} is a finite process.

  <u>Step 4</u>: If $\|t - z_{i+1}\| = \|t - z_i\|$, stop; else, set $i = i+1$ and go to Step 1.

To see that algorithm (4.11) does indeed correspond to the model (4.2), we set $c(z) = \|t - z\|$, and we define $A(j, z)$ by (2.23), with $W(z)$ replaced by the set $W(j, z) = \{\omega(z, y, j) \mid y \in T\}$. It is now reasonably easy to show that the assumptions of theorem (4.3) are satisfied.[+]

[+]To obtain this result, we begin by showing that for every z and y in T, there exists an $\varepsilon(z, y) > 0$ and a sequence $\{\delta_j(z, y)\}_{j=0}^{\infty}$, with $\delta_j(z, y) > 0$ and $\delta_j \to 0$ as $j \to \infty$, such that $<t-z'$, $\omega'> - <t-z', \omega(z', y', j)> \leq \delta_j(z, y)$ for all z', y', ω' in T satisfying $\|z'-z\| \leq \varepsilon(z, y)$, $\|y'-y\| \leq \varepsilon(z, y)$, and $<t-z', \omega'> = \max\{<t-z', \overline{\omega}> \mid \overline{\omega} \in T\}$. (For this purpose one needs some of the details appearing in the proofs in [2].) Once this fact has been established, it is reasonably easy to show that our claim is correct.

5. Conclusion

The methods for conceptual algorithm implementation described in this paper represent only a small fraction of possibilities. A number of other algorithm models which can guide us in our invention of implementation schemes can be found in [3], [4]. The book [3] treats a large number of algorithms in terms of the ideas highlighted in this paper.

REFERENCES

1. E. Polak, "Computational Methods in Discrete Optimal Control and Nonlinear Programming: A Unified Approach, " Univ. of Calif. Berkeley, Electronics Res. Lab., Memo No. ERL-M261, Feb. 1969.

2. _____, "On the Convergence of Optimization Algorithms, " RIRO, No. R1, pp. 17-34, 1969.

3. _____, "Computational Methods in Optimization: A Unified Approach, " Academic Press, 1970 (in Press).

4. G. Meyer and E. Polak, "Abstract Models for the Synthesis of Optimization Algorithms, " Univ. of Calif., Berkeley, Electronics Res. Lab. Memo No. ERL-M269, Oct. 1969.

5. P. Huard, "Programmation Mathematique Convex, " RIRO, R7, pp. 43-59, 1968.

6. M. Franke and P. Wolfe, "An Algorithm for Quadratic Programming, " Naval Log. Quart., Vol. 3, pp. 95-110, 1956.

Some Convex Programs Whose Duals Are Linearly Constrained

R. TYRRELL ROCKAFELLAR

ABSTRACT

Let $f_0, f_1, \ldots, f_m$ be convex functions on R^n, and let (P) denote the problem of minimizing $f_0(x)$ subject to $f_1(x) \leq 0, \ldots, f_m(x) \leq 0$. According to the theory of conjugate functions, many different dual problems can be associated with (P), each one corresponding to a particular class of perturbations of (P). Thus, in developing dual methods of solutions of (P), one has considerable flexibility in the choice of the dual problem, and the choice can be made in view of its suitability for a given purpose.

This paper treats some simple possibilities in the important case where each of the functions f_i satisfies the following condition: f_i is not affine along any line segment, unless it is affine along the entire line extending the segment (The latter holds, for example, if f_i is analytic.) It is shown that the perturbations can be chosen in this case so that the corresponding dual problem (P*) consists essentially of maximizing a differentiable concave function subject to linear constraints. The duality theorems applicable to (P) and (P*) are then somewhat more refined than those in the general theory; e.g. the infimum in (P) and the supremum in (P*) are necessarily equal, if (P) is consistent.

The duality theory for the geometric programs of Duffin, Peterson and Zener, and the quadratic and ℓ^p programs of Peterson and Ecker, is derived as an illustration.

1. Introduction

Let $f_0, f_1, \ldots, f_m$ be convex functions on R^n, and let (P) denote the problem of minimizing $f_0(x)$ subject to the constraints $f_1(x) \leq 0, \ldots, f_m(x) \leq 0$. According to the theory of conjugate functions, many different dual problems can be associated with (P), each one corresponding to a particular class of perturbations of the objective function and constraints. Thus, in developing dual methods of solution of (P), one has considerable flexibility in the choice of the dual problem, and the choice can be made in view of its suitability for a given purpose.

In this paper we describe cases where the dual can be regarded essentially as a problem of maximizing a differentiable concave function subject to linear constraints. Actually, it is possible to find many such cases, and the approaches to them are quite diverse. Rather than attempting a general survey, however, we concentrate here on presenting a few especially sharp results for problems (P) of a restricted type.

To this end, we assume that the functions f_i are everywhere differentiable and satisfy the following condition, called <u>faithful</u> convexity: f_i is not affine (simultaneously convex and concave) along any line segment, unless f_i is affine along the entire line extending the line segment. The class of faithfully convex functions obviously includes all strictly convex functions and all affine or quadratic convex functions. In fact, it includes all <u>analytic</u> convex functions. Thus the results in this paper are applicable in particular to convex programming problems with analytic objective and analytic constraints.

We begin with a discussion of the "ordinary" dual (D_0) of (P), where the constraint functions are perturbed by subtracting constants. Further perturbations are then introduced, leading to a dual problem (D_1) which involves more variables, but which can be handled more directly, provided that the Legendre transformation can be carried out. This extended dual, originally presented in a less elaborate form in 1964 [11], has hitherto not been exploited for its computational properties.

Our assumptions on the nature of the functions f_i allow certain refinements of the general theorems in [12]. These results encompass the duality theorems of Duffin, Peterson and Zener for geometric programming [1, 2, 3] and those of Peterson and Ecker for (quadratic and) ℓ^p-programming [7, 8, 9, 10].

We show that, after an optimal solution has been determined for (D_1), an optimal solution may be obtained for (P), if not immediately, then by solving the dual (D_1') of just one further problem (P') of the same type as (P). This may be contrasted with the procedure given by Peterson and Ecker in ℓ^p-programming, where a sequence of subsidiary dual problems, numbering perhaps as many m+1, might have to be solved in order to determine an optimal solution to (P). When the simplified procedure given here is applied to an ℓ^p-program (P), the problem (P') is another ℓ^p-program.

2. Dual problems

In the terminology of [12], the problem (P) is not actually called a <u>convex program</u> until a suitable class of perturbations has been singled out. Unless otherwise specified, however, it is assumed that the perturbations are the following, in which event one speaks of an <u>ordinary</u> convex program: for each vector $u = (u_1, \ldots, u_m)$ in R^m one considers the problem of minimizing $f_0(x)$ subject to the constraints

$$f_i(x) - u_i \leq 0, \quad i = 1, \ldots, m. \tag{2.1}$$

The dual problem corresponding to this class of perturbations is that of maximizing the concave function

$$g(y) = \inf\{f_0(x) + y_1 f_1(x) + \cdots + y_m f_m(x) \mid x \in R^n\} \tag{2.2}$$

over the convex set

$$C = \{y = (y_1, \ldots, y_m) \in R^m \mid y \geq 0, \; g(y) > -\infty\} \tag{2.3}$$

We call this the <u>ordinary</u> dual of (P) and denote it by (D_0).

Problems closely related to (D_0) have, of course, been studied by many authors. Concerning computation, the papers of Falk [4] and Geoffrion [5] are especially noteworthy. Roughly speaking, (D_0) may be expected to be useful computationally in solving (P) if the minimization in (2.2) is relatively easy to carry out for any $y \geq 0$, as for example if the functions f_i are all separable, or all quadratic. In the separable case, solving (P) by way of (D_0) is an application of the decomposition principle (for a general discussion see [12, pp. 285-290]).

Let C' denote the subset of C consisting of the vectors y such that the infimum in (2.2) is attained. Results of Falk [4] show that, if f_0 is strictly convex, then C' is convex and open relative to the orthant R_+^m, and g is continuously differentiable relative to C' with relative gradient $(f_1(x), \ldots, f_m(x))$, where the x corresponding to a given $y \in C'$ is the unique element of R^n for which the infimum in (2.2) is attained. Furthermore, if (P) is strictly consistent (i.e. satisfies the Slater condition) and has an optimal solution, or if C' is nonempty and the supremum of g over C' is attained, then

$$\sup_{y \in C'} g(y) = \sup(D_0) = \inf(P).$$

The main restriction in applying Falk's results is the requirement that f_0 be strictly convex. Thus the case where (P) is a linear programming problem is not covered, even though this is the case on which the results are patterned. On the other hand, if f_0 is not strictly convex it can be made so by adding a strictly convex term. For instance one can add $\varepsilon |x - \bar{x}|^2$, where $\bar{x}$ is an estimate of an optimal solution to (P) and $|\cdot|$ denotes the Euclidean norm, and one then has $C = C' = R_+^n$ [14, p. 136]. This device has computational uses, but one disadvantage could be an increase in the dimensionality of the dual problem. In the altered (D_0) the convex set C is m-dimensional in R^m, whereas in the original (D_0) it might be of smaller dimension, corresponding to the fact that the dual variables had to satisfy certain linear relations as in linear programming.

Under the conditions we have imposed on the functions f_i, Falk's results can be generalized in a rather thorough way to the case where f_0 is not strictly convex, and the relationship between C and C' can also be described in greater detail. Instead of doing this here, however, we develop related results for a different dual problem in which linear relations among the dual variables appear explicitly.

Let each f_i be expressed in the form

$$f_i(x) = h_i(A_i x + a_i) + b_i x + c_i, \qquad (2.5)$$

where h_i is a differentiable, faithfully convex function on R^{n_i}, A_i is a matrix of dimension $n_i \times n$, $a_i \in R^{n_i}$, $b_i \in R^n$ and $c_i \in R^1$. Certainly such an expression (2.5) is possible, since one can always take $n_i = n$, $A_i = I$, $a_i = 0$, $b_i = 0$ and $c_i = 0$. It is easily seen from the theory of lineality vectors of convex functions [12, pp. 70-71] that the faithful convexity property of f_i is equivalent to the existence of a representation (2.5) with h_i strictly convex and $n_i =$ rank f_i. (As an extreme case of (2.5), we allow $n_i = 0$; then the term $h_i(A_i x + a_i)$ is omitted.) In what follows we do not assume, however, that h_i is strictly convex, since that would make it awkward to treat certain examples such as geometric programs.

The dual problem we want to discuss is the one corresponding to the following class of perturbations of (P), as explained in [12, pp. 324-325]. With each vector

$$(u, v_0, \ldots, v_m) \in R^m \times R^{n_0} \times \ldots \times R^{n_m} \quad (2.6)$$

one associates the problem of minimizing

$$h_0(A_0 x + a_0 - v_0) + b_0 x + c_0 \quad (2.7)$$

subject to the constraints

$$h_i(A_i x + a_i - v_i) + b_i x + c_i - u_i \leq 0, \quad i = 1, \ldots, m. \quad (2.8)$$

The dual problem, denoted by (D_1), consists of maximizing

$$c_0 + a_0 z_0 - h_0^*(z_0) + \sum_{i=1}^{m} [c_i y_i + a_i z_i - y_i h_i^*(y_i^{-1} z_i)] \quad (2.9)$$

subject to the constraints

$$b_0 + \sum_{i=1}^{m} y_i b_i + \sum_{i=0}^{m} A_i^* z_i = 0, \quad (2.10)$$

$$z_0 \in C_0, \quad z_i \in y_i C_i \text{ and } 0 \leq y_i \in R^1 \text{ for } i = 1, \ldots, m, \quad (2.11)$$

where A_i^* is the transpose of A_i, h_i^* is the convex function conjugate to h_i, that is,

$$h_i^*(z_i) = \sup\{z_i v_i - h_i(v_i) \mid v_i \in R^{n_i}\}, \quad (2.12)$$

and C_i is the (convex) effective domain of h_i^*,

$$C_i = \{z_i \in R^{n_i} \mid h_i^*(z_i) < +\infty\}. \tag{2.13}$$

We use the convention in (2.9) that

$$y_i h_i^* (y_i^{-1} z_i) = 0 \text{ if } y_i = 0 \text{ and } z_i = 0. \tag{2.14}$$

The circumstances in which (D_1) is likely to be more useful computationally than (D_0) are those in which it is comparatively easy to determine a solution v_i (if it exists, not necessarily uniquely) to any equation of the form $\nabla h_i(v_i) = z_i$. Then, as we explain in the next section, the values of the objective function (2.9) and its directional derivatives are readily available, and the possibly nonlinear aspects of the constraints $z_i \in y_i C_i$ can, in a sense, be ignored.

3. The nature of problem (D_1)

We now state some general facts about the functions h_i^* and sets C_i and how they may be determined from h_i, particularly in light of our assumptions of differentiability and faithful convexity.

Let L_i denote the lineality space of h_i [12, p. 70]. Thus L_i is a subspace of R^{n_i}, and a vector s_i belongs to L_i if and only if the difference quotient

$$[h_i(v_i + \lambda s_i) - h_i(v_i)]/\lambda, \quad \lambda \neq 0,$$

is a constant independent of λ and v_i. Faithful convexity implies that $v_i' - v_i \in L_i$ if h_i is affine on the line segment joining v_i and v_i'; cf. [12, Theorem 8.8].

The affine hull M_i of C_i can be obtained from L_i as follows [12, Cor. 13.3.4(d)]. Let the vectors s_{ik} ($k = 1, \ldots, \ell_i$) generate L_i, and let r_{ik} be the constant (3.1) corresponding to s_{ik}. Then M_i is the set of vectors $z_i \in R^{n_i}$ satisfying the linear equation

$$S_i z_i = r_i, \qquad (3.2)$$

where S_i is the matrix of dimension $\ell_i \times n_i$ whose k^{th} row is s_{ik}, and r_i is the vector in R^{ℓ_i} with components r_{ik}. Therefore, the linear equation

$$y_i r_i - S_i z_i = 0 \qquad (3.3)$$

is a constraint implicit in the condition $z_i \in y_i C_i$ in (2.11).

The affine set M_i is all of R^{n_i} if and only if $L_i = \{0\}$, which means, because of faithful convexity, that h_i is strictly convex. Thus it could be arranged by appropriate choice of the representations (2.5) that $M_i = R^{n_i}$ for $i = 0, \ldots, m$, and then every C_i would have a nonempty interior in R^{n_i}.

In general C_i need not have a nonempty interior in R^{n_i}, but it has a nonempty interior relative to M_i, which we denote here by C_i'. Of course, C_i' is convex and has the same closure as C_i. The following facts are elementary generalizations to faithful convexity of facts derived in [12, §26] for differentiable, strictly convex functions.

(a) The set C_i' is the range of the gradient mapping ∇h_i. Thus z_i belongs to C_i' if and only if the supremum in (2.12) is attained by some v_i. Moreover, if z_i belongs to C_i' and v_i is any vector such that $\nabla h_i(v_i) = z_i$, one has

$$h_i^*(z_i) = z_i v_i - h_i(v_i) . \tag{3.4}$$

(b) If z_i and z_i' are elements of R^{n_i} such that

$$z_i + \lambda z_i' \in C_i', \quad 0 < \lambda < \lambda_0 , \tag{3.5}$$

then

$$h_i^*(z_i) = \lim_{\lambda \downarrow 0} h_i^*(z_i + \lambda z_i') . \tag{3.6}$$

Thus the values of h_i^* on the closure of C_i can be obtained as simple limits of the values of h_i^* on C_i'.

(c) h_i^* is a continuously differentiable, strictly convex function relative to C_i'. Indeed, suppose that z_i and z_i' satisfy (3.6), and let

$$\varphi(\lambda) = h_i^*(z_i + \lambda z_i'), \quad \lambda \geq 0 . \tag{3.7}$$

If z_i belongs to C_i', and v_i is any vector such that $\nabla h_i(v_i) = z_i$, then $\varphi'(\lambda)$ decreases to $\varphi'(0)$ as λ tends to 0, and one has $\varphi'(0) = z_i' v_i$. (Thus, in particular, v_i gives the directional derivatives of h_i^* at z_i; in fact, $v_i = \nabla h_i^*(z_i)$ if C' is full-dimensional.) On the other hand, if z_i does not belong to C_i', then the derivative $\varphi'(\lambda)$ decreases to $-\infty$ as λ tends to 0. (In other words, h_i^* becomes "infinitely steep" as one approaches the relative boundary of C_i'.)

(d) One has $C_i = C_i' = M_i$ if and only if

$$\lim_{\lambda \to +\infty} h_i(\lambda s_i)/\lambda = +\infty \text{ for every } s_i \notin L_i . \tag{3.8}$$

These facts yield much information about the nature of (D). For notational convenience, let us set $y = (y_1, \ldots, y_m)$ in R^m, $z = (z_0, \ldots, z_m)$ in $R^N (N = n_0 + \ldots + n_m)$,

$$G(y, z) = k_0(z_0) + k_1(y_1, z_1) + \ldots + k_m(y_m, z_m), \quad (3.9)$$

$$k_0(z_0) = c_0 + a_0 z_0 - h_0^*(z_0), \quad (3.10)$$

$$k_i(y_i, z_i) = c_i y_i + a_i z_i - y_i h_i^*(y_i^{-1} z_i) \text{ if } y_i > 0,$$

$$= 0 \text{ if } y_i = 0 \text{ and } z_i = 0, \quad (3.11)$$

$$= -\infty \text{ otherwise } (i = 1, \ldots, m).$$

Note that for $i \neq 0$ one has

$$k_i(\lambda y_i, \lambda z_i) = \lambda k_i(y_i, z_i), \quad \lambda \geq 0. \quad (3.12)$$

In (D_1), G is to be maximized subject to (2.10). The functions k_i are concave and upper semicontinuous [12, p. 67 and Theorem 13.3], and therefore G is concave and upper semicontinuous. Thus for every real number α, the set of feasible solutions (y, z) to (D_1) giving a value $\geq \alpha$ to the objective function in (D_1) is a closed convex set.

The differential properties of G can be derived from those of the functions k_i, which are apparent from (c) above. In particular, for $i = 1, \ldots, m$ let

$$F_i = \{(y_i, z_i) | k_i(y_i, z_i) > -\infty\} = \{(y_i, z_i) | y_i \geq 0, z_i \in y_i C_i\}.$$
$$(3.13)$$

Then F_i is a convex cone whose relative interior is

$$\{(y_i, z_i) | y_i > 0, \ z_i \in y_i C_i'\} \qquad (3.14)$$

[12, Theorem 6.8], and k_i is continuously differentiable relative to this relative interior. Furthermore, k_i becomes "infinitely steep" as one approaches a relative boundary point of F_i, unless the boundary point is the origin. At the origin, k_i is linear on every ray in F_i by (3.12), and the directional derivatives of k_i are therefore trivial to calculate.

It is possible, in view of all this, to regard (D_1) essentially as a problem of maximizing a differentiable concave function subject to only linear constraints. The exact sense is explained by the theorem which follows.

Let F denote the set of all feasible solutions (y, z) to (D_1), and let F' be the modification of F obtained by substituting C_i' for C_i in (2.11). Let F" be the modification obtained by not only substituting C_i' for C_i, but also strengthening the constraint $y_i \geq 0$ to $y_i > 0$, except for indices i such that $n_i = 0$ (f_i affine). Of course, the sets F, F' and F" are convex, but they need not be closed. Their closures coincide, however, if F" $\neq \emptyset$.

Theorem 1. <u>Suppose that</u> F" $\neq \emptyset$. <u>Then the objective function in</u> (D_i) <u>has the same supremum over</u> F' <u>as it has over</u> F, <u>and the optimal solutions to</u> (D_1), <u>if any, all belong to</u> F'.

<u>Furthermore, let</u> (y, z) <u>and</u> (y', z') <u>be such that</u>

$$(y + \lambda y', z + \lambda z') \in F', \quad 0 < \lambda < \lambda_0. \qquad (3.15)$$

Then the concave function

$$\varphi(\lambda) = G(y + \lambda y', z + \lambda z') \qquad (3.16)$$

is continuous for $0 < \lambda < \lambda_0$ and continuously differentiable for $0 < \lambda < \lambda_0$. The derivative $\varphi'(\lambda)$ increases to $+\infty$ as λ tends to 0, unless $(y, z) \in \bar{F}'$, in which event $\varphi'(\lambda)$ increases to the (finite) right derivative of φ at $\lambda = 0$.

Proof. The hypothesis implies that F lies between F' and the closure of F'. Therefore G, being concave, has the same supremum over F' as over F [12, Cor. 7.3.1] The differential properties of G described in the theorem are immediate from the properties of the functions k_i noted above, and they imply in particular that the supremum cannot occur at a point of F not in F'.

Theorem 1 asserts that, if $F'' \neq \emptyset$, the constraints $z_0 \in C_0$ and $z_i \in y_i C_i$ ($i = 1, \ldots, m$) can be replaced in (D_1) by $z_i \in y_i C_i'$. Moreover, latter constraints are automatically taken care of, in the sense that, as one approaches a point (y, z) which is excluded by these constraints but not by the other constraints, G becomes "infinitely steep." Thus the only constraints in (D_1) which have a practical effect, in terms of gradient projections and related computational techniques, are the linear constraints (2.10) and

$$r_0 - S_0 z_0 = 0, \quad y_i r_i - S_i z_i = 0 \text{ and } y_i \geq 0 \quad (3.17)$$

$$\text{for } i = 1, \ldots, m.$$

Of course, if the closures of the convex sets C_i (and C_i') are all polyhedral, then the closure of F (and F') is polyhedral and hence describable entirely by a system of linear equations and inequalities. In particular, suppose that every h_i satisfies condition (3.8). Then one has $F' = F$, and the closure of F is described simply by (2.10) and (3.17). As one approaches boundary points of F not in F itself, the objective function tends to $-\infty$.

4. Examples

The following pairs of problems (P) and (D_1) illustrate the facts described in §3, as well as indicate areas of application of the duality results to be developed in §5.

Example 1. (Geometric programming [1, 2, 3]). Let

$$h_i(v_i) = \log\left(\sum_{k=1}^{n_i} e^{v_{ik}}\right) \quad i = 0, \ldots, m, \quad (4.1)$$

where v_{ik} is the kth component of $v_i \in R^{n_i}$. Then h_i is an analytic convex function (hence a faithfully convex function) such that

$$C_i = \{z_i \in R^{n_i} \mid z_{ik} \geq 0, \sum_{k=1}^{n_i} z_{ik} = 1\}, \quad (4.2)$$

$$h_i^*(z_i) = \sum_{k=1}^{n_i} z_{ik} \log z_{ik}, \quad z_i \in C_i, \quad (4.3)$$

(with $0 \log 0 = 0$). Setting $x_i = \log t_i$, one sees that (P) is equivalent to a typical geometric program. Let $b_i = 0$ and $c_i = 0$, and let the components of A_i and a_i be denoted by a_{kj}^i and a_{k0}^i, respectively. The dual problem (D_1) consists of maximizing the concave function

$$\sum_{i=0}^{m} \sum_{k=1}^{n_i} z_{ik}(a_{k0}^i - \log z_{ik}) + \sum_{i=1}^{m} y_i \log y_i \quad (4.4)$$

subject to the linear constraints

$z_{ik} \geq 0$ for $i = 0, \ldots, m$ and $k = 1, \ldots, n_i$,

$$\sum_{k=1}^{n_0} z_{0k} = 1, \text{ and } \sum_{k=1}^{n_i} z_{ik} = y_i \text{ for } i = 1, \ldots, m, \quad (4.5)$$

$$\sum_{i=0}^{m} \sum_{k=1}^{n_i} a_{kj}^i z_{ik} = 0 \text{ for } j = 1, \ldots, m.$$

Here the feasible set F is actually polyhedral, because the sets C_i are polyhedral and bounded.

Example 2. (Quadratically constrained quadratic programming; cf. [7, 8, 9, 10]). If each of the convex functions f_i is quadratic, it is simple to write down representations of the form (2.5) with

$$h_i(v_i) = \frac{1}{2} |v_i|^2 = \frac{1}{2} \sum_{k=1}^{n_i} v_{ik}^2. \quad (4.6)$$

One then has $C_i = R^{n_i}$ and

$$h_i^*(z_i) = \frac{1}{2} |z_i|^2 = \frac{1}{2} \sum_{k=1}^{n_i} z_{ik}^2, \quad (4.7)$$

so that in (2.9) one has

$$y_i h_i^*(y_i^{-1} z_i) = |z_i|^2/2y_i \text{ if } y_i > 0 \ (z_i \text{ arb.}),$$

$$= 0 \text{ if } y_i = 0 \text{ and } z_i = 0, \quad (4.8)$$

$$= +\infty \text{ otherwise.}$$

Observe that the last remarks of §3 are applicable to this example; h_i satisfies (3.8).

The more general ℓ^p-programs of Peterson and Ecker may be obtained by letting h_i be of the form

$$h_i(v_i) = \sum_{k=1}^{n_i} (1/p_{ik})|v_{ik}|^{p_{ik}}, \quad 1 < p_{ik} < +\infty, \qquad (4.9)$$

in which event one has

$$h_i^*(z_i) = \sum_{k=1}^{n_i} (1/q_{ik})|z_{ik}|^{q_{ik}}, \quad 1 < q_{ik} < +\infty, \qquad (4.10)$$

where $(1/p_{ik}) + (1/q_{ik}) = 1$. The next example shows that a much broader class of problems can actually be handled just as easily.

Example 3. (Quasiseparable Programming). The convex function f_i is said to be <u>quasiseparable</u> if it can be expressed as a sum of functions, each of which is a linear function on R^n_1 composed with a convex function (possibly infinite) on R^1. Thus f_i is quasiseparable if and only if f_i can be represented as in (2.5) with h_i separable:

$$h_i(v_i) = \sum_{k=1}^{n_i} h_{ik}(v_{ik}). \qquad (4.11)$$

Assuming that every f_i has this property, in addition to the properties already specified, we can actually get representations in which the functions h_{ik} on R^1 are all differentiable and strictly convex. The conjugate functions h_{ik}^* are then, of course, relatively simple to determine, and one has

$$h_i^*(z_i) = \sum_{k=1}^{n_i} h_{ik}^*(z_{ik}). \qquad (4.12)$$

Furthermore, C_i is the product of the (nondegenerate) intervals

$$C_{ik} = \{z_{ik} \in R^1 \mid h_{ik}^*(z_{ik}) < +\infty\}. \qquad (4.13)$$

Thus C_i has a polyhedral closure and a nonempty interior ($M_i = R^{n_i}$). It follows that the feasible set F has a polyhedral closure.

Example 4. (Convex programming with linear constraints). Suppose that f_i is affine for $i = 1, \ldots, m$, so that the term $h_i(A_i x + a_i)$ can be omitted in (2.5). Then in (D_1) one maximizes

$$c_0 + \sum_{i=1}^{m} c_i y_i + a_0 z_0 - h_0^*(z_0) \tag{4.14}$$

subject to the constraints

$$b_0 + \sum_{i=1}^{m} y_i b_i + A_0^* z_0 = 0, \quad y_i \geq 0, \quad z_0 \in C_0. \tag{4.15}$$

If these constraints can be satisfied with $z_0 \in C_0'$, then, as explained in §3, the condition $z_0 \in C_0$ reduces for practical purposes to the linear constraint $S_0 z_0 = r_0$ (which is vacuous if C_0 has a nonempty interior). If h_0 is strictly convex and satisfies the growth condition (3.8), then $C_0 = R^{n_0}$.

If f_0 itself is affine, so that (P) is a linear programming problem, everything concerning z_0 can be omitted from (4.14) and (4.15), and (D_1) is the usual dual linear programming problem, coinciding with (D_0).

Example 5. This miscellaneous, but specific example illustrates some useful tricks, as well as a particular computation of the conjugate functions h_i^*. We consider the problem of minimizing

$$(1/4)(1 + 3x_1)^4 - 7x_2 + \exp|x_1 - x_2| \tag{4.16}$$

over all $(x_1, x_2) \in R^2$ satisfying the constraints

$$(x_1^5 + x_2^5)^{1/3} + 2x_2^2 \le 9 , \qquad (4.17)$$

$$x_1 \ge 0, \quad x_2 \ge 0, \quad x_1 x_2 \ge 1 . \qquad (4.18)$$

In the given formulation, this problem does not satisfy our assumptions, because the absolute value term in (4.16) spoils the differentiability of the objective function, and the quantity $1 - x_1 x_2$ in (4.18) is not convex as a function of x_1 and x_2. Also, the first term in (4.17) is not globally convex as a function of x_1 and x_2, although it is convex for $x_1 \ge 0$ and $x_2 \ge 0$. Note, however, that the objective function is the pointwise maximum of two analytic convex functions, and that the set of points satisfying (4.18) is indeed convex.

We may transform this problem into the desired form as follows. First, where the expression $|x_1 - x_2|$ occurs, we replace it by a new variable x_3, which is required to satisfy

$$x_3 \ge x_1 - x_2 \quad \text{and} \quad x_3 \ge x_2 - x_1 .$$

(A similar device can be used whenever the given objective function is the maximum of several functions which are differentiable and faithfully convex.) We next replace x_1 and x_2 by $|x_1|$ and $|x_2|$ in (4.17) to get a globally convex function (see below); this involves no loss of generality, because of the nonnegativity in (4.18). Finally, we replace (4.18) by an equivalent convex constraint:

$$(4 + (x_1 - x_2)^2)^{1/2} - x_1 - x_2 \le 0 . \qquad (4.20)$$

(The trick used to obtain this constraint is the following one, which is applicable under quite general circumstances. Let the set of all (x_1, x_2) satisfying (4.18) be denoted by H.

We observe that for each $(x_1, x_2) \in R^2$ there exists a unique smallest real number λ such that $(x_1 + \lambda, x_2 + \lambda)$ belongs to H. Denoting this λ by $r(x_1, x_2)$, we have

$$H = \{(x_1, x_2) \mid r(x_1, x_2) \leq 0\},$$

and r is convex. The function r is easily computed in this case, and the left side of (4.20) is $2r(x_1, x_2)$.)
The given problem is thus equivalent to minimizing

$$f_0(x) = (1/4)(1 + 3x_1)^4 - 7x_2 + \exp x_3$$

subject to the constraints

$$f_1(x) = (|x_1|^5 + |x_2|^5)^{1/3} + 2x_2^2 - 9 \leq 0,$$

$$f_2(x) = (4 + (x_1 - x_2)^2)^{1/2} - x_1 - x_2 \leq 0,$$

$$f_3(x) = x_1 - x_2 - x_3 \leq 0,$$

$$f_4(x) = -x_1 + x_2 - x_3 \leq 0,$$

where $x = (x_1, x_2, x_3) \in R^3$. The functions f_i are all differentiable and faithfully convex, so this is a problem (P) of the desired type.

The next step is to choose suitable representations of the form (2.5). Here convenience in computing the conjugate functions is the chief guide, and this is dependent on one's knowledge of general rules and examples, such as those in [12] (cf. the situation in computing indefinite integrals). We take

$$h_0(v_{01}, v_{02}) = (1/4)v_{01}^4 + \exp v_{02},$$

$$h_1(v_{11}, v_{12}, v_{13}) = (|v_{11}|^5 + |v_{12}|^5)^{1/3} + 2v_{13}^2,$$

$$h_2(v_{21}) = (4 + v_{21}^2)^{1/2},$$

$$A_0 = \begin{bmatrix} 3 & 0 & 0 \\ 0 & 0 & 1 \end{bmatrix}, \quad A_1 = \begin{bmatrix} 1 & 0 & 0 \\ 0 & 1 & 0 \\ 0 & 1 & 0 \end{bmatrix}, \quad A_2 = [1, -1, 0]$$

$$a_0 = (1, 0), \quad a_1 = (0, 0, 0), \quad a_2 = 0$$

$$b_0 = (0, -7, 0), \quad b_1 = (0, 0, 0), \quad b_2 = (-1, -1, 0),$$

$$b_3 = (1, -1, -1), \quad b_4 = (-1, 1, -1)$$

$$c_0 = c_2 = c_3 = c_4 = 0, \quad c_1 = -9.$$

We do not define h_i, A_i or a_i for $i = 3, 4$, because f_3 and t_4 are linear. The only trick here that deserves special mention is the introduction of the variable v_{13}, where v_{12} could apparently have been used just as well. This makes it easier to compute h_1^*. (The same trick can be used whenever a function h_i is a sum of convex functions, each of which has a known conjugate.)

The conjugate of h_0 is easy to determine directly (cf. [12, p. 106]):

$$h_0^*(z_{01}, z_{02}) = (3/4)z_{01}^{4/3} + z_{02}(\log z_{02} - 1),$$

$$C_0 = \{(z_{01}, z_{02}) \in R^2 \mid z_{02} \geq 0\}.$$

The conjugate of h_1 is obtained by a fancier method, although in principle the Legendre transformation would give the global answer [12, Theorem 26.6]. We write

$$h_1(v_{11}, v_{12}, v_{13}) = (5/3)w_1(v_{11}, v_{12}) + 4w_2(v_{13}),$$

where

$$w_1(v_{11}, v_{12}) = (3/5)[(|v_{11}|^5 + |v_{12}|^5)^{1/5}]^{5/3}$$

$$w_2(v_{13}) = (1/2)v_{13}^2.$$

(The convexity of w_1 follows from [12, Theorem 5.1] and the convexity of ℓ^p norms; this provides in particular one way of verifying the convexity of h_1.) We then have

$$h_1^*(z_{11}, z_{12}, z_{13}) = (5/3)w_1^*((3/5)z_{11}, (3/5)z_{12}) +$$
$$+ 4w_2^*((1/4)z_{13})$$

with $C_1 = R^3$, where, using [12, Theorem 15.3],

$$w_1^*(z_{11}, z_{12}) = (2/5)[(|z_{11}|^{5/4} + |z_{12}|^{5/4})^{4/5}]^{5/2},$$

$$w_2^*(z_{13}) = (1/2)z_{13}^2.$$

Finally, we compute from the definition that

$$h_2^*(z_{21}) = -2(1-z_{21}^2)^{1/2}, \quad C_2 = \{z_{21} \mid -1 \leq z_{21} \leq 1\}.$$

Substituting these expressions in (2.9), (2.10) and (2.11), we obtain the following dual problem (D_1) in the variables y_i ($i = 1, \ldots, 4$), z_{0k} ($k = 1, 2$), z_{1k} ($k = 1, 2, 3$) and z_{21}: maximize

$$z_{01} + z_{02} - (3/4)z_{01}^{4/3} - z_{02}\log z_{02} - 9y_1$$
$$- (2/3)(3/5)^{5/2} y_1^{-3/2}(|z_{11}|^{5/4} + |z_{12}|^{5/4})^2 \quad (4.21)$$
$$-(1/8)y_1^{-1} z_{13}^2 + 2(y_2^2 - z_{21}^2)^{1/2}$$

subject to the linear constraints

$$-y_2 + y_3 - y_4 + 3z_{01} + z_{11} + z_{21} = 0,$$
$$-y_2 - y_3 + y_4 + z_{12} + z_{13} - z_{21} = 7,$$
$$-y_3 - y_4 + z_{02} = 0,$$
$$(4.22)$$
$$y_1 \geq 0, \; y_2 \geq 0, \; y_3 \geq 0, \; y_4 \geq 0, \; z_{02} \geq 0, \; -y_2 \leq z_{21} \leq y_2.$$

5. Relationships between (P), (D_0) and (D_1)

The general duality results in [12, §30] are, of course, applicable to (P), (D_0) and (D_1). It would be repetitious to state these results here, so we only present certain refinements which follow from the assumption that the functions f_i are faithfully convex. The main goal is to indicate the extent to which (P) can be solved by solving (D_1).

We denote the optimal value in (P) by inf (P); this is $+\infty$ by convention if (P) has no feasible solutions. The expressions sup (D_0) and sup (D_1) have an analogous meaning.

Before giving a duality theorem concerning these optimal values, we describe the basic relationship between problems (D_0) and (D_1). Let g_0 and g_1 be the upper semicontinuous, concave functions defined by

$$g_0(y) = g(y) \text{ if } y \geq 0, \\ = -\infty \text{ if } y \ngeq 0, \tag{5.1}$$

$$g_1(y, z) = G(y, z) \text{ if } (2.10) \text{ holds}, \\ = -\infty \text{ if } (2.10) \text{ does not hold}, \tag{5.2}$$

where g is given by (2.2) and G by (3.9). Then

$$\sup(D_0) = \sup\{g_0(y) \mid y \in R^m\}, \tag{5.3}$$

$$\sup(D_1) = \sup\{g_1(y, z) \mid y \in R^m, z \in R^N\}. \tag{5.4}$$

Theorem 2. *For each* $y \in R^m$ *one has*

$$g_0(y) = \sup\{g_1(y, z) \mid z \in R^N\}, \tag{5.5}$$

where the supremum is attained by some z. *Thus*

$$\sup(D_0) = \sup(D_1), \tag{5.6}$$

and (D_0) *has an optimal solution if and only if* (D_1) *has an optimal solution.*

Proof. The proof of this result does not depend on faithful convexity, but it uses the finiteness of the functions h_i in applying Fenchel's Duality Theorem [12, Theorem 31.1]. Let p_0 and p_1 be the perturbation functions associated with (D_0) and (D_1), respectively. Thus $p_0(u)$ is for each $u \in R^m$ the infimum of $f_0(x)$ subject to (2.1), while $p_1(u, v)$ is for each $u \in R^m$ and $v \in R^N$ the infimum of (2.7) subject to (2.8). The functions p_0 and p_1 are convex, and we have the conjugacy relations

$$g_0(y) = \inf\{yu + p_0(u) \mid u \in R^m\} = (-p_0)^*(y), \quad (5.7)$$

$$g_1(y, z) = \inf\{yu + zv + p_1(u, v) \mid u \in R^m, v \in R^N\} = \\ = (-p_1)^*(y, z) \quad (5.8)$$

(see [12, Theorem 30.2]). Also

$$p_0(u) = p_1(u, 0). \quad (5.9)$$

Therefore, assuming for the moment that $p_1(u, v)$ is never $-\infty$, we have for every $\bar{y} \in R^m$

$$g_0(\bar{y}) = \inf\{q(u, v) + p_1(u, v) \mid u \in R^m, v \in R^N\}, \quad (5.10)$$

where q is the convex function on $R^m \times R^N$ whose value at (u, v) is $\bar{y}u$ if $v = 0$ and $+\infty$ if $v \neq 0$. From Fenchel's Duality Theorem, we then have

$$g_0(\bar{y}) = \sup\{(-p_1)^*(y, z) - q^*(y, z) \mid y \in R^m, z \in R^N\}, \quad (5.11)$$

where the supremum is attained, because the interior of the convex set $\{(u, v) \mid p_1(u, v) < +\infty\}$ meets the subspace $\{(u, v) \mid q(u, v) < +\infty\}$. Of course, the conjugate function q^* vanishes at points of the form $(\bar{y}, z)$ and has the value $+\infty$ at all other points. Thus (5.11) is equivalent to (5.5) by (5.8). If p_1 takes on $-\infty$, the expression $\infty - \infty$ could occur in (5.10), and a different argument must be given. In this degenerate case, p_1 takes on $-\infty$ throughout the interior of its effective domain $\{(u, v) \mid p_1(u, v) < +\infty\}$ [12, Theorem 7.2], so that p_0 also takes on $-\infty$ by (5.9). But then g_0 and g_1 are identically $+\infty$ by (5.7) and (5.8), and (5.5) holds trivially.

Theorem 3. (Duality) If (P) has a feasible solution, then

$$\inf(P) = \sup(D_1) . \qquad (5.12)$$

Moreover, if (P) has a feasible solution x such that $f_i(x) < 0$ for every index i such that f_i is not affine, and if the common extremum in (5.12) is not $-\infty$, then (D_1) has an optimal solution.

On the other hand, suppose (D_1) has a feasible solution (y, z) such that $y_i > 0$ and $z_i \in C_i'$ for every index i such that $n_i \neq 0$ (that is, the set F" is nonempty, as in Theorem 1). Then again (5.12) necessarily holds, and if the common extremum is not $+\infty$, (P) has an optimal solution.

Proof. The assertions in the first paragraph are immediate, in view of Theorem 2, from the corresponding assertions for (D_0) (see [13] and [12, Theorem 28.2]). To prove the assertions in the second paragraph, we express g_1 as $-(K_0 + \ldots + K_{m+1})$, where the convex functions K_i on $R^m \times R^N$ are defined from the functions k_i in (3.10) and (3.11) by

$$K_0(y, z) = -k_0(z_0) , \qquad (5.13)$$

$$K_i(y, z) = -k_i(y_i, z_i) \text{ for } i = 1, \ldots, m, \qquad (5.14)$$

$$K_{m+1}(y, z) = 0 \text{ if (2.10) is satisfied}, \qquad (5.15)$$

$$= +\infty \text{ if (2.10) is not satisfied} .$$

We then have

$$\sup(D_1) = -\inf(K_0 + \ldots + K_{m+1}) = (K_0 + \ldots + K_{m+1}) * (0,0)$$

$$= (K_0^* \square \ldots \square K_{m+1}^*)(0,0) \qquad (5.16)$$

by Theorem 20.1 of [12], where $\square$ denotes infimal convolution, and the infimum in the definition of $\square$ is attained. (The hypothesis of the cited theorem, for the functions K_i, is the condition that $F'' \neq \emptyset$.) The conjugates K_i^* are easily computed, and one sees thereby that the infimum symbolized by the final expression in (5.16) is $\inf(P)$. We leave the straightforward details of this to the reader.

Remark 1. If $\inf(P)$ is finite and the functions h_i all satisfy (3.8), as in Example 2, then (P) has an optimal solution. This follows from [12, Cor. 27.3.3].

Remark 2. Theorem 2, applied to Examples 1 and 2, yields the duality theorems of Duffin, Peterson and Zener in geometric programming [1, 2, 3] and Peterson and Ecker in ℓ^p-programming [7, 8, 9, 10], except for those results involving the subinfimum in (P). The latter results are covered by the general theorems stated in [12, §30], as has already been pointed out in the case of geometric programming by Hamala [6].

It is clear from Theorem 3 that the optimal value in (P) can usually be obtained by solving (D_1), but further analysis is needed to see how optimal solutions to (P) can likewise be obtained by a dual approach. In this analysis, it is convenient to represent each of the lineality spaces L_i introduced in §3 as

$$L_i = \{w \in R^{n_i} \mid B_i w = 0\}, \qquad (5.17)$$

where B_i is some matrix. Such a representation is of course, easy to obtain from the matrix S_i by elementary linear algebra. In particular, if h_i is strictly convex, one can take B_i to be the $n_i \times n_i$ identity matrix. Note that the subspace

$$L'_i = \{w \in R^n \mid B_i A_i w = 0\} \tag{5.18}$$

is the lineality space of f_i. Thus $f_i(x + \lambda w)$ is an affine function of λ for all x, if w satisfies $B_i A_i w = 0$.

Theorem 4. Let (y, z) be an optimal solution to (D_1), and let I be the index set consisting of 0 and all the indices $i \in \{1, \ldots, m\}$ such that $y_i \neq 0$. Then (P) has an optimal solution if and only if $y_i^{-1} z_i \in C'_i$ for every $i \in I$.

Furthermore, suppose that the latter condition is satisfied, and for each $i \in I$ let v_i be an element of R^{n_i} such that $\nabla h_i(v_i) = y_i^{-1} z_i$ (with the factor y_i^{-1} omitted if $i = 0$). Let M be the affine subset of R^n consisting of the vectors x such that

$$B_i A_i x = B_i(v_i - a_i) \text{ for every } i \in I. \tag{5.19}$$

The functions f_i for $i \in I$ are then affine on M, and the optimal solutions to (P) are the vectors x such that

$$x \in M \text{ and } f_i(x) = 0 \text{ for every } i \in I, \; i \neq 0, \tag{5.20}$$

$$f_i(x) \leq 0 \text{ for every } i \notin I. \tag{5.21}$$

Proof. It follows from the first assertion of Theorem 3 that x is an optimal solution to (P) if and only if x is a feasible solution to (P) such that

$$f_0(x) = g_1(y, z). \tag{5.22}$$

Using the defintion of the conjugate functions h_i^*, one can easily verify that (5.22) is equivalent to (5.20) and (5.21), if M is taken to be the set of all $x \in R^n$ such that

$$\nabla h_i(A_i x_i + a_i) = y_i^{-1} z_i \quad \text{for every } i \in I \qquad (5.23)$$

(with y_i^{-1} omitted if $i = 0$). The equivalence of this description of M with the one in the theorem follows from the fact that h_i is faithfully convex. Indeed, assuming that v_i satisfies $\nabla h_i(v_i) = y_i^{-1} z_i$, one has $\nabla h_i(w) = y_i^{-1} z_i$, if and only if $w - v_i \in L_i$. Thus (5.23) holds if and only if

$$B_i(A_i x_i + a_i - v_i) = 0 \quad \text{for every } i \in I,$$

or in other words (5.19). This completes the proof of Theorem 4.

The affine set M in Theorem 4, if nonempty, is a translate of the subspace L equal to the intersection of the L_i' for $i \in I$, and hence one has

$$\dim M = \dim L \leq \min_{i \in I} \dim L_i'. \qquad (5.24)$$

Therefore in particular, if one of the functions f_i for $i \in I$ (e.g. the objective function f_0) is strictly convex, so that $\dim L_i' = 0$, there is only one element x in M. This x must automatically satisfy the equations and inequalities in (5.20) and (5.21) and thus be the unique optimal solution in (P).

More generally, if M is not zero-dimensional, we may look at the set M' consisting of the vectors x which satisfy (5.20). Since the functions f_i, $i \in I$, are affine on M, this involves solving a further system of <u>linear</u> equations, and M' is an affine set. If M' is zero-dimensional, its unique element x is the unique optimal solution to (P), as in the case just considered. Otherwise, the problem is reduced to the following: find an $x \in M'$ satisfying (5.21). This can be solved by linear programming if the functions f_i for $i \notin I$, like those for $i \in I$, are affine on M', as is true certainly if $L'_i \supset L$ for every $i \neq I$. (Observe that this holds in Example 4. It also holds if $L'_i \supset L'_0$ for $i = 1, \ldots, m$.)

At all events, if none of these shortcuts can be used, one can pass to the following convex programming problem, in order to obtain a solution to (5.20) and (5.21) and thereby an optimal solution to (P):

(P')   minimize $\frac{1}{2} |x|^2$ over all $x \in M'$ satisfying (5.21).

This problem is in fact of the same type as (P). (The linear equations expressing the condition $x \in M'$ could be represented as linear inequalities, but they could also be used to eliminate some of the variables x_j and thus transform (P') into a similar problem of reduced dimensionality.) Thus (P') can be attacked via its dual (D'_1), just like (P). Since the objective function in (P') is strictly convex, an optimal solution to (D'_1) immediately yields an optimal solution to (P'), as just explained.

Thus, having computed an optimal solution to (D_1), one can determine an optimal solution to (P) by solving at most one more problem (D'_1), which is similar to (D_1) but probably of a vastly lower dimension.

REFERENCES

1. R. J. Duffin, "Linearizing geometric programs," SIAM Review, 1970.

2. R. J. Duffin and E. L. Peterson, "Duality theory for geometric programming," SIAM J. App. Math. 14 (1966), 1307-1349.

3. R. J. Duffin, E. L. Peterson and C. Zener, Geometric Programming, Wiley, New York, 1967.

4. J. E. Falk, "Lagrange multipliers and nonlinear programming," J. Math. Anal. Appl. 19 (1967), 141-159.

5. A. Geoffrion, "Duality in nonlinear programming: a simplified applications-oriented development," Working Paper no. 150, Western Management Science Institute, U. of California, Los Angeles, 1969.

6. M. Hamala, "Geometric programming in terms of conjugate functions," discussion paper no. 6811, Center for Operations Research and Econometrics, Louvain, 1968.

7. E. L. Peterson and J. G. Ecker, "A unified duality theory for quadratically constrained quadratic programs and ℓ_p-constrained ℓ_p-approximation problems," Bull. Amer. Math. Soc. 74 (1968), 316-321.

8. E. L. Peterson and J. G. Ecker, "Geometric programming: duality in quadratic programming and ℓ_p-approximation I," Proc. Internat. Symp. on Math. Programming, Princeton, 1967, H. W. Kuhn and A. W. Tucker, eds. (to appear).

9. E. L. Peterson and J. G. Ecker, "Geometric programming: duality in quadratic programming and ℓ_p-approximation II (canonical programs)," SIAM J. Appl. Math. 17 (1969), 317-340.

10. E. L. Peterson and J. G. Ecker, "Geometric programming: duality in quadratic programming and ℓ_p-approximation III (degenerate programs)," J. Math. Anal. App. 29 (1970), 365-383.

11. R. T. Rockafellar, "An extension of Fenchel's duality theorem," Duke Math. J. (1966), 81-90.

12. R. T. Rockafellar, Convex Analysis, Princeton Univ. Press, 1969.

13. R. T. Rockafellar, "Ordinary convex programs without a duality gap," to appear.

14. A. V. Fiacco and G. P. McCormick, *Nonlinear Programming: Sequential Unconstrained Minimization Techniques*, Wiley, 1968.

Sufficiency Conditions and a Duality Theory for Mathematical Programming Problems in Arbitrary Linear Spaces

LUCIEN W. NEUSTADT

ABSTRACT

This paper is devoted to an investigation of mathematical programming problems in arbitrary linear vector spaces. The constraints of the problem are assumed to be of three types: (a) the point must belong to a given (arbitrary) convex set in the underlying linear space, (b) a finite-dimensional equality constraint must be satisfied, (c) a generalized (possibly infinite-dimensional) inequality constraint, defined in terms of a closed convex cone with non-empty interior in a linear topological space, must be satisfied.

Assuming that the equality constraints are affine, that the "inequality" constraints are, in a certain generalized sense, convex, and that the problem is "well-posed", Kuhn Tucker type conditions which are both necessary and sufficient for optimality are obtained.

A duality theory for obtaining the "multipliers" in the generalized Kuhn-Tucker conditions is presented.

An application to optimal control theory is also presented.

1. Introduction

In this paper we shall investigate mathematical programming problems in arbitrary linear spaces, where the scalar-valued function to be minimized is assumed to be convex, and the functional constraints are in the form of a finite-dimensional equality constraint and a not-necessarily-finite-dimensional inequality constraint, the inequality being defined in terms of a closed convex cone with a non-empty interior in a linear topological vector space. The function that defines the equality constraint is assumed to be affine, and the function that defines the inequality constraint is assumed to be convex (in a generalized sense).

After stating the problem precisely in Section 2, we derive in Section 3 necessary conditions that solutions of this problem must satisfy. Sufficient conditions, which almost coincide (and indeed do coincide if the problem is what we call well-posed) with the necessary conditions, are also obtained. These conditions are given in two forms which are shown to be equivalent. The first form is in terms only of the functions that characterize the problem, whereas the second (or differential) form is in terms of the differentials of these functions, evaluated at the solution point. These conditions may be looked upon as generalized Lagrange multiplier rules, or as generalization of the Kuhn-Tucker conditions.

In section 4 we present a duality theory by constructing a dual of the original problem, the solutions of which coincide with the "Lagrange multipliers" that appear in the necessary and sufficient conditions obtained in Section 3.

In Section 5, we apply our sufficiency theorem in differential form to develop sufficient conditions for optimal control problems in which the differential equation is linear in the state variable (but is arbitrary in the control variable), in the presence of a convex-type phase inequality constraint. These sufficient conditions "almost" coincide with the necessary conditions.

2. Mathematical preliminaries and problem statement

Let us first generalize the concept of a convex function to functions whose range is contained in an arbitrary topological linear vector space. Thus, let Z be topological linear vector space. A set $\overline{Z} \subset Z$ will be called a <u>convex cone</u> if $\lambda_1 z_1 + \lambda_2 z_2 \in \overline{Z}$ whenever $z_1, z_2 \in \overline{Z}$ and $\lambda_1, \lambda_2 \geq 0$. Clearly, a convex cone is always a convex set. If $\overline{Z}$ is a closed convex cone in Z and $z_1, z_2 \in Z$, then we shall write $z_1 \leq_{\overline{Z}} z_2$ to mean that $z_1 - z_2 \in \overline{Z}$. It is trivial to show that $\leq_{\overline{Z}}$ is a transitive and reflexive relation.

If $\mathcal{Y}$ is a linear vector space and ϕ is a function from $\mathcal{Y}$ into Z, then we shall say that ϕ is $\overline{Z}$-convex (where $\overline{Z}$ is a closed convex cone in Z) if $\phi(\lambda_1 y_1 + \lambda_2 y_2) \leq_{\overline{Z}} \lambda_1 \phi(y_1) + \lambda_2 \phi(y_2)$ whenever $y_1, y_2 \in \mathcal{Y}$, $\lambda_1 \geq 0$, $\lambda_2 \geq 0$, and $\lambda_1 + \lambda_2 = 1$.

This generalization of the concept of a convex function was apparently first made by Hurwicz [1, page 68].

Note that if $Z = R$ and $\overline{Z} = \{\xi : \xi \leq 0\}$, then $\phi: \mathcal{Y} \to Z$ is $\overline{Z}$-convex if and only if it is convex in the ordinary sense. If $Z = R^\mu$ (Euclidean μ-space), and $\overline{Z}$ is the closure of the negative orthant in Z, then $\phi: \mathcal{Y} \to Z$ is $\overline{Z}$-convex if and only if each component ϕ^i of ϕ is convex in the ordinary sense. Further, if $\phi: \mathcal{Y} \to Z$ is $\overline{Z}$-convex, then (for all positive integers ν)

$$\phi \sum_{i=1}^{\nu} \lambda_i y_i \leq_Z \sum_{i=1}^{\nu} \lambda_i \phi(y_i) \text{ whenever } y_1, \ldots, y_\nu \in \mathcal{Y},$$

$$\lambda_i \geq 0 \text{ for each } i \text{ and } \sum_{i=1}^{\nu} \lambda_i = 1.$$

Henceforth in this paper, $\mathcal{Y}$ will denote a linear vector space, Z will denote a topological linear vector space, and $\mathcal{Z}$ will denote a closed convex cone in Z with a non-empty interior. We shall denote by Z^* the conjugate space of Z, i.e., the linear vector space of all continuous linear functionals defined on Z. We shall say that a function φ from $\mathcal{Y}$ into Z is affine if there are a linear function $\varphi_L : \mathcal{Y} \to Z$ and an element $z_0 \in Z$ such that $\varphi(y) = \varphi_L(y) + z_0$ for all $y \in \mathcal{Y}$. We shall in this case refer to φ_L as the <u>linear part</u> of φ. Clearly, every linear function $\varphi : \mathcal{Y} \to Z$ is affine, and every affine function $\varphi : \mathcal{Y} \to Z$ is $\mathcal{Z}$-convex (no matter what the closed convex cone $\mathcal{Z}$).

A function ϕ from $\mathcal{Y}$ into Z will be said to be <u>directionally differentiable</u> at a point $y_0 \in \mathcal{Y}$ if there is a function $D\phi : \mathcal{Y} \to Z$ such that

$$\frac{\phi(y_0 + \varepsilon y) - \phi(y_0)}{\varepsilon} \xrightarrow[\varepsilon \to 0^+]{} D\phi(y) \text{ for all } y \in \mathcal{Y}.$$

In this case, $D\phi$ will be called the <u>directional differential</u> of ϕ at y_0. If ϕ is directionally differentiable at each point $y_0 \in \mathcal{Y}$, then we shall simply say that ϕ is directionally differentiable.

Note that we do not require $D\phi$ to be linear in the preceding definition.

It is easy to see that if $\mathcal{Y}$ and Z are Banach spaces and if ϕ has a Fréchet differential $D\phi$ at y_0, then ϕ is directionally differentiable at y_0 with directional differential $D\phi$. If $\varphi : \mathcal{Y} \to Z$ is affine, then φ is directionally differentiable, and $D\varphi$ at any $y_0 \in \mathcal{Y}$ coincides with the linear part of φ. Further, if $Z = R$ and ϕ is convex in the ordinary sense, then ϕ is directionally differentiable (see [2, page 39]). Finally, we point out that (as it trivially verified) if $\phi: \mathcal{Y} \to Z$ is Z-convex and directionally differentiable at $y_0 \in \mathcal{Y}$, then $D\phi$ is also Z-convex.

We now consider the following convex mathematical programming problem.

Problem 1. Given a convex set $\mathcal{E}$ in $\mathcal{Y}$, a Z-convex function $\phi: \mathcal{Y} \to Z$, an affine function $\varphi: \mathcal{Y} \to R^m$, and a convex function $\phi^0: \mathcal{Y} \to R$, find an element $y_0 \in \mathcal{E}$ that achieves a minimum for ϕ^0 (on $\mathcal{E}$) subject to the constraints $\varphi(y) = 0$, $\phi(y) \leq_Z 0$.

We remind the reader that Z is a closed, convex cone in the topological linear vector space Z and that Z has a non-empty interior.

An important special case of Problem 1 is the following one.

Problem 1a. Given a convex set $\mathcal{E}$ in $\mathcal{Y}$, real-valued convex functions $\phi^0, \phi^1, \ldots, \phi^\mu$ defined on $\mathcal{Y}$, and real-valued affine functions $\varphi^1, \ldots, \varphi^m$ defined on $\mathcal{Y}$, find an element $y_0 \in \mathcal{E}$ that achieves a minimum for ϕ^0 (on $\mathcal{E}$) subject to the constraints $\varphi^i(y) = 0$ for $i = 1, \ldots, m$, and $\phi^i(y) \leq 0$ for $i = 1, \ldots, \mu$.

If we set $\varphi = (\varphi^1, \ldots, \varphi^m)$, $\phi = (\phi^1, \ldots, \phi^\mu)$, $Z = R^\mu$ and $Z = \{(\xi^1, \ldots, \xi^\mu) : \xi^i \leq 0 \text{ for } i = 1, \ldots, \mu\}$, then Problem 1a clearly falls under the category of Problem 1.

We shall say that Problem 1 is well-posed if (1) there is a vector $y_1 \in \mathcal{E}$ such that $\varphi(y_1) = 0$ and $\phi(y_1) \in$ interior of Z, and (2) 0 is an interior point of $\varphi(\mathcal{E})$.

Roughly speaking, Problem 1 is well-posed if the equality and inequality constraints are consistent on the constraint set $\mathcal{E}$, and if the equality constraints are independent on $\mathcal{E}$. (Note that $\varphi(\mathcal{E})$ is convex).

Also of interest is the problem that differs from Problem 1 in that the equality constraint $\varphi(y) = 0$ is omitted. (Such a problem is well-posed if $\phi(y_1) \in$ interior of Z for some $y_1 \in \mathcal{E}$). Although we shall not discuss this problem in the sequel, all the results that we shall obtain for Problem 1 carry over in an obvious way (with evident modifications) to the problem without equality constraints. A similar statement holds for problems without inequality constraints.

3. Necessary conditions and sufficient conditions

The basic necessary conditions for Problem 1 are given by the following theorem.

Theorem 3.1. Let $y_0 \in \mathcal{E}$ be a solution of Problem 1. Then there exist a vector $\alpha \in R^m$, a number $\beta^0 \leq 0$, and a functional $\ell \in Z^*$, not all zero, such that

$$(\alpha \cdot \varphi + \beta^0 \phi^0 + \ell \circ \phi)(y) \leq (\alpha \cdot \varphi + \beta^0 \phi^0 + \ell \circ \phi)(y_0) =$$
$$= \beta^0 \phi^0(y_0) \text{ for all } y \in \mathcal{E}, \quad (3.1)$$

$$\ell(z) \geq 0 \text{ for all } z \in Z. \quad (3.2)$$

Although Theorem 3.1 follows directly from the general results of [3], we shall present a proof of this theorem which is far simpler than the proof of the very general necessary conditions presented in [3].

Proof. Let $\mathcal{E}_1 = \{y : y \in \mathcal{E}, \phi^0(y) < \phi^0(y_0), \phi(y) \leq_Z 0\}$. It follows at once from our hypotheses that $\mathcal{E}_1$ is convex and that $0 \notin \varphi(\mathcal{E}_1)$ in R^m. Since $\varphi(\mathcal{E}_1)$, as the image of a convex set under an affine map, is a convex set in R^m, we can therefore conclude (on the basis of the main separation theorem for convex sets in R^m -- see, e.g., [4, p. 162, Lemma 2] -- that there is a non-zero vector $\tilde{\alpha} \in R^m$ such that $\tilde{\alpha} \cdot \varphi(y) \leq 0$ for all $y \in \mathcal{E}_1$. But this means that the set

$$B_1 = \{(-\tilde{\alpha} \cdot \varphi(y), \theta, z) : y \in \mathcal{E}, z \in Z, \theta \geq \phi^0(y) -$$
$$- \phi^0(y_0), \phi(y) \leq_Z z\}$$

does not meet the set

$$B_2 = \{(\theta^1, \theta^2) : \theta^1 < 0, \theta^2 < 0\} \times Z.$$

It is easily verified that B_1 is a non-empty convex set in $R^2 \times Z$, and that B_2 is an open convex cone in $R^2 \times Z$. By the principal separation theorem in topological linear vector spaces [4, p. 244], there are a non-zero linear, continuous functional $\tilde{\ell}$ defined on $R^2 \times Z$ and a real number γ_0 such that

$$\tilde{\ell}(\zeta_1) \leq \gamma_0 \leq \tilde{\ell}(\zeta_2) \quad \text{whenever} \quad \zeta_1 \in B_1 \text{ and } \zeta_2 \in B_2.$$

This is easily seen to imply (if we take into account that B_2 is a convex cone) that there are a functional $\ell \in Z^*$ and numbers $\beta^0 \leq 0$ and $\beta^1 \leq 0$, not all zero, such that

$$-\beta^1 \tilde{\alpha} \cdot \varphi(y) + \beta^0 \phi^0(y) - \beta^0 \phi^0(y_0) + \ell \circ \phi(y) \leq 0 \tag{3.3}$$

$$\text{for all } y \in \mathcal{E},$$

$$\ell(z) \geq 0 \quad \text{for all} \quad z \in Z. \tag{3.4}$$

Since $y_0 \in \mathcal{E}$ and $\varphi(y_0) = 0$, it follows from (3.3) that $\ell \circ \phi(y_0) \leq 0$. But, since $\phi(y_0) \in Z$, (3.4) implies that $\ell \circ \phi(y_0) \geq 0$, i.e., $\ell \circ \phi(y_0) = 0$. Setting $\alpha = -\beta^1 \tilde{\alpha}$, our desired conclusion now follows at once.

Sufficient conditions for Problem 1 are given by the following theorem.

Theorem 3.2. Suppose that $y_0 \in \mathcal{E}$, $\alpha \in R^m$, $\beta^0 \leq 0$ and $\ell \in Z^*$ satisfy (3.1) and (3.2) as well as the relations

$$\beta^0 \neq 0, \quad \varphi(y_0) = 0, \quad \phi(y_0) \leq_Z 0. \tag{3.5}$$

Then y_0 is a solution of Problem 1.

Proof. Without loss of generality, we shall assume that $\beta^0 = -1$. Then (3.1) implies that

$$\phi^0(y_0) \le \phi^0(y) - \alpha \cdot \varphi(y) - \ell \circ \phi(y) \text{ for all } y \in \mathcal{E}. \quad (3.6)$$

Let $y_1 \in \mathcal{E}$ be such that $\varphi(y_1) = 0$ and $\phi(y_1) \le_Z 0$. Then it follows from (3.6) and (3.2) that $\phi^0(y_0) \le \phi^0(y_1)$, which means that y_0 is a solution of Problem 1.

Note that Theorem 3.2 is valid even if, in the statement of Problem 1, we dispense with the requirements that $\mathcal{E}$ is a convex set, that φ is an affine function, that ϕ^0 is a convex function and that ϕ is a Z-convex function.

If we examine the two preceding theorems, we see that the only difference between the necessary conditions of Theorem 3.1 and the sufficient conditions of Theorem 3.2 is that, in the latter, β^0 must not vanish. If Problem 1 is well-posed, it turns out that the number β^0 in Theorem 3.1 cannot be zero, so that we obtain conditions which are both necessary and sufficient for y_0 to be a solution of the problem. In other words, the following theorem holds.

Theorem 3.3. Suppose that Problem 1 is well-posed. Then $y_0 \in \mathcal{E}$ is a solution of this problem if and only if there exist a vector $\alpha_0 \in R^m$ and a functional $\ell_0 \in Z^*$ such that

$$\varphi(y_0) = 0 \text{ and } \phi(y_0) \le_Z 0, \quad (3.7)$$

$$(\phi^0 + \alpha_0 \cdot \varphi + \ell_0 \circ \phi)(y) \ge (\phi^0 + \alpha_0 \cdot \varphi + \ell_0 \circ \phi)(y_0) = \quad (3.8)$$

$$= \phi^0(y_0) \text{ for all } y \in \mathcal{E},$$

$$\ell_0(z) \le 0 \text{ for all } z \in Z. \quad (3.9)$$

Proof. By Theorems 3.1 and 3.2, it suffices to prove that relations (3.1) and (3.2) (with α, β^0, and ℓ not all zero) imply that $\beta^0 \ne 0$, if Problem 1 is well-posed. Let $y_1 \in \mathcal{E}$ be such that $\varphi(y_1) = 0$ and $\phi(y_1) \in$ interior of Z. We argue by contradiction, and suppose that $\beta^0 = 0$. Then

(3.1) implies that $\ell \circ \phi(y_1) \leq 0$, which, in view of (3.2), means that $\ell \circ \phi(y_1) = 0$, which is possible only if $\ell = 0$. Consequently, $\alpha \neq 0$ and (3.1) takes the form $\alpha \cdot \varphi(y) \leq 0$ for all $y \in \mathcal{E}$, contradicting the hypothesis that 0 is an interior point of $\varphi(\mathcal{E})$, and completing the proof.

If the function ϕ is directionally differentiable at y_0, then Theorems 3.1 - 3.3 may also be rewritten in "differential" form as follows.

Theorem 3.1 (Differential form). Let $y_0 \in \mathcal{E}$ be a solution of Problem 1, and suppose that ϕ is directionally differentiable at y_0. Then there exist a vector $\alpha \in R^m$, a number $\beta^0 \leq 0$, and a functional $\ell \in Z^*$, not all zero, such that (3.2) holds and such that

$$\alpha \cdot \varphi_L(y - y_0) + \beta^0 D\phi^0(y - y_0) + \ell \circ D\phi(y - y_0) \leq 0 \tag{3.10}$$

for all $y \in \mathcal{E}$,

$$\ell \circ \phi(y_0) = 0, \tag{3.11}$$

where $D\phi^0$ and $D\phi$ denote the directional differentials of ϕ^0 and ϕ, respectively, at y_0, and φ_L denotes the linear part of φ.

Theorem 3.2 (Differential form). Suppose that $y_0 \in \mathcal{E}$, $\alpha \in R^m$, $\beta^0 \leq 0$ and $\ell \in Z^*$ satisfy (3.5), (3.2), and (3.11), that ϕ is directionally differentiable at y_0, and that (3.10) holds, where φ_L, $D\phi^0$ and $D\phi$ are as in the preceding theorem. Then y_0 is a solution of Problem 1.

Theorem 3.3 (Differential form). Suppose that Problem 1 is well-posed and that ϕ is directionally differentiable. Then $y_0 \in \mathcal{E}$ is a solution of this problem if and only if there exist a vector $\alpha_0 \in R^m$ and a functional $\ell_0 \in Z^*$ such that (3.7), (3.9), and (3.11) -- with $\ell = \ell_0$ -- hold and such that

$$\alpha_0 \cdot \varphi_L(y - y_0) + D\phi^0(y - y_0) + \ell_0 \circ D\phi(y - y_0) \geq 0 \tag{3.12}$$

for all $y \in \mathcal{E}$

(where φ_L, $D\phi^0$ and $D\phi$ are as in the preceding theorems).

The differential forms of Theorems 3.1 - 3.3 follow at once from their conventional forms and the following lemma.

Lemma 3.1. Suppose that (for Problem 1) ϕ is directionally differentiable at some $y_0 \in \mathcal{E}$, and that $\varphi(y_0) = 0$. Further, suppose that $\alpha \in R^m$, $\beta^0 \leq 0$, and $\ell \in Z^*$ are such that (3.2) holds. Then (3.1) holds if and only if (3.10) and (3.11) hold, where $D\phi^0$ and $D\phi$ are the directional differentials of ϕ^0 and ϕ, respectively, at y_0, and φ_L is the linear part of φ.

Proof. First suppose that (3.1) and (3.2) hold, so that (since $\varphi(y_0) = 0$) (3.11) certainly is satisfied. Since $\mathcal{E}$ is convex, $y_0 + \varepsilon y_1 = (1-\varepsilon)y_0 + \varepsilon(y_0 + y_1) \in \mathcal{E}$ whenever $(y_0 + y_1) \in \mathcal{E}$ and $0 < \varepsilon < 1$, and therefore, by virtue of (3.1),

$$\alpha \cdot \frac{\varphi(y_0 + \varepsilon y_1) - \varphi(y_0)}{\varepsilon} + \beta^0 \frac{\phi^0(y_0 + \varepsilon y_1) - \phi^0(y_0)}{\varepsilon} +$$

$$+ \ell\left(\frac{\phi(y_0 + \varepsilon y_1) - \phi(y_0)}{\varepsilon}\right) \leq 0 \quad \text{whenever } (y_0 + y_1) \in \mathcal{E}$$

and $\varepsilon \in (0, 1)$.

Passing to the limit in the preceding relation as $\varepsilon \to 0^+$, we at once arrive at (3.10).

Conversely, suppose that (3.10), (3.11) and (3.2) hold. Since $\varphi(y_0) = 0$, the equality in (3.1) follows at once. Since ϕ^0 is convex and ϕ is Z-convex,

$$\frac{\phi^0(y_0 + \varepsilon(y - y_0)) - \phi^0(y_0)}{\varepsilon} \leq \phi^0(y) - \phi^0(y_0),$$

$$\frac{\phi(y_0 + \varepsilon(y - y_0)) - \phi(y_0)}{\varepsilon} \leq_Z \phi(y) - \phi(y_0)$$

for all $y \in \mathcal{U}$ and $\varepsilon \in (0, 1)$. Passing to the limit as $\varepsilon \to 0^+$, we conclude that $D\phi^0(y - y_0) \leq \phi^0(y) - \phi^0(y_0)$ and $D\phi(y - y_0) \leq Z\phi(y) - \phi(y_0)$ for all $y \in \mathcal{U}$, which, by virtue of (3.2) and (3.10), implies that the inequality in (3.1) holds, completing our proof.

Note 3.1. It is important to observe, for the purpose of applications to optimal control (see Section 5), that the differential form of Theorem 3.2 remains in force if we remove the requirements in Problem 1 that the set $\mathcal{E}$ is convex. However, in contrast to the ordinary form of the theorem, we cannot dispense with the convexity hypotheses for ϕ^0 and ϕ or the hypothesis that φ is affine. This is because Lemma 3.1 requires that φ, ϕ^0 and ϕ satisfy these hypotheses.

Note 3.2. The differential form of the necessary conditions (Theorem 3.1) can also be derived even if the hypotheses on the data of Problem 1 are considerably relaxed. For example, one need not assume that φ is affine or that ϕ^0 and ϕ are convex and Z-convex, respectively, so long as they are suitably differentiable. (In this case, φ_L must be replaced by $D\varphi$ in (3.10).) Further, the hypothesis that $\mathcal{E}$ is convex can be considerably weakened. For details, the reader is referred to [3].

The necessary conditions and sufficient conditions of Theorems 3.1 - 3.3 have been presented in two forms. In addition, they can be presented in a third form. Namely, if $y_0 \in \mathcal{E}$, $\alpha \in R^m$, $\beta^0 \leq 0$, and $\ell \in Z^*$ satisfy (3.2) and (3.7), then (3.1) holds if and only if there is a <u>linear</u> functional $\Lambda^0 : \mathcal{U} \to R$ such that (3.11) holds and such that

$$(\alpha \cdot \varphi + \beta^0 \Lambda^0 + \ell \circ \phi)(y) \leq (\alpha \cdot \varphi + \beta^0 \Lambda^0 + \ell \circ \phi)(y_0) \quad (3.13)$$

for all $y \in \mathcal{E}$,

$$\Lambda^0(y) \leq \phi^0(y_0 + y) - \phi^0(y_0) \text{ for all } y \in \mathcal{U}. \quad (3.14)$$

In other words, relations (3.11), (3.13) and (3.14) may be used in place of (3.1) in Theorems 3.1 and 3.2. It is obvious that, with y_0, α, β^0 and ℓ as indicated and Λ^0 linear, (3.11), (3.13), and (3.14) imply that (3.1) holds. The converse, namely that (3.1) and (3.7) imply the existence of a linear map $\Lambda^0: \mathcal{Y} \to R$ such that (3.11), (3.13), and (3.14) hold is a somewhat deeper result which follows from Lemma 5.1 in [3]. (Also, see [5].)

A linear functional $\Lambda^0: \mathcal{Y} \to R$ that satisfies (3.14) is sometimes referred to as a support functional to ϕ^0 at y_0. If ϕ^0 has a <u>linear</u> directional differential $D\phi^0$ at y_0, then (3.14) evidently implies that $\Lambda^0 = D\phi^0$.

Of course, in a similar way, Relation (3.8) in Theorem 3.3 may be replaced by (3.11) together with the inequalities (3.14) and $(\Lambda^0 + \alpha_0 \cdot \varphi + \ell_0 \circ \phi)(y) \geq (\Lambda^0 + \alpha_0 \cdot \varphi + \ell_0 \circ \phi)(y_0)$ for all $y \in \mathcal{E}$.

In the case of Problem 1a, the term $\ell \circ D\phi$ in (3.1) can also be "replaced" by a linear term. In fact, Theorems 3.1 - 3.3 can be summarized as follows for this problem.

<u>Theorem 3.4.</u> Let $y_0 \in \mathcal{E}$ be a solution of Problem 1a. Then there exist vectors $\alpha = (\alpha^1, \ldots, \alpha^m) \in R^m$ and $\beta = (\beta^0, \beta^1, \ldots, \beta^\mu) \in R^{\mu+1}$, not both zero, such that

$$\left(\sum_{i=1}^{m} \alpha^i \varphi^i + \sum_{i=0}^{\mu} \beta^i \phi^i\right)(y) \leq \left(\sum_{i=1}^{m} \alpha^i \varphi^i + \sum_{i=0}^{\mu} \beta^i \phi^i\right)(y_0) =$$

$$= \beta^0 \phi^0(y_0) \text{ for all } y \in \mathcal{E},$$

$$\beta^i \leq 0 \text{ for } i = 0, 1, \ldots, \mu+1.$$

(3.15)

If $\beta^0 \neq 0$ (as must be the case if the problem is well-posed), then the preceding necessary conditions, together with the relations

$$\varphi^i(y_0) = 0 \text{ for } i = 1, \ldots, m, \text{ and}$$

$$\phi^i(y_0) \leq 0 \text{ for } i = 1, \ldots, \mu,$$

are also sufficient for y_0 to be a solution of Problem 1a.

Corollary 1. In Theorem 3.4, (3.15) may be replaced by the relations

$$\sum_{i=1}^{m} \alpha^i \varphi_L^i(y - y_0) + \sum_{i=0}^{\mu} \beta^i D\phi^i(y - y_0) \leq 0 \text{ for all } y \in \mathcal{E},$$

$$\sum_{i=1}^{\mu} \beta^i \phi^i(y_0) = 0, \qquad (3.16)$$

where (for each i) $D\phi^i$ is the directional differential of ϕ^i at y_0 and φ_L^i is the linear part of φ^i.

Corollary 2. In Theorem 3.4, (3.15) may be replaced by the relations

$$\left(\sum_{i=1}^{m} \alpha^i \varphi_L^i + \sum_{i=0}^{\mu} \beta^i \Lambda^i\right)(y) \leq \left(\sum_{i=1}^{m} \alpha^i \varphi_L^i + \sum_{i=0}^{\mu} \beta^i \Lambda^i\right)(y_0)$$

for all $y \in \mathcal{E}$,

$$\Lambda^i(y) \leq \phi^i(y_0 + y) - \phi^i(y_0) \text{ for all } y \in \mathcal{Y} \text{ and each}$$

$$i = 0, 1, \ldots, \mu,$$

together with (3.16), for some suitably chosen linear functionals Λ^i (i = 0, ..., μ) defined on $\mathcal{Y}$.

Theorem 3.4 follows at once from Theorems 3.1 - 3.3, and its Corollary 1 follows directly from the differential form of these theorems. Corollary 2 follows from a repeated application of Lemma 5.1 in [3]. (Also see [5].)

The results of this section are closely related to some of the classical theorems of Kuhn and Tucker [6] (which were stated in finite-dimensional spaces). The results of Kuhn and Tucker have been extended in various ways, in a large number of papers, to Banach spaces. (In particular, see [7], whose Theorems 1.1 and 1.6 essentially coincide with our Theorem 3.4 and its Corollary 2 in case $\mathcal{Y}$ is a Banach space.) Hurwicz in [1, Theorem V.3.1] considered a problem which included our Problem 1 as a special case (except that Hurwicz did not consider equality-type constraints), and obtained necessary conditions which essentially include the necessity part of our Theorem 3.3 (in ordinary form).

The differential form of Theorems 3.2 and 3.3 are, to the best of the author's knowledge, novel, and the former (taking into account Note 3.1) is particularly useful for obtaining sufficient conditions for optimal control problems (See Section 5).

4. Duality

In this section, we shall be concerned with the problem that is the dual to Problem 1, i.e. (under the assumption that this problem is well-posed), we shall construct an auxiliary mathematical programming problem the solution of which yields a pair $(\ell_0, \alpha_0) \in Z^* \times R^m$ with the properties described in Theorem 3.3.

With this aim in mind, we define the function $\hat{\phi}$ on $Z^* \times R^m$ as follows:

$$\hat{\phi}(\ell, \alpha) = \inf_{y \in \mathcal{E}} [(\phi^0 + \alpha \cdot \varphi + \ell \circ \phi)(y)], \quad \ell \in Z^*, \quad \alpha \in R^m.$$

Note that $\hat{\phi}$ may take on the value $-\infty$. Let

$$\mathcal{E}^* = \{(\ell, \alpha) : \ell \in Z^*, \alpha \in R^m, \hat{\phi}(\ell, \alpha) > -\infty\}.$$

It is trivial to verify that $\mathcal{E}^*$ is a convex set in $Z^* \times R^m$. Further, $\hat{\phi}$, as the infimum of a class of linear functionals, is concave (i.e., $-\hat{\phi}$ is convex) on $\mathcal{E}^*$.

Let us denote the dual cone of Z by Z^*, i.e.,

$$Z^* = \{\ell : \ell \in Z^*, \ell(z) \leq 0 \text{ for all } z \in Z\}.$$

The mathematical programming problem which consists in finding an element $(\ell_0, \alpha_0) \in Z^* \times R^m$ which achieves a maximum on the convex set $\mathcal{E}^*$ for the concave functional $\hat{\phi}$, subject to the constraint $\ell \leq_{Z^*} 0$, will be called the dual of Problem 1.

It turns out that any solution of the dual problem may be used for the functional ℓ_0 and the vector α_0 in Theorem 3.3. Namely, we have the following result.

Theorem 4.1. Suppose that Problem 1 is well-posed and that $y_0 \in \mathcal{E}$ is a solution thereof. Then the dual problem has at least one solution. Further, $(\ell_0, \alpha_0) \in Z^* \times R^m$ is a solution of the dual problem if and only if (3.8) and (3.9) are satisfied, and, in this case,

$$\hat{\phi}(\ell_0, \alpha_0) = \phi^0(y_0) . \qquad (4.1)$$

Thus, according to Theorem 4.1, Problem 1, if it is well-posed and has a solution, may be solved as follows: First, find any solution (ℓ_0, α_0) of the dual problem (this often turns out to be a relatively easy task). Then the solution points of Problem 1 will be a subset of the set of solution points of the problem which consists in minimizing the function $(\phi^0 + \alpha_0 \cdot \varphi + \ell_0 \circ \phi)$ on $\mathcal{E}$ (subject neither to equality nor to inequality constraints). Finally, the original problem and its dual have the same optimum "payoff".

Proof of Theorem 4.1. We shall first show that if $(\ell_0, \alpha_0) \in Z^* \times R^m$ satisfies (3.8) and (3.9), then (ℓ_0, α_0) is a solution of the dual problem, and (4.1) holds. Since, by Theorem 3.3, a pair $(\ell_0, \alpha_0) \in Z^* \times R^m$ satisfying (3.8) and (3.9) does exist, we shall be able to conclude that the dual problem has at least one solution.

But if (3.8) and (3.9) hold, then $\ell_0 \leq_{Z^*} 0$, and

$$\hat{\phi}(\ell_0, \alpha_0) = \inf_{y \in \mathcal{E}} [(\phi^0 + \alpha_0 \cdot \varphi + \ell_0 \circ \phi)(y)] = \phi^0(y_0) > -\infty ,$$

i.e., $(\ell_0, \alpha_0) \in \mathcal{E}^*$. Further, if $(\ell_1, \alpha_1) \in \mathcal{E}^*$ and $\ell_1 \leq_{Z^*} 0$, then, since (3.7) holds,

$$\hat{\phi}(\ell_0, \alpha_0) = \phi^0(y_0) \geq (\phi^0 + \alpha_1 \cdot \varphi + \ell_1 \circ \phi)(y_0) \geq$$

$$\geq \inf_{y \in \mathcal{E}} \ [(\phi^0 + \alpha_1 \cdot \varphi + \ell_1 \circ \phi)(y)] = \hat{\phi}(\ell_1, \alpha_1),$$

so that (ℓ_0, α_0) is a solution of the dual problem, and (4.1) is satisfied.

Conversely, suppose that (ℓ_0, α_0) is a solution of the dual problem, so that (3.9) holds. By what has already been proved, (4.1) is satisfied. Further, since $\ell_0 \leq_{Z^*} 0$ and (3.7) holds,

$$(\phi^0 + \alpha_0 \cdot \varphi + \ell_0 \circ \phi)(y_0) \leq \phi^0(y_0) = \hat{\phi}(\ell_0, \alpha_0) =$$

$$= \inf_{y \in \mathcal{E}} \ [(\phi^0 + \alpha_0 \cdot \varphi + \ell_0 \circ \phi)(y)] \leq (\phi^0 + \alpha_0 \cdot \varphi + \ell_0 \circ \phi)(y_0),$$

which implies that (3.8) holds, completing our proof.

Corollary 1. Suppose that Problem 1 is well-posed and that $y_0 \in \mathcal{E}$ satisfies (3.7). Then y_0 is a solution of Problem 1 if and only if there is a solution (ℓ_0, α_0) of the dual problem such that (4.1) holds. Further, the "if" part holds even if Problem 1 is not well-posed.

Proof. The necessity follows at once from Theorem 4.1. Conversely, if $(\ell_0, \alpha_0) \in \mathcal{E}^*$ is a solution of the dual problem and $\hat{\phi}(\ell_0, \alpha_0) = \phi^0(y_0)$, where $y_0 \in \mathcal{E}$ satisfies (3.7), then we can show, as in the proof of Theorem 4.1, that (3.8) and (3.9) hold, and consequently conclude, on the basis of Theorem 3.2, that y_0 is a solution of Problem 1.

Corollary 2. If $y_0 \in \mathcal{E}$ and $(\ell_0, \alpha_0) \in \mathcal{E}^*$ satisfy (3.7) and (4.1), then y_0 is a solution of Problem 1, and (ℓ_0, α_0) is a solution of its dual.

Proof. Arguing as in the proof of Theorem 4.1, we can show that the hypotheses of the corollary imply that (ℓ_0, α_0) is a solution of the dual problem, and we can then conclude, on the basis of Corollary 1, that y_0 is a solution of Problem 1.

We point out that if $Z = R^\mu$ (e.g., in the case of Problem 1a), then $\mathcal{E}^*$ is a subset of a finite-dimensional space, i.e., the dual problem is finite-dimensional (whether or not $\mathcal{U}$ is).

If the data of Problem 1 satisfy certain hypotheses (as is the case in many applications), then the dual problem has a particularly simple form. These hypotheses are as follows:

1. $\mathcal{E}$ is a convex cone in $\mathcal{U}$,
2. $\phi^0(\gamma y) = \gamma \phi^0(y)$ for all $\gamma \geq 0$ and $y \in \mathcal{U}$,
3. there are a function $\bar{\phi}: \mathcal{U} \to Z$ and an element $z_0 \in Z$ such that (i) $\phi(y) = \bar{\phi}(y) + z_0$ for all $y \in \mathcal{U}$ and (ii) $\bar{\phi}(\gamma y) = \gamma \bar{\phi}(y)$ for all $\gamma \geq 0$ and $y \in \mathcal{U}$.

If the preceding three hypotheses are satisfied, then the set $\mathcal{E}^*$ and the function $\hat{\phi}$ of the dual problem have the following simple forms:

$$\mathcal{E}^* = \{(\ell, \alpha) : \ell \in Z^*, \alpha \in R^m, (\phi^0 + \alpha \cdot \varphi_L + \ell \circ \bar{\phi})(y) \geq 0$$

$$\text{for all } y \in \mathcal{E}\}, \tag{4.2}$$

$$\hat{\phi}(\ell, \alpha) = \alpha \cdot \xi_0 + \ell(z_0) \text{ for all } (\ell, \alpha) \in \mathcal{E}^*, \tag{4.3}$$

where φ_L is the linear part of φ and $\xi_0 = \varphi(y) - \varphi_L(y)$ for any $y \in \mathcal{U}$.

Indeed, by definition,

$$\hat{\phi}(\ell, \alpha) = \inf_{y \in \mathcal{E}} [(\phi^0 + \alpha \cdot \varphi_L + \ell \circ \bar{\phi})(y) + \alpha \cdot \xi_0 + \ell(z_0)] \tag{4.4}$$

for all $(\ell, \alpha) \in Z^* \times R^m$. If (ℓ, α) is such that $(\phi^0 + \alpha \cdot \varphi_L + \ell \circ \bar{\phi})(y_1) < 0$ for some $y_1 \in \mathcal{E}$, then, by

Hypotheses 1-3 and the linearity of φ_L, $\hat{\phi}(\ell, \alpha) \le$ $\inf_{\gamma > 0} [\gamma(\phi^0 + \alpha \cdot \varphi_L + \ell \circ \bar{\phi})(y_1) + \alpha \cdot \xi_0 + \ell(z_0)] = -\infty$, i.e., $(\ell, \alpha) \notin \mathcal{E}^*$. On the other hand, if $(\phi^0 + \alpha \cdot \varphi_L + \ell \circ \bar{\phi})(y) \ge 0$ for all $y \in \mathcal{E}$, then the infimum in (4.4) is attained at $y = 0 \in \mathcal{E}$ and is equal to $[\alpha \cdot \xi_0 + \ell(z_0)] > -\infty$, so that $(\ell, \alpha) \in \mathcal{E}^*$, and (4.2) and (4.3) hold.

Thus, if Hypotheses 1-3 are satisfied, then the dual problem consists in finding a maximum of the <u>linear</u> function $\hat{\phi}(\ell, \alpha)$ given by (4.3) on $Z^* \times R^m$, subject to the constraints $\ell \le_{Z^*} 0$ and $(\alpha \cdot \varphi_L + \ell \circ \bar{\phi})(y) \ge -\phi^0(y)$ for all $y \in \mathcal{E}$.

For the special case where $Z = R^\mu$ and $Z = \{(\xi^1, \ldots, \xi^\mu) : \xi^i \le 0 \text{ for all } i\}$ (i.e., the case of Problem 1a), the dual problem consists in finding a maximum for the linear function

$$\hat{\phi}(\beta^1, \ldots, \beta^\mu, \alpha^1, \ldots, \alpha^m) = \hat{\phi}(\beta, \alpha) = \alpha \cdot \xi_0 + \beta \cdot z_0$$

(4.5)

subject to the constraints

$$\beta^i \ge 0 \text{ for } i = 1, \ldots, \mu, \quad (\alpha \cdot \varphi_L + \beta \cdot \bar{\phi})(y) \ge$$

(4.6)

$$\ge -\phi^0(y) \text{ for all } y \in \mathcal{E}$$

(provided, of course, that Hypotheses 1-3 are satisfied).

A duality theory for convex programming problems, in which the dual problem is defined in the same way as was done at the beginning of this section, has been developed by Pshenichnyi [2, pages 111-115] -- for the special case where $\mathcal{Y}$ is a Banach space and Problem 1 has the form of Problem 1a -- and by Gol'shteyn [8] -- in the absence of equality constraints and for the case of $\mathcal{Y}$ a Banach space. Pshenichnyi's principal result [2, Theorem 5.7 on page 115] is a spacial case of our Theorem 4.1, and Gol'shteyn's results, though in a somewhat different direction, bear some similarity to ours.

In case our problem is linear (i.e., the functions ϕ^0 and ϕ are linear and affine, respectively) and finite dimensional (i.e., $\mathcal{Y} = R^n$ and $Z = R^\mu$ for some positive integers n and μ), and if $Z = \{(\xi^1, \ldots, \xi^\mu) : \xi^i \leq 0 \text{ for all } i\}$ and $\mathcal{E} = \{(\varepsilon^1, \ldots, \varepsilon^n) : \varepsilon^i \geq 0 \text{ for all } i\}$ (so that our Problem has the form of Problem 1a, and Hypotheses 1-3 are satisfied), then the problem of maximizing the function (4.5) subject to the constraints (4.6) is equivalent to the classical dual problem for linear programming.

In closing this section, we remark that the duality theory developed in this section may be applied to obtain some useful computational algorithms for a number of optimal control problems.

5. An application to the theory of optimal control

In this section we shall illustrate our results by applying the differential form of Theorem 3.2 to obtain sufficient conditions for an optimal control problem with restricted phase coordinates. As we shall we, these sufficient conditions "almost" coincide with well-known necessary conditions.

The problem we consider is the following: Consider the ordinary differential equation:

$$\dot{x}(t) = A(t)x(t) + h(u(t), t), \quad t_1 \leq t \leq t_2, \qquad (5.1)$$

where x is an absolutely continuous function from $I = [t_1, t_2]$ into R^n, A is a given continuous $(n \times n)$-matrix valued function defined on I, h is a given continuous n-vector valued function defined on $U \times I$, U being a given set in R^r, and $u(\cdot)$ is some function in $\mathcal{U}$, where $\mathcal{U}$ is a given class of measurable functions from I into U with the property that $h(u(\cdot), \cdot)$ is integrable over I for each $u \in \mathcal{U}$. Further, let there be given vectors $\zeta_1, \ldots, \zeta_m$, $\eta_1, \ldots, \eta_m$ in R^n; real numbers $\nu_1, \ldots, \nu_m$; real-valued, continuously differentiable functions g and f defined on $R^n \times I$ such that, for each $t \in I$, $f(\cdot, t)$ and $g(\cdot, t)$ are convex functions, and $D_x g(\cdot, \cdot)$ is continuously differentiable; and a real-valued, convex, continuously differentiable function χ defined on R^n.

Then our problem consists in finding a pair $(x(\cdot), u(\cdot))$ that satisfies the constraints (1) $u \in \mathcal{U}$, (2) Eq. (5.1) is satisfied for almost all $t \in I$ with x absolutely continuous on I,

  (3) $\zeta_i \cdot x(t_1) + \eta_i \cdot x(t_2) = \nu_i$ for $i = 1, \ldots, m$,
  (4) $g(x(t), t) \leq 0$ for all $t \in I$,

and in so doing achieves a minimum for the functional

$$\chi(x(t_2)) + \int_{t_1}^{t_2} f(x(t), t) dt .$$

Let us reduce this problem to the form of Problem 1. Let $\mathcal{Y}$ denote the linear vector space of all absolutely continuous functions from I into R^n, let $\mathcal{E}$ denote the set of all $x \in \mathcal{Y}$ such that (5.1) holds a.e. on I for some $u \in \mathcal{U}$, let $\mathcal{Z}$ denote the linear vector space of all real-valued continuous functions defined on I with the usual sup norm topology, and let

$$Z = \{z : z \in \mathcal{Z}, z(t) \leq 0 \text{ for all } t \in I\} .$$

Further, let φ^i ($i = 1, \ldots, m$) and ϕ^0 be real-valued functions defined on $\mathcal{Y}$ through the formulas

$$\varphi^i(x(\cdot)) = \zeta_i \cdot x(t_1) + \eta_i \cdot x(t_2) - \nu_i, \quad i = 1, \ldots, m,$$
$$x(\cdot) \in \mathcal{Y},$$

$$\phi^0(x(\cdot)) = \chi(x(t_2)) + \int_{t_1}^{t_2} f(x(t), t) dt, \quad x(\cdot) \in \mathcal{Y},$$

let $\varphi = (\varphi^1, \ldots, \varphi^m)$, and let ϕ be the function from $\mathcal{Y}$ into $\mathcal{Z}$ given by the formula

$(\phi(x(\cdot)))(t) = g(x(t), t)$ for all $t \in I$, $x(\cdot) \in \mathcal{Y}$.

With $\mathcal{Y}$, $\mathcal{E}$, Z, $\mathcal{Z}$, φ, ϕ^0, and ϕ defined in this manner, it follows at once that our optimal control problem is equivalent to Problem 1. Further, it is trivial to verify that our problem data satisfy the requirements that we imposed in this problem, with the exception that $\mathcal{E}$ is not necessarily a convex set. In addition, ϕ^0 and ϕ have linear directional differentials at any $x_0 \in \mathcal{Y}$ given by the formulas

$$D\phi^0(\delta x) = D\chi(x_0(t_2)) \cdot \delta x(t_2) + \int_{t_1}^{t_2} D_x f(x_0(t), t) \cdot \delta x(t) dt, \quad \delta x \in \mathcal{Y},$$

$$(D\phi(\delta x))(t) = D_x g(x_0(t), t) \cdot \delta x(t) \text{ for all } t \in I, \quad \delta x \in \mathcal{Y},$$

where $D_x f$, $D_x g$ and $D\chi$ are the obvious gradients of f, g, and χ with respect to x.

Let us now appeal to the differential form of Theorem 3.2 (See Note 3.1). According to this theorem (making use of the well-known representation theorem for Z^*), if $x_0 \in \mathcal{Y}$, $u_0 \in \mathcal{U}$, $\alpha = (\alpha^1, \ldots, \alpha^m) \in R^m$, $\beta^0 \leq 0$, and the real-valued, non-increasing function λ defined on I satisfy the conditions

$$\dot{x}_0(t) = A(t) x_0(t) + h(u_0(t), t) \text{ for almost all } t \in I, \quad (5.2)$$

$$\zeta_i \cdot x_0(t_1) + \eta_i \cdot x_0(t_2) = \nu_i \text{ for } i = 1, \ldots, m, \quad (5.3)$$

$$g(x_0(t), t) \leq 0 \text{ for all } t \in I, \quad (5.4)$$

$$\beta^0 \neq 0, \quad (5.5)$$

$\lambda(t_2) = 0$, λ is continuous from the right on (t_1, t_2) and is constant on subintervals of I on which

$$g(x_0(t), t) < 0, \tag{5.6}$$

$$\sum_{i=1}^{m} \alpha^i \zeta_i \cdot \delta x(t_1) + \left[\sum_{i=1}^{m} \alpha^i \eta_i + \beta^0 D\chi(x_0(t_2)) \right] \cdot \delta x(t_2) +$$

$$+ \beta^0 \int_{t_1}^{t_2} D_x f(x_0(t), t) \cdot \delta x(t) dt +$$

$$+ \int_{t_1}^{t_2} D_x g(x_0(t), t) \cdot \delta x(t) d\lambda(t) \leq 0 \text{ for all } \delta x(\cdot) \in \mathcal{E} - x_0, \tag{5.7}$$

then x_0 is a solution of our problem.

By the well-known variations of parameters formula, the solution of Eq. (5.1) is given by the formula

$$x(t) = \Phi(t) \left[x(t_1) + \int_{t_1}^{t} \Phi^{-1}(s) h(u(s), s) ds \right], \quad t \in I,$$

where $\Phi(\cdot)$ is the fundamental matrix of the equation $\dot{x} = A(t)x$ such that $\Phi(t_1)$ is the identity matrix. Hence, (5.7) holds if

$$\sum_{i=1}^{m} \alpha^i \zeta_i + \left[\sum_{i=1}^{m} \alpha^i \eta_i + \beta^0 D\chi(x_0(t_2)) \right] \Phi(t_2) +$$

$$+ \beta^0 \int_{t_1}^{t_2} D_x f(x_0(t), t) \Phi(t) dt + \int_{t_1}^{t_2} D_x g(x_0(t), t) \Phi(t) d\lambda(t) = 0 \tag{5.8}$$

(where the ζ_i and η_i, $D\chi$, $D_x f$, and $D_x g$ are to be considered row-vectors), and if

$$\left[\sum_{i=1}^{m}\alpha^i\eta_i + \beta^0 D\chi(x_0(t_2))\right]\Phi(t_2)\int_{t_1}^{t_2}\Phi^{-1}(s)[h(u(s),s) -$$

$$- h(u_0(s),s)]ds + \beta^0\int_{t_1}^{t_2}D_x f(x_0(t),t)\Phi(t)\int_{t_1}^{t}\Phi^{-1}(s)[h(u(s),s) -$$

$$- h(u_0(s),s)]ds\,dt + \int_{t_1}^{t_2}D_x g(x_0(t),t)\Phi(t)\int_{t_1}^{t}\Phi^{-1}(s)[h(u(s),s) -$$

$$- h(u_0(s),s)]ds\,d\lambda(t) \le 0 \quad \text{for all} \ u \in \mathcal{U}.$$

If we interchange the order of integration in each of the double integrals in the preceding expression, and integrate by parts in the latter of these, we conclude that (5.7) holds if (5.8) does and if

$$\int_{t_1}^{t_2}[\psi(s) - \lambda(s)D_x g(x_0(s),s)]h(u(s),s)ds \le$$

$$\le \int_{t_1}^{t_2}[\psi(s) - \lambda(s)D_x g(x_0(s),s)]h(u_0(s),s)ds \quad \text{for all} \ u \in \mathcal{U}, \tag{5.9}$$

where

$$\psi(s) = \left\{\left[\sum_{i=1}^{m}\alpha^i\eta_i + \beta^0 D\chi(x_0(t_2))\right]\Phi(t_2) + \beta^0\int_{s}^{t_2}D_x f(x_0(t),t)\Phi(t)dt - \int_{s}^{t_2}\lambda(t)\frac{d}{dt}[D_x g(x_0(t),t)\Phi(t)]dt\right\}\Phi^{-1}(s), \quad t_1 \le s \le t_2. \tag{5.10}$$

Taking into account the equation $\frac{d}{ds}(\Phi^{-1}(s)) = -\Phi^{-1}(s)A(s)$ (which follows at once from the definition of Φ), we easily conclude that ψ satisfies (5.10) if and only if ψ is a solution of the differential equation

$$\dot{\psi}(s) = -\psi(s)A(s) - \beta^0 D_x f(x_0(s), s) +$$
$$+ \lambda(s) D_x p(x_0(s), s), \quad t_1 \leq s \leq t_2, \tag{5.11}$$

where

$$p(x, t) = D_x g(x, t)(A(t)x + h(u_0(t), t) + D_t g(x, t), \tag{5.12}$$
$$x \in R^n, \quad t \in I,$$

(ψ as well as $D_x p$ are to be considered row-vectors), with boundary value

$$\psi(t_2) = \sum_{i=1}^{m} \alpha^i \eta_i + \beta^0 D\chi(x_0(t_2)). \tag{5.13}$$

Further, (5.8) holds if and only if

$$\psi(t_1) = -\sum_{i=1}^{m} \alpha^i \zeta_i + \lambda(t_1) D_x g(x_0(t_1), t_1). \tag{5.14}$$

Thus, we have shown the following: If $x_0 \in \mathcal{U}$, $u_0 \in \mathcal{U}$, $\alpha = (\alpha^1, \ldots, \alpha^m) \in R^m$, $\beta^0 \leq 0$, the real-valued, non-increasing function λ defined on I and the n-row-vector valued absolutely continuous function ψ defined on I satisfy the relations (5.2) - (5.6), (5.9), and (5.11) - (5.14), then (x_0, u_0) is a solution of our problem.

If $\mathcal{U}$ consists of all measurable functions u from I into U such that $h(u(\cdot), \cdot)$ is integrable over I, then the preceding sufficient conditions -- with the exception of (5.5) -- are also necessary (see, e.g., [9, Section 8]), and the maximum condition (5.9) is equivalent to the relation

$$\max_{v \in U} [\psi(s) - \lambda(s)D_x g(x_0(s), s)]h(v, s) = [\psi(s) -$$

$$- \lambda(s)D_x g(x_0(s), s)]h(u_0(s), s) \quad \text{for almost all} \quad s \in I.$$

Similar sufficient conditions, for optimal control problems which are related to the one described in this section, have previously been obtained by Mangasarian [10], Pshenichnyi [7] and Gilbert and Funk [11]. Those in the latter paper are particularly close to ours. Further, Pshenichnyi's approach had much in common with the one presented in this section.

REFERENCES

1. L. Hurwicz, Programming in linear spaces, in Studies in Linear and Nonlinear Programming by K. J. Arrow, L. Hurwicz and H. Uzawa, Stanford University Press, Stanford, Calif., 1958.

2. B. N. Pshenichnyi, Necessary Conditions for an Extremum, Izdat. Nauka, Moscow, 1969 (in Russian).

3. L. W. Neustadt, A general theory of extremals, J. Comput. System Sci., 3 (1969), pp. 57-92.

4. C. Berge, Topological Spaces, Macmillan, New York, 1963.

5. H. Halkin and L. W. Neustadt, General necessary conditions for optimization problems, Proc. Nat. Acad. U.S.A., 56 (1966), pp. 1066-1071.

6. H. W. Kuhn and A. W. Tucker, Nonlinear programming, in Proceedings of the Second Berkeley Symposium on Mathematical Statistics and Probability, Univ. of California Press, Berkeley, 1951, pp. 481-492.

7. B. N. Pshenichnyi, Linear optimal control problems, SIAM J. Control, 4 (1966), pp. 577-593.

8. E. G. Gol'shteyn, Dual problems of convex and fractionally-convex programming in functional spaces, Societ Math. -Dokl., 8 (1967), pp. 212-216.

9. L. W. Neustadt, An abstract variational theory with applications to a broad class of optimization problems. II. Applications, SIAM J. Control 5 (1967), pp. 90-134.

10. O. L. Mangasarian, Sufficient conditions for the optimal control of nonlinear systems, SIAM J. Control, 4 (1966), pp. 139-152.

11. E. G. Gilbert and J. E. Funk, Some sufficient conditions for optimality in control problems with state space constraints, SIAM J. Control, 8 (1970), to appear.

This research was supported in part by the United States Air Force Office of Scientific Research under grant AF-AFOSR-1029-67, and in part by the National Science Foundation under grant GK-15787.

Recent Results on Complementarity Problems

C. E. LEMKE

ABSTRACT

In the last few years a number of important results have taken place dealing with the Complementarity Problem, linear and nonlinear. Mainly, there have been results dealing with existence and uniqueness; with classes of matrices in the linear case; and with generalizations of the problem.

In this paper, an attempt is made to summarize these results, and to bring them together into a comprehensive whole in a manner as to suggest areas of further concentration.

Introduction

This paper is largely a summarization of recent results relating to the <u>Complementarity Problem</u> (CP):
Given a column $f(z)$ of n real functions $f_i(z)$ find columns z which are feasible solutions (f.s.); that is:

$$f(z) \geq 0; \quad z \geq 0, \qquad (1)$$

and are complementary solutions (c.s.); that is, are feasible and satisfy:

$$z_i f_i(z) = 0; \text{ all } i \text{ (equivalently: } z'f(z) = 0). \qquad (2)$$

Prime denotes matric transposition. Most of the pages are concerned with the <u>linear</u> (CP); namely the case:

$$f(z) = q + Mz; \quad (M \text{ is of order } n \text{ by } n). \qquad (3)$$

We try to show how the results tie together, and tend to suggest further research and provide further unification than before. The list of references shows that we have not been able to acknowledge all of the work of recent years. Also, to minimize the list, we often refer to lists of references in prior papers.

Regarding (CP), the usual questions arise concerning existence and uniqueness of solutions, and computing solutions. It is to be expected that the linear case (LCP) will have received more attention than the nonlinear case (NLCP), not only because of the relative manageability of linearity, but because of the diversity of applications, such as in the area of approximations to other problems; in furnishing proofs; etc. For example, (LCP) has generated much additional interest in certain classes of matrices and their characterizations. Also, the combinatoric aspects of (LCP) have proven useful in other areas, and have placed focus on the use of 'constructive proofs'.

In this paper, when proofs are instructive and easy to expound they are given. Some proofs are relegated to Appendices, so as not to detract from the flow of the presentation.

Section I deals with (LCP); Section II considers some results in (NLCP).

I. The Linear Complementarity Problem

We relegate the inequality constraints to nonnegativity of all variables by introducing slack variables:

$$w = q + Mz; \quad w, z, \geq 0; \quad \text{and} \quad w'z = 0. \quad (4)$$

Throughout, e refers to a nonnegative column, which, unless otherwise indicated, is a column of 1's of appropriate order (used without loss of generality). z_i and w_i are <u>complements</u> of one another.

In <u>I.1</u> classes of matrices are so introduced as to suggest further areas of investigation. In <u>I.2</u>, the present state with regard to existence and uniqueness of solutions is considered, serving also to characterize some of the classes of matrices considered. <u>I.3</u> considers some of the more interesting special cases.

I.1. Classes of Matrices

One form of the basic result involving inequalities for an arbitrary rectangular matrix is that credited to J. Ville (see Ref. [8]):

There is $x \geq 0$ such that $Ax > 0$, or there is (5) $0 \neq y \geq 0$ such that $A'y \leq 0$. (In the spirit of the paper this is demonstrated in an appendix.)

The following classes are defined and considered extensively by Fiedler and Ptak, [10]: (where for brevity in (6) <u>a</u>, for example, we mean that S_0 is the class of A (any order) satisfying the condition following. We also write, for example, $A \in -S_0$ to mean $-A \in S_0$. Let us write: $-A' = A^*$.):

<u>a</u>. $S_0 : \exists\, 0 \neq x \geq 0 \ni Ax \geq 0$

<u>b</u>. $S : \exists\ x \geq 0 \ni Ax > 0$ (equivalently, (6)

$\exists x > 0 \ni Ax > 0$).

Thus, Ville's Theorem asserts that for any A:

$A \in S$ <u>or</u> $A \in S_0^*$ (clearly not both, See also App. 1) (7)

We restrict attention now mainly to square matrices, which we label M. We define some properties: let $\emptyset \neq a \subset (1, 2, \ldots, n)$. Let $\underline{a} = (1, 2, \ldots, n) - a$. Let M_{aa} denote the matrix obtained from M by deleting all rows and columns corresponding to indices in <u>a</u>. M_{aa} is a <u>principal sub-matrix</u> (p.s.m.) of M. Determinants of p.s.m.'s are called <u>principal minors</u> (p.m.). We may partition M (using the same permutation on the rows as on the columns when necessary) as:

$$M = \begin{pmatrix} M_{aa}, & M_{a\underline{a}} \\ M_{\underline{a}a}, & M_{\underline{a}\underline{a}} \end{pmatrix}. \qquad (8)$$

When M_{aa} is non-singular, a __principal transform__ (p. t.), with block pivot M_{aa} is the matrix:

$$T_a(M) = \begin{pmatrix} -M_{aa}, & 0 \\ -M_{\underline{a}a}, & I \end{pmatrix}^{-1} \begin{pmatrix} -I, & M_{a\underline{a}} \\ 0, & M_{\underline{a}\underline{a}} \end{pmatrix} = $$

$$\begin{pmatrix} M_{aa}^{-1}, & -M_{aa}^{-1} M_{a\underline{a}} \\ M_{\underline{a}a} M_{aa}^{-1}, & M_{\underline{a}\underline{a}} - M_{\underline{a}a} M_{aa}^{-1} M_{a\underline{a}} \end{pmatrix}$$

(9)

We say that a class T of matrices is closed under p.s.m.'s, for example, if whenever $M \in T$ any p.s.m. of M is in T.
For our purposes the next most basic classes are:

<u>a</u>. E_0: $\forall a$: $M_{aa} \in S_0'$ (i.e., $M_{aa} \not\in -S$)

(10)

<u>b</u>. E: $\forall a$: $M_{aa} \in S'$ (i.e., $M_{aa} \not\in -S_0$),

and:

<u>a</u>. P_0: nonnegative p.m.'s (n.n.p.m.'s)

(11)

<u>b</u>. P: positive p.m.'s (called P-matrices).

These three classes may be characterized in terms of components. For (6) and (11) see, for example, [3]. In Reference [9] Eaves notes the characterization for E. (Unless otherwise noted, Eaves' results are to be found in the comprehensive and nicely developed paper above.) We shall note class inclusions. It turns out that all inclusions noted are proper.

S_0: $\forall\ x \geq 0\quad \exists\ i \ni (M'x)_i \geq 0$

S: $\forall\ 0 \neq x \geq 0\quad \exists\ i \ni (M'x)_i > 0\quad (S \subset S_0)$

E_0: $\forall\ 0 \neq x \geq 0\quad \exists\ i \ni x_i > 0$ and $(Mx)_i \geq 0$ (see App. 1)

E: $\forall\ 0 \neq x \geq 0\quad \exists\ i \ni x_i > 0$ and $(Mx)_i > 0\quad (E \subset E_0)$

P_0: $\forall\ x \neq 0\quad \exists\ i \ni x_i \neq 0$ and $x_i(Mx)_i \geq 0$

P: $\forall\ x \neq 0\quad \exists\ i \ni x_i(Mx)_i > 0.\quad (P \subset P_0)$.

(12)

One has: $P_0 \subset E_0 \subset S_0'$ and $P \subset E \subset S'$.

Clearly, E_0, E, P_0 and P are closed under p.s.m.'s P_0 and P, since they involve determinants, are closed under transposition: $P_0 = P_0'$; $P = P'$. Thus: $P_0 \subset S_0$; $P \subset S$. There are two other classes which are familiar and which share the just-mentioned properties, but which do not appear to lend themselves to a 'component-type' characterization:

The co-positive classes:

C_0: $x \geq 0$ implies $x'Mx \geq 0\quad (C_0 \subset E_0)$

C: $0 \neq x \geq 0$ implies $x'Mx > 0\quad (C \subset C_0;\ C \subset E)$,

(13)

and the nonnegative definite classes:

D_0: $x'Mx \geq 0\quad\quad (D_0 \subset C_0;\ D_0 \subset P_0)$

(14)

D: $x \neq 0$ implies $x'Mx > 0\ (D \subset D_0;\ D \subset C;\ D \subset P)$.

For $D_0 \subset P_0$ see, for example, [6].

Of interest is the property of D_0: $x'Mx = 0$ implies $(M + M')x = 0$ $(= \frac{1}{2}$ Grad. $x'Mx)$. We define the co-positive plus class:

C^+: $C^+ \subset C_0$, and $x \geq 0$; $x'Mx = 0$ implies

implies $(M + M')x = 0$. (15)

One has: C, D_0, $D \subset C^+$; the first and third because $x \geq 0$ and $x'Mx = 0$ imply $x = 0$. The classes C_0 and C have been studied extensively. In this regard see [6], and [7]. In [6] the following characterizations of C^+ and D_0 are given:

C^+: $x \geq 0$ and $x'Mx = 0$ imply $(M+M')x = 0$, and, when $M \neq 0$ there exists $x \geq 0$ such that $x'Mx > 0$.
(16)
D_0: $x'Mx = 0$ implies $(M + M')x = 0$; and, when $M \neq 0$ there exists x such that $x'Mx > 0$.

Also, see [7] for characterizations of C_0, C and C^+.
With an eye toward (LCP), Eaves defines the class:

E_1: $x \geq 0$; $Mx \geq 0$; and $x'Mx = 0$ implies there are nonnegative diagonal matrices D_1 and D such (17) that: when $x \neq 0$: $Dx \neq 0$ and $(D_1 M + M'D)x = 0$,

and the class:

L: $L = E_0 \cap E_1$ $(E, C^+ \subset L)$. (18)

The class L is quite extensive. Eaves points out, for example, that matrices of the form:

$$\begin{pmatrix} p & n & n & n & n \\ a & p & n & n & n \\ a & a & p & n & n \\ a & a & a & p & n \\ a & a & a & a & p \end{pmatrix} \quad \text{are in } L;$$

where an 'n' denotes any nonnegative number; a 'p' denotes any positive number; and an 'a' denotes an arbitrary real.

A final class of interest is:

Z: $M_{i,j} \leq 0$; for all i and j such that $i \neq j$. (19)

Z has been studied extensively by Fiedler and Ptak [11]. They show, for example, that $Z \cap S \subset P$.

I.2. Relative to the Linear Complementarity Problem

We start this section by paraphrasing somewhat some of Eaves' results relative to (LCP). One has the characterizations:

E_0: (LCP) has a unique c.s. for all $q > 0$ (namely $z = 0$)

(20)

E: (LCP) has a unique c.s. for all $q \geq 0$ (namely $z = 0$)

Next, consider (LCP) for given M, q, where $e+q > 0$:

$w = (q+z_0 e) + Mz; \quad w, z \geq 0; \quad z_0 \geq 0; \quad \text{and} \quad w'z = 0.$ (21)

Call solutions: "z_0-complementary solutions" (z_0 - c.s.). We wish to consider "z_0-complementary rays" (z_0 - c.r.). These consist of z_0 - c.s.'s: $<w^1 + \theta\bar{w}, z^1 + \theta\bar{z}, z_0^1 + \theta\bar{z}_0>$ for all $\theta \geq 0$; where $<\bar{z}, \bar{z}_0> \neq 0$:

$$w^1 + \theta\bar{w} = q + (z_0^1 + \theta\bar{z}_0)e + M(z^1 + \theta\bar{z}). \qquad (22)$$

That these constitute a z_0 - c.r. implies:

a. $w^1, \bar{w}, z^1, \bar{z} \geq 0$; $z_0^1, \bar{z}_0 \geq 0$;

b. $\bar{w} = \bar{z}_0 e + M\bar{z}$; $\qquad (23)$

c. $w^{1\prime}z^1 = w^{1\prime}\bar{z} = \bar{w}'z^1 = \bar{w}'\bar{z} = 0$.

Case 1: z_0 unbounded on the z_0 - c.r.
  Then, for $\theta \geq 1$, $q + z_0 e > 0$. Now if $M \in E_0$, (20) implies $z = 0$, which implies $z^1 = \bar{z} = 0$. Then the z_0 - c.r. has the unique form: $z = 0$; $w = q + z_0 e$, for all z_0 with $w \geq 0$, and we call the ray the <u>primary ray</u>.

Case 2: z_0 bounded on the z_0 - c.r. Thus $\bar{z}_0 = 0$. We call such a ray a secondary ray. Now suppose $M \in E_1$. Since $\bar{z} \neq 0$, (23) implies $M\bar{z} \geq 0$; $\bar{z}'M\bar{z} = 0$. Regarding (17), let $y = D\bar{z} \neq 0$, so that $M'y = -D_1 M\bar{z} \leq 0$. Further:

$$y'M\bar{z} = \bar{z}'DM\bar{z} = 0, \text{ because } \bar{z}'M\bar{z} = 0, \text{ and}$$

$$y'Mz^1 = -z^{1\prime}D_1 M\bar{z} = -z^{1\prime}D_1\bar{w} = 0, \text{ because } z^{1\prime}\bar{w} = 0.$$

Hence, regarding (22): $y'w = 0 = y'(q + z_0^1 e)$. But $q + e > 0$ implies $y'(q + e) > 0$. Hence, $z_0^1 < 1$. Hence: $(1-z_0^1)y'q < 0$.
  We thus have: $y \geq 0$; $M'y \leq 0$; and $q'y < 0$. Then, for any $z \geq 0$, with $w = q + Mz$, one has $y'w < 0$, which implies $w \not\geq 0$. Thus, for the given (M, q), (LCP) is infeasible. Summarizing: on a z_0 - c.r.:

If $M \in E_0$ and z_0 is unbounded, the ray is the primary ray.
If $M \in E_1$ and z_0 is bounded, (LCP) is already infeasible.
  Note that in the latter case, if $M \in E$, then by (20), $\overline{w} = M\overline{z}$ implies $\overline{z} = 0$, implying that there is no ray. Hence if $M \in E$ there can be no secondary rays (due to Cottle and Dantzig [3]).

  We next recall a pivot scheme (Scheme 1 in [18]) applied to (21) which, after an initial pivot, works wholly with z_0 - c.s.'s. Assuming non-degeneracy, the sequence of pivots necessarily terminates, and either in a c.s. to (4) (when $z_0 = 0$), or in a z_0 - c.r. which then necessarily is not the primary ray; and this holds for any M and q. In the general case, when (21) is not necessarily non-degenerate for the purposes of proof one may set up a perturbed or lexicographical system corresponding to (21) which is inherently non-degenerate, and whose corresponding sequence of pivots terminates with the above-mentioned results for (4). Eaves gives a thorough exposition of this.

  We may therefore conclude:
  If $M \in L$ then a c.s. exists for all feasible q, and if, further, $M \in E$ then there exists a c.s. for all q.

  This proof is constructive in the sense that for $M \in L$, Scheme 1 will yield a c.s. or demonstrate infeasibility. (Note that if $M \in S$, (4) is feasible for any q. (10) shows $E \subset S'$. We now have that $E \subset S$ because (4) is feasible for $q = -e$.)

  The Preceding suggests the definition of two classes relative to (LCP):

$$\begin{aligned} K&: \text{a c.s. exists for all feasible } q \\ Q&: \text{a c.s. exists for all } q \quad (Q \subset K). \end{aligned} \quad (24)$$

  Thus: $L \subset K$, and $E \subset Q$. Chandrasekaran [1] shows that also $Z \subset K$, although $Z \not\subset L$. As indicated in the next sub-section, K is not exhausted thereby.
  In particular, $P \subset E$, and P has the <u>characterization</u>:

P: a unique c.s. exists for all q (see, e.g., [21]).
(25)

Murty has refined these results. See App. 2.

If M is anti-symmetric ($M = M^*$), then $M \in C^+$. In this regard, see App. 1.

Space permits only the mention of the results relating to the <u>number</u> of c.s. in the non-degenerate case, and to some consequences of these results.

I.3. Some Special Problems and Pivot Schemes

We first consider a 'parametric cast' of Scheme 1; then discuss Chandrasekaran's result for the class Z; then consider a special scheme employed by Howson; and finally consider some problems which may be put into (LCP)-form, sometimes for special q, and contrast questions of existence of c.s.'s with the foregoing.

A. Referring to the z_0-problem (21), McCammon [20] has developed a Parametric Pivot Scheme, whose name refers to the treatment of z_0 as a <u>parameter</u> rather than as a variable. The scheme tends to make clearer the relationship between Scheme 1 and the well-known Principal Pivot Methods developed by Cottle and Dantzig [4]. We paraphrase McCammon's scheme briefly.

We first identify a sub-scheme, applied to the problem:

$w = q + Mz$; $w, z \geq 0$; $w'z = 0$; where $q \geq 0$;
(26)

and some unique component of q is 0: $q_r = 0$.

The sub-scheme has interest in its own right. We are assuming non-degeneracy. (We will be thinking of $q = q(z_0) = q^0 + z_0 e^0$; for some fixed value of z_0, where e^0 is a given column.)

Associate with a form such as (26) the <u>basic</u> f.s. obtained by setting non-basic variables to zero: $z = 0$;

$w = q$. w_r is the 'designated variable' (des. v.). The intent of the scheme is to get w_r non-basic, but to preserve feasibility and $w'z = 0$ throughout. After some k pivots the form is:

$$w^k = q^k + M^k z^k; \quad q^k \geq 0; \quad q_r^k = 0. \tag{27}$$

A pivot will involve the exchange of the pivot pair, written generically (w_t^k, z_s^k). Initially, $(k = 0)$, $z_s^k = z_r$, the complement of des. v. w_r. Subsequently, z_s^k is the complement of the variable which became non-basic on the previous pivot.

Having z_s^k one seeks to complete a pivot pair as follows: If $M_{r,s}^k \neq 0$, $(w_t^k, z_s^k) = (w_r, z_s^k)$ is pivot pair, in which case, completing the pivot terminates the sub-scheme. If $M_{r,s}^k = 0$, then, if $M_{\cdot,s}^k \geq 0$ (indicating a ray), the sub-scheme is terminated. Otherwise, the pivot pair is determined (as usual) so as to preserve feasibility. If no ray is found, the sub-scheme will terminate in a form such as (26).

The overall scheme consists of applications of the sub-scheme. After some t such applications one has a form:

$$w^t = (q^t + z_0^t e^t) + M^t z^t; \quad (q^t + z_0^t e^t)_i > 0; \text{ for } i \neq r_t$$
$$\text{and } (q^t + z_0^t e^t)_{r_t} = 0; \tag{28}$$

such that (with a permutation of rows if necessary) M^t is a p.t. of $M = M^0$ (that is, the complement of w_i^t is some z_j^t); $w_{r_t}^t$ is the current des.v., and the sub-scheme may be applied.

Now, in place of the fixed z_0^t there is an interval $a_t \leq z_0 \leq b_t$ for which, setting $z^t = 0$ yields a z_0 - c.s. It turns out that z_0^{t-1} is either a_t or b_t. Then z_0^t is selected as a_t if z_0^{t-1} is b_t; as b_t if z_0^{t-1} is a_t.

(Initially, $q = q^0$, $a_0 = \max_i . - q_i$ -- assumed positive --
and $b_0 = \infty$.

The total scheme is: perform a sequence of sub-schemes. Terminate if either $z_0^t = 0$ (in a c.s. for (4)); or in a z_0 - c.r.

The execution of a sub-scheme is analogous to Cottle and Dantzig's "major cycle".

The relationship between the scheme and Scheme 1 is given by McCammon:

Let L_m denote the form after the first m pivots according to Scheme 1, and M_m the form after the first m pivots according to McCammon's scheme. Then, having the form M_m, with $w_{r_k}^k$ the des.v., executing the pivot with pivot pair $(w_{r_k}^k, z_0)$ yields (except for permutations) L_{m+1}.

In this sense McCammon's scheme gives an interpretation of Scheme 1 which brings out the resemblance to the Principal Pivot Scheme of Cottle and Dantzig (and appears also to be a more natural approach than Scheme 1 to (LCP)).

B. As shown by Chandrasekaran [1], $Z \subset K$ provides an example of a class of matrices for which the resolution of the (LCP) (for any q) is extremely neat and instructive. We alter his scholarly demonstration slightly, so as to fit into the discussion so far:

By a principal permutation, if necessary, permute the system (4) so as to have negative q_i on top: In partitioned form:

$$\begin{pmatrix} w_1 \\ w_2 \end{pmatrix} = \begin{pmatrix} q_1 \\ q_2 \end{pmatrix} + \begin{pmatrix} M_{11}, & M_{12} \\ M_{21}, & M_{22} \end{pmatrix} \begin{pmatrix} z_1 \\ z_2 \end{pmatrix} ; \qquad (29)$$

where $q_1 < 0$; $q_2 \geq 0$.

In particular, for feasibility:

$$M_{11}z_1 = w_1 - q_1 - M_{12}z_2 > 0, \text{ (since } M_{12} \leq 0). \tag{30}$$

Clearly, $M_{11} \in Z$ requires $z_1 > 0$ for any f.s. In turn this implies $M_{11} \in S$; hence (see statement following (19)) $M_{11} \in P$; which implies (Fiedler and Ptak [11]) $M_{11}^{-1} \geq 0$.

Assuming feasibility one may thus perform a p.t. yielding:

$$\begin{pmatrix} z_1 \\ w_2 \end{pmatrix} = \begin{pmatrix} q_1^1 \\ q_2^1 \end{pmatrix} + \begin{pmatrix} M_{11}^1, & M_{12}^1 \\ M_{21}^1, & M_{22}^1 \end{pmatrix} \begin{pmatrix} w_1 \\ z_2 \end{pmatrix} ; \tag{31}$$

and one has: $q_1^1 > 0$; $(M_{11}^1, M_{12}^1) \geq 0$; $M_{21}^1 \leq 0$; and, again, $M_{22}^1 \in Z$.

One may then apply the same analysis to:

$$w_2 = (q_2^1 + M_{21}^1 w_1) + M_{22}^1 z_2; \text{ (for any fixed } w_1 \geq 0). \tag{32}$$

Therefore, assuming feasibility, one may successively perform the above 'reductions'; terminating whenever (generically speaking) a 'q_2^1' is found which is nonnegative. Then, for a c.s., one sets to 0 each component w_i for which $z_i > 0$ has been seen to be necessary for feasibility, and otherwise $w_2 = q_2^1 \geq 0$; $z_2 = 0$ for the final 'q_2^1'. Thus, $Z \in K$, and of course, in the event of infeasibility, this must show up.

Chandrasekaran remarks that the sequence of principal (block) pivots may be viewed as a sequence of scalar pivots; more precisely, that a direct variant of the Principal Pivot Method of Cottle and Dantzig for P-matrices may also be used (the latter may be viewed as one manner of performing the inversions). At the present time, it is not clear that Scheme I will resolve the problem.

C. A Special Pivot Scheme

The following pivot scheme enunciated by Howson [14] tends to suggest either the possibility of an efficient general scheme for (LCP), or at least for special classes of problems:

Visualize n players in a game; player i has m_i pure strategies. Set $m = \Sigma_i m_i$. One is given n^2 matrices $M^{i,j}$ of order m_i by m_j satisfying: $M_{ii} = 0$; $M_{ij} > 0$, for $i \neq j$. Write: $x = <x^1, x^2, \ldots, x^n>$. Define:

$$f^i(x) = \sum_j^n M^{i,j} x^j, \quad \text{('partial payoff' to i)}; \qquad (33)$$

for mixed strategies x^i:

$$x^i \geq 0; \quad e'x^i = 1. \qquad (34)$$

One seeks x (equilibrium point) such that:

$$x^{i'} f^i(x) = \min_k f_k^i(x) \ (= v_i); \quad i = 1, 2, \ldots, n; \qquad (35)$$

or equivalently (introducing slacks $y^i \geq 0$)

$$y^i = f^i(x) - v_i e; \quad x^{i'} y^i = 0. \qquad (36)$$

Introducing (artificially) complements for v_i:

$$u_i = -1 + e'x^i, \qquad (37)$$

one has the (LCP) which, for $n = 3$, is:

$$\begin{pmatrix} y^1 \\ y^2 \\ y^3 \\ u^1 \\ u^2 \\ u^3 \end{pmatrix} = \begin{pmatrix} 0 \\ 0 \\ 0 \\ -1 \\ -1 \\ -1 \end{pmatrix} + \begin{pmatrix} 0 & M^{12} & M^{13} & -e & 0 & 0 \\ M^{21} & 0 & M^{23} & 0 & -e & 0 \\ M^{31} & M^{32} & 0 & 0 & 0 & -e \\ e' & 0 & 0 & 0 & 0 & 0 \\ 0 & e' & 0 & 0 & 0 & 0 \\ 0 & 0 & e' & 0 & 0 & 0 \end{pmatrix} \begin{pmatrix} x^1 \\ x^2 \\ x^3 \\ v^1 \\ v^2 \\ v^3 \end{pmatrix} \quad (38)$$

Let "P" denote a f.s. Let $r_i = x^{i\prime}y^i$; $r = \langle r_1, r_2, \ldots, r_n \rangle$. One starts by taking (somewhat arbitrarily) $x_1^i = 1$; i.e., performing the n pivots with pivot pairs (u_i, x_1^i).

Define a Permissible Feasible Point (PFP), as one satisfying:

a. $r_i = x_1^{i\prime}y_1^i$
b. $u_i = 0$; $i = 1, 2, \ldots, n$.
c. $v_i = 0$ implies $v_{i+1} = 0$
d. $r_{i+1} = $ implies $r_i = 0$; $i = 1, 2, \ldots, n-1$.
e. $r'v = r_p v_p$; some p.

The pivot scheme stays in the set of PFP's. We note:

    i. The start yields an initial PFP, characterized by $v = 0$; and one seeks an equilibrium point characterized by $r = 0$.

    ii. Insisting on <u>b</u> implies, see (36), that the set of PFP's is bounded.

    iii. Since $u = 0$ requires $e'x^i = 1$: $v_i = 0$ implies $r_i > 0$. Hence $r_i = 0$ implies $v_i > 0$. In fact, if $v_i = 0$ (Player i inactive), <u>a</u> and <u>b</u> additionally imply (see (36)) that for $j \geq i$, $x_1^j = 1$; $y^j > 0$; thus, in the computation, no attention need be paid to constraints (36), (37) for $j \geq i$.

Now, for a PFP P, let q be least for which $v_i = 0$; let t be greatest for which $r_i = 0$. One sees that this implies two cases:

<u>Case I:</u> $t = q-1$. Take $p = q$. Thus $r'v = r_p v_p = 0$.

<u>Case II:</u> $t = q-2$. Take $p = q-1$. Thus $r'v = r_p v_p > 0$.

Assuming non-degeneracy, for Case I, if $q = 1$ one has the initial PFP (take $t = 0$), from which the only pivot to make to remain in the PFP set is given by pivot pair (y_j^1, v_1); some unique j. Otherwise, there are exactly two possible pivots:

<u>Case Ia:</u> Pivot on pivot pair which uses v_p (make $v_p > 0$).

<u>Case Ib:</u> Since $r_{p-1} = 0$, it is readily seen that just one of the complementary pair (y_1^{p-1}, x_1^{p-1}) is non-basic. Pivot on pivot pair which uses the non-basic of the pair (make $r_{p-1} > 0$).

In Case II, $v_p > 0$ (<u>1st</u> p players 'active'); and $r_p > 0$ or:

$$\sum_1^p x^{i\prime} y^i = x_1^p y_1^p ; \qquad (39)$$

and, regarding note <u>iii</u>, one is in a 'well-known' almost-complementary pivoting situation, and thus knows that there are just two pivots possible which keep one in the set of PFP's.

Howson's scheme is to set up the start, and then to pivot so as to remain in the set of PFP's, starting by making $v_1 > 0$. By our remarks, the path thus generated is 'reversible', which implies that no basic set repeats. Since the set of PFP's is bounded, one must terminate, and can only terminate in the desired $r = 0$ (setting $q = n+1$ in that case).

The matrix M in (38) is not in L.

D.1. The Cottle-Dantzig Generalization

In [5], Cottle and Dantzig nicely extend some of the (LCP) concepts considered above. It is reasonable to expect that there will be valuable applications of their results forthcoming.

As before, let z have n components. However, here each z_i has some m_i complements, and we may speak of the $\underline{i\text{th}}$ $\underline{\text{complementary set}}$ $(z_i, w_1^i, w_2^i, \ldots, w_{m_i}^i)$. Now $\underline{\text{complementarity}}$ requires that $\underline{\text{some}}$ variable in the set be equal to 0. Let $m = \sum_i m_i$. The system has the form:

$$w = q + Nz; \quad w \geq 0; \quad z \geq 0, \quad (N \text{ is } m \text{ by } n); \tag{40}$$

and $w = \langle w^1, w^2, \ldots, w^n \rangle$; where $w^i = \langle w_1^i, w_2^i, \ldots, w_{m_i}^i \rangle$, and the problem is to find a f.s. which is complementary in the generalized sense. As in Scheme 1, they work with the system:

$$w = (q + z_0 e) + Nz, \tag{41}$$

and z_0 - c.s.'s. In this regard, after each non-terminal pivot the basic set of variables has the property that, for each i except one, precisely m_i of the m_i+1 variables of the $\underline{i\text{th}}$ complementary set are basic! Again, in this way termination is either in a c.s. ($z_0 = 0$), or in a ray.

To ensure favorable results analogous to those given above, they generalize to 'vertical block matrices' such as N some of the classes of matrices considered above, as follows: an n by n matrix M is a $\underline{\text{representative submatrix}}$ (r.s.m.) of N, whenever the $\underline{i\text{th}}$ row of M is one of the rows from the $\underline{i\text{th}}$ block N^i of N. Then, for example, a vertical block matrix N is called a P-matrix ($N \in P$) if and only if each of its r.s.m.'s are in P. (Defining a p.s.m. of N as a p.s.m. of some r.s.m., note that all p.s.m.'s of N are in P also.) We list some of their results:

If either $N \in C$, or $N \in P$ then a c.s. exists for all q. If $N \in C+$, then a c.s. solution exists for all feasible q. (This could no doubt be readily extended to L.)

Let U, of order n by m have the form: (for n = 3)

$$U = \begin{pmatrix} u^1 & 0 & 0 \\ 0 & u^2 & 0 \\ 0 & 0 & u^3 \end{pmatrix} \geq 0; \text{ where row } u^i \neq 0.$$

Then: $N \in P$ iff for all such U: $UN \in P$. Also, $N \in S$.

It is of interest to observe that one may 'put' the Cottle-Dantzig Generalization in square form. Let us define:

$$F = \begin{pmatrix} e, & 0, & \ldots, & 0 \\ 0, & e, & \ldots, & 0 \\ \vdots & \vdots & & \vdots \\ 0, & 0, & \ldots, & e \end{pmatrix} \qquad (42)$$

Observing that the i-requirement for complementarity is that:

$$\min.(z_i, w^i_1, w^i_2, \ldots, w^i_{m_i}) = 0;$$

leads to an (LCP):

$$\begin{pmatrix} w \\ v \end{pmatrix} = \begin{pmatrix} q \\ e \end{pmatrix} + \begin{pmatrix} 0, & N \\ -F', & 0 \end{pmatrix} \begin{pmatrix} u \\ z \end{pmatrix}; \begin{pmatrix} w \\ v \end{pmatrix}, \begin{pmatrix} u \\ z \end{pmatrix} \geq 0; \begin{pmatrix} w \\ v \end{pmatrix}' \begin{pmatrix} u \\ z \end{pmatrix} = 0.$$

(43)

(Write: $u = <u^1, u^2, \ldots, u^n>$. When written out it is seen that: $z_i > 0$ implies $v_i = 0$ implies $u^i \neq 0$ implies $w^i \not> 0$; and $w^i > 0$ implies $u^i = 0$ implies $v_i = 1$ implies $z_i = 0$.)

D.2. A Class of Problems of Scarf

Using the notation of the preceding sub-section, let N be a vertical block matrix of order m by n, and b a column with m components. Scarf [22] considers the 'complementary' problem: Find $x \geq 0$ for which:

$$r_i(x) = \underset{j}{\text{Max}}(N^i x - q^i)_j \geq 0; \text{ and } \sum_i x_i r_i(x) = 0. \qquad (44)$$

Scarf proves that a solution exists for all q if N has the property:

$$x \geq 0; \text{ and } \sum_i x_i \underset{j}{\text{Max}} (N^i x)_j \leq 0 \text{ implies } x = 0. \quad (45)$$

He gives a constructive proof, similar to Scheme 1. We shall note that this problem again may be put into (LCP) form:

For given x, to say that scalar $r_i = \underset{j}{\text{Max}}(N^i x - q^i)_j$ is equivalent to asserting that one has:

$$y^i = q^i - N^i x + r_i e; \ y^i \geq 0; \text{ and } y^i \not> 0, \qquad (46)$$

and $x'r = 0$ is the desired complementarity condition.

Problem (44) may be expressed in (LCP) as:

$$\begin{pmatrix} y \\ v \\ z \end{pmatrix} = \begin{pmatrix} q \\ -e \\ 2e \end{pmatrix} + \begin{pmatrix} 0, & -N, & F \\ F', & 0, & 0 \\ -F', & 0, & 0 \end{pmatrix} \begin{pmatrix} w \\ x \\ r \end{pmatrix} + z_0 \begin{pmatrix} -e \\ 0 \\ 0 \end{pmatrix} ; \quad (47)$$

when $z_0 = 0$. (To see this, note that complementarity yields -- since always $v_i + z_i = 1$ -- <u>i</u>. $0 \neq w^i$ implies $y^i \not> 0$. <u>ii</u>. $r'x = 0$ since: $r_i > 0$ implies $z_i = 0$ implies $v_i = 1$ implies $x_i = 0$; and $x_i > 0$ implies $v_i = 0$ implies $z_i = 1$ implies $r_i = 0$.)

Paraphrasing Scarf, he considers z_0 - c.s.'s to the problem (47). It is of interest to prove Scarf's result in an Appendix, where we also display the "Scarf Start".

Observe that there are similarities in the matrices M so far considered in this sub-section.

We may here observe that for a z_0 - c.s., since one always has $r'x = 0$; and $y^i \neq 0$, one has:

$$r_i = \operatorname*{Max}_{j}(N^i x - q^i)_j + z_0 . \qquad (48)$$

D.3. Scarf's Constructive Fixed-Point Proof

Since Scarf's initial papers on the subject, he and others (see Ref. [23]) have made potentially significant strides in the application of his basic discovery. Space does not permit an exhaustive summary of the results. We shall content ourselves with discussing his main theorem and making note of some of the applications.

Consider a fixed matrix $C = (J, D)$, of order n by m, where m is thought of as being very large (for the applications considered), and J is n by n. C has the properties:

$$C > 0; \quad J_{ii} = \min_{j} C_{i,j} \quad \text{(row min)}, \quad \text{whereas:} \qquad (49)$$
$$\text{for } i \neq j: \quad J_{ij} \geq D_{ik}; \text{ all } k.$$

(the columns of D may be considered as points interior to the simplex: $x \geq 0$; $e'x = 1$; thus $e'D = e'$).

We are concerned with n by n sub-matrices B of C, and the column $u = u(B)$ defined by:

$$u_i = \operatorname*{Min}_{j} B_{i,j} \quad \text{(row min.)} \qquad (50)$$

and with sub-matrices B, called <u>Primitive</u>, satisfying:

$$C^i \not\geq u = u(B); \text{ for all columns } C^i \text{ of } C. \tag{51}$$

Alternatively, Primitive sub-matrices have the property:

$$r_i = \max_j (u - C^i)_j \geq 0; \text{ for all columns } C^i \text{ of } C. \tag{52}$$

Note that if C^i is a column of B: $C^i \geq u(B)$, and by (52) for some j: $C^i_j = u_j$. Scarf makes the 'non-degeneracy assumption that in any row of C the components are all different. One effect of this is clearly that $r_i = 0$ if and only if C^i is a column of B. There are many Primitive matrices.

Now let n by m A and b be given, and satisfy:

$$\text{The set: } b = Ax; \; x \geq 0 \text{ is bounded.} \tag{53}$$

<u>Theorem</u>: (Scarf) There is a b.f.s. x for (53), and a primitive sub-matrix B of C such that C^i is a column of B if and only if $x_i > 0$.

The particular application of the Theorem determines the form of (A, b). The theorem has been used to prove Brouwer's Fixed-point Theorem; Kakutani's Fixed-point Theorem; for computing approximate solutions to convex programming problems; and for proof and computation in some other problems. In this regard, Kuhn [17] has developed an efficient procedure for utilizing the implied algorithm.

Again in the spirit of this paper we point out that the Theorem asserts the existence of solutions for some (LCP). (As in the two previous examples, there are other formulations).

Equivalent to (50) is the requirement:

$$y^i = c^i - u + r_i e; \; u \geq 0; \; y^i \geq 0; \; r_i \geq 0; \; y \not\geq 0; \tag{54}$$

If in addition some n of the r_i are 0, then the n corresponding C^i form a primitive sub-matrix, (and more than n r_i's 0 would violate Scarf's non-degeneracy assumption). Constraints (53) and (54) together with the complementary condition:

$$r'x = 0 \qquad (55)$$

yield a problem whose solution is guaranteed by Scarf's Theorem.

To put this in (LCP) form, we employ the previously used tricks, artificially introducing complements for u and t = e + r. This yields the (LCP) with equation:

$$\begin{pmatrix} r \\ y \\ w \\ v \end{pmatrix} = \begin{pmatrix} -e \\ c-e \\ -b \\ -e \end{pmatrix} + \begin{pmatrix} 0 & 0 & 0 & I \\ 0 & 0 & -G & F \\ A & 0 & 0 & 0 \\ 0 & F' & 0 & 0 \end{pmatrix} \begin{pmatrix} x \\ s \\ u \\ t \end{pmatrix} ; \qquad (56)$$

where: $c = \langle C^1, C^2, \ldots, C^m \rangle$; and $G = \begin{pmatrix} I \\ I \\ \vdots \\ I \end{pmatrix}$.

II. Nonlinear Complementarity Problems

Section A deals with some Nash-type Applications of Brouwer's Fixed-Point Theorem. Section B considers some general theorems. C considers some 'pivot-type' schemes, and D deals with a generalization of (NLCP).

A. Some Nash-type Applications of Brouwer's Theorem

If h(x) is a real function, we may consider its 'positive' and 'negative' parts, w(x) and u(x):

$$h = w - u; \quad w, u \geq 0, \quad \text{and} \quad wu = 0.$$ (57)

If h is continuous, so are w and u. This extends immediately to columns $h(x)$. We start with:

Lemma: Let $S_p = \{x: x \geq 0; e'x = p\}$; for p positive; a simplex. Let $g(x)$ be continuous on S_p and satisfy:

$$x'g(x) = 0; \quad \text{for all } x \text{ in } S_p.$$ (58)

Then there is an x in S_p for which $g(x) \geq 0$.

Proof: With regard to (57) for $h = g$, the continuous function:

$$\phi(x) = p(x + u)/(p + e'u)$$ (59)

is from S_p to S_p. Brouwer's theorem yields a fixed point x_0; so that $(e'u_0)x_0 = pu_0$. Since $x_0'w_0 = x_0'u_0$:

$$pu_0'u_0 = (e'u_0)(x_0'u_0) = (e'u_0)(x_0'w_0) = pu_0'w_0 = 0.$$ (60)

Hence $u_0 = 0$, and x_0 is the desired point.

For some applications, consider, for $f(x)$ continuous:

$$g(x) = pf(x) - x'f(x)e.$$ (61)

By the Lemma, for all $p > 0$, there is an x_p satisfying:

$$pf(x) \geq x'f(x)e.$$ (62)

(This is Karamardian's Lemma 1 in [15], which he proves by invoking Kakutani's Theorem.). Setting $p = 1$, if, for example: $f(x) > 0$ for x in S_1, in (NLCP) form: where $z = \langle x, v \rangle$; v a scalar:

$h(z) \geq 0$; $z \geq 0$; $z'h(z) = 0$; where $h(z) = \begin{pmatrix} f(x) - ve \\ -1 + e'x \end{pmatrix}$, (63)

and, by the Lemma, a fixed point yields a solution to the (NLCP). Generalizing the latter, consider the compact set:

$$S = \{x = <x^1, x^2, \ldots, x^n>; \; x^i \geq p; \; e'x^i = 1\}. \quad (64)$$

In a game setting, consider the problem of finding x in S satisfying:

$$x^{i'}f^i(x)e \leq f^i(x); \quad i = 1, 2, \ldots, n; \quad (65)$$

where $f(x)$ is a continuous column on S. Here:

$$g^i(x) = f^i(x) - x^{i'}f^i(x)e = w^i - u^i; \; w^i, u^i \geq 0; \; w^{i'}u^i = 0. \quad (66)$$

The function:

$$\emptyset(x) = \begin{pmatrix} (x^1 + u^1)/(1 + e'u^1) \\ (x^2 + u^2)/(1 + e'u^2) \\ \vdots \\ (x^n + u^n)/(1 + e'u^n) \end{pmatrix} \quad (67)$$

is continuous from S to S, and again Brouwer's theorem yields a fixed point satisfying (66), with $u^i = 0$, all i.
As another example, consider (LCP) for $q < 0$: take

$$g(x) = Mx + (x'Mx)q = w - u; \quad (68)$$

373

where $x \in S = \{x: x \geq 0; -q'x = 1\}$. The obvious $\phi(x)$ yields a fixed point with $u = 0$. Suppose $M \in C^+$.
  Case 1: $x'Mx \neq 0$. Then $x_0 = x/x'Mx$ solves (LCP).
  Case 2: $x'Mx = 0$. Then $Mx \geq 0$ implies $M'x \leq 0$.
Since $x'q = -1 < 0$, (LCP) is infeasible.
  The author does not see how to extend this kind of proof to L; or for general q.

B. Some General Existence and Uniqueness Results

Consider the convex programming problem: (see [19])

$$g_0(x) \text{ (min); subject to: } x \geq 0; \ g(x) \leq 0; \qquad (69)$$

where g_0 and column g are convex. The by-now-classical Kuhn-Tucker necessary and sufficient conditions (with reasonable constraint qualifications -- See [19]) that x yield a minimum are that x be part of a c.s. $z = \langle x, y \rangle$ for the (NLCP):

$$f(z) \geq 0; \ z \geq 0; \text{ and } z'f(z) = 0; \qquad (70)$$

where in this instance:

$$f(z) = \begin{pmatrix} Dg_0 + y'Dg \\ -g \end{pmatrix} \qquad (71)$$

where Dg_0 denotes gradient g_0; Dg the Jacobian of g.
  We content ourselves with stating results of Cottle [2] and Karamardian [15].
  <u>Definition</u> (Cottle): f has positively bounded Jacobians, if there is a $k > 0$ such that, for all x, the p.m.'s of $Df(x)$ lie between k and $1/k$.

__Theorem__ (Cottle): If f has positively bounded Jacobians, (70) has a solution.
(The condition that Jacobians have positive p.m.'s is not, as Cottle shows by example, sufficient.)

__Definition__ (Minty, see [15]): f is strongly monotone if there is a $k > 0$ such that for all x and y: $(x-y)'f(x) - f(y)) \geq k|x-y|^2$.

__Theorem__ (Karamardian): If f is continuous and is strongly monotone then (70) has a unique solution.
(In the differentiable case, Karamardian defines a square matrix $M(x)$ to be strongly positive definite if there is a $k > 0$ such that for all x: $y'M(x)y \geq k|y|^2$, for all y. Then, comparing with the Cottle result, if Df is strongly positive definite, then f is strongly monotone.)

In the linear case: $f = Mx + q$; positively bounded Jacobian reduces to $M \in P$; strongly monotone to $M \in D$.

For example, both results are applied to certain non-linear programming problems, and saddle-point problems. In a second paper [16] Karamardian resolves the existence problem for certain games.

C. Some Generalized Pivot Schemes

The prospect of 'pivoting' in the nonlinear case has interest from the standpoint of potential computational schemes as well as of proofs.

For positively bounded Jacobians, Cottle proves his result by showing that the Prinicpal Pivot Method for $M \in P$ may be directly generalized to this case: that for a principal pivot performed on $w = f(z)$, leading to $w^* = f^*(w^*)$, f^* again has positively bounded Jacobians.

As another example, referring to the game situation in I.C. in (33), if, for example, $f^i(x) > 0$ for $e'x^i = 1$, with sufficient conditions assumed for the constraint set (roughly, that it be homeomorphic to a convex polyhedron) the same scheme as there outlined leads constructively to a c.s. Wilson [24] gives a constructive scheme for finding equilibria for n-person games which is exactly the case in point. In his case the $f^i(x)$ are multilinear (f^i_j is a sum of terms, each of which is a constant times the product of n-1 variables), and $f^i(x)$ is independent of x^i.

In fact, Howson's scheme was inspired by Wilson's, and may be considered to be a variant of it.

As Cottle has generalized the Principal Pivot Method for P-matrices, of great interest would be a generalization to nonlinear cases of Scheme 1, say for classes of functions which would be, in some sense, generalizations of the class L of Eaves.

D. A Generalization of (LCP) and (NLCP)

In this sub-section we shall summarize some of the results of Habetler and Price [12], [13], who consider the generalization of (NLCP) from the nonnegative orthant of n-space to arbitrary convex cones K.

The <u>polar cone</u> K^* of K is the set:

$$K^* = \{w: w'z \geq 0 \text{ for all } z \text{ in } K\}. \qquad (72)$$

Since K defines an ordering on n-space, to preserve the analogies we shall write:

$$z(\geq K) \, 0 \text{ to mean } z \in K. \qquad (73)$$

The K-(NLCP) is: find z satisfying:

$$f(z) \, (\geq K^*) \, 0; \, z \, (\geq K) \, 0; \text{ and } z'f(z) = 0. \qquad (74)$$

Analogous to (57), given $h = h(z)$ there are <u>unique</u> w and u satisfying:

$$h = w-u; \text{ where } w \, (\geq K^*) \, 0; \, z \, (\geq K) \, 0; \text{ and } w'u = 0.$$
$$(75)$$

Comparing with Minty's definition above:

Definition: U is a domain. f(z) is strongly K-monotone on U if there exists a $k > 0$ such that for z_1, z_2 in U with $z_1 - z_2 \;(\geq K)\; 0$:

$$(z_1 - z_2)'(f(z_1) - f(z_2)) \geq k|z_1 - z_2|^2.$$

Definition: U is a domain. f(z) is strongly K-copositive on U if there exists a $k > 0$ such that for all y in U:

$$z'Df(y)z \geq k|z|^2; \quad \text{for all } z \;(\geq K)\; 0;$$

where DM(y) is the Gateaux derivative with respect to K (assumed to exist on U).

We write: $q = f(0)$. K is <u>solid</u> if Int K $\neq \emptyset$. K is <u>sub-polar</u> if $K \subset K^*$.

The next two theorems hold for continuous f which is either Strongly K-monotone <u>on K</u>, or Strongly K-copositive <u>on K</u>.

Theorem (Habetler-Price): If K is solid and $q \;(< K^*)\; 0$, then K-(NLCP) has a c.s.

Theorem (Habetler-Price): If K is solid and sub-polar, then K-(NLCP) has a c.s.

The proofs are based on (75) and utilize the Brouwer Theorem in a manner similar to the use in (68).

Regarding Karamardian's <u>uniqueness</u> theorem, the authors observe that if the condition: 'f strongly K-monotone on K' is replaced by the stronger: 'f strongly monotone on K' (i.e.: $(z_1 - z_2)'(f(z_1) - f(z_2)) \geq k|z_1 - z_2|^2$; for all z_1, z_2 $(\geq K)\; 0$), then there is a <u>unique</u> c.s. to the K-(NLCP).

The authors suggest an iterative scheme for some cases, which we describe for a special K-(LCP):

$$Mz + q \;(\geq K^*)\; 0; \quad z \;(\geq K)\; 0; \quad z'(Mz + q) = 0. \qquad (76)$$

(In the linear case M is strictly K-copositive: for all $0 \neq z \; (\geq K) \; 0$: $z'Mz > 0$; true if and only if M is strongly copositive on K.)

The iterative scheme is valid for M <u>symmetric</u> and strictly K-copositive, and $q \; (\not\geq K^*) \; 0$. Define:

$$W = \{z \; (\geq K) \; 0 : z'(Mz + q) = 0\} . \tag{77}$$

W is shown to be homeomorphic to a compact convex set. Fix on scalar k satisfying: $0 < k < 2/\rho$; where ρ is the spectral radius of M. Set $\bar{z} = z + ku$; where u is as in (75). Then $z \; (\geq K) \; 0$ implies $\bar{z} \; (\geq K) \; 0$.

Defining the map:

$$\phi(z) = (-\bar{z}'q/\bar{z}'M\bar{z})\bar{z}, \tag{78}$$

it is verified that $\phi(W) \subset W$ (and hence, since ϕ is continuous on W, the fixed-point furnished by Brouwer's Theorem yields a c.s. for K-(LCP).) If one defines:

$$F(z) = \tfrac{1}{2} z'Mz + z'q, \text{ and } z_{n+1} = F(z_n) , \tag{79}$$

it is shown that:

$$F(z_{n+1}) \leq F(z_n); \text{ equality if and only if } u_n = 0. \tag{80}$$

The authors develop conditions which ensure that K-(LCP) has a <u>finite</u> number of c.s.'s.

They show that if there are only a finite number of c.s.'s, the sequence $F(z_n)$ converges; hence to a c.s.

APPENDIX

1. Applying Ville's Theorem to A^* yields the refinement:

$$A \in S \quad \text{or} \quad A \in S^* \quad \text{or} \quad A \in V; \tag{A.1}$$

where $V = S_0 \cap S_0^*$, and the three classes are pairwise disjoint.

For a further refinement involving $-A$ and A', there are the nine non-empty, exhaustive, pairwise disjoint classes:

$$\begin{array}{ccc} S \cap -S & S \cap S' & S \cap -V \\ S^* \cap -S & S^* \cap S' & S^* \cap -V \\ V \cap -S & V \cap S' & V \cap -V \end{array} \tag{A.2}$$

For the class of symmetric matrices the only possibilities are S, $-S$, and V.

The class of anti-symmetric matrices: $M = M^*$ has the distinction that for (LCP):

$$w = Mz; \quad w, z \geq 0 \quad \text{and} \quad w'z = 0, \tag{A.3}$$

any f.s. is a c.s.

Regarding the 'component characterization' of E_0, the negation of (10) <u>a</u>. asserts that there exists an a such that $M_{aa} \in -S$, by Ville's Theorem. Thus there is $x_a > 0$ such that $M_{aa}x_a < 0$; or let $x = \langle x_a, 0 \rangle$. Thus

$$\exists 0 \neq x \geq 0 \colon \forall i, \ x_i > 0 \ \text{implies} \ (Mx)_i < 0,$$

the negation of which is the statement (12) E_0.

As an illustration of the results in I.2, we prove Ville's Theorem. Consider the z_0-problem:

$$\begin{pmatrix} u \\ v \end{pmatrix} = \begin{pmatrix} -e \\ 0 \end{pmatrix} + z_0 \begin{pmatrix} e \\ e \end{pmatrix} + \begin{pmatrix} 0, & A \\ A^*, & 0 \end{pmatrix} \begin{pmatrix} y \\ x \end{pmatrix}$$

(so that $M = M^*$) (A.4)

Thus $M \in C^+$. For a f.s. with $z_0 = 0$, one has $Ax \geq e$, so that $A \in S$. Otherwise, Scheme 1 uncovers a secondary ray: $\langle y^1 + \theta \bar{y}, x^1 + \theta \bar{x} \rangle \neq 0$. Using (A.4) this implies:

$$z_0^1(e'(x^1 + \theta \bar{x}) + e'(y^1 + \theta \bar{y})) = e'(y^1 + \theta \bar{y});$$

$$z_0^1 > 0,$$ (A.5)

for all $\theta \geq 0$. Equating coefficients of θ:

$$z_0^1(e'\bar{x} + e'\bar{y}) = e'\bar{y}.$$ (A.6)

Now $\bar{y} \neq 0$, since this would imply by (A.6) that $\bar{x} = 0$, implying no ray. Hence $\bar{y} \geq 0$. Hence, referring to (23) b, with $\bar{z}_0 = 0$, we have $0 \neq \bar{y} \geq 0$, and $A^*\bar{y} \geq 0$; i.e., $A \in S_0^*$.

2. Murty has refined the result (25) (see [21]) as follows:
    (i) if there is a unique c.s. whenever q is a column of $(-I, M, e)$, then $M \in Q$.
    (ii) if $M \in Q$, and there is a unique c.s. whenever q is a column of $(I, -M)$, then $M \in P$.
From these he deduces the characterization of P:
    (iii) $M \in P$ if and only if there is a unique c.s. whenever q is one of the $4n+1$ columns of $(I, -M, -I, M, e)$.

3. Consider Scarf's problem, D.2. in Section I. We assume without loss of generality that i. q_1^i = Min$_j$ q_j^i, for each i, and that ii. q_1^1 = Max q_1^i.
Then $q_1^1 \leq 0$ is seen to be the assertion that $x = 0$ yields a c.s. We now assume that $q_1^1 > 0$, and set up the initial z_0 - c.s. as Scarf does. This is seen to be accomplished by performing the following pivots, indicated by pivot pairs: (y_1^1, z_0); (y_1^i, r_i); for $i = 2, 3, \ldots, n$; and (z_i, w_1^i); for $i = 1, 2, \ldots, n$, (in any sequence!).

To examine the possibilities of z_0 - c.r.'s, we write out the constraints as:

$$y^i = q^i - N^i x + r_i e - z_0 e; \quad i = 1, 2, \ldots, n; \tag{A.7}$$

$$v_i + z_i = 1; \text{ and } 1 + v_i = e'w^i.$$

Clearly, from the latter equations, v, z, and y are bounded on the set of f.s.'s. Referring to (48), since initially $x = 0$, we may consider the <u>initial ray</u> characterized by:

$$r_i = \text{Max}_j (-q_j^i) + z_0. \tag{A.8}$$

Now on a secondary z_0 - c.r. (using the 'bar' notation adopted in I.2), it is readily seen that we have:

$$\bar{r}_i = \text{Max}_j (N^i \bar{x})_j + \bar{z}_0; \text{ and } \bar{r}'\bar{x} = 0; \tag{A.9}$$

which conditions yield:

$$0 = \bar{r}'\bar{x} = \sum_{i=1}^{n} \bar{x}_i \text{ Max}_j (N^i \bar{x})_j + \bar{z}_0 e'\bar{x}; \tag{A.10}$$

Then Scarf's assumption implies that $\bar{x} = 0$. Then if $\bar{z}_0 = 0$ one would have $\bar{r} = 0$, indicating no ray, whereas $\bar{z}_0 > 0$ implies that $\bar{r} = \bar{z}_0 e > 0$, and $\bar{r}'x = 0$ would imply that $x = 0$ on the ray, so that one has the initial ray, which is impossible.

REFERENCES

1. Chandrasekaran, R., "A Special Case of the Complementary Pivot Problem", to appear in Opsearch, 1970.

2. Cottle, R. W., "Nonlinear Programs with Positively Bounded Jacobians", SIAM App. Math. Vol. 14, No. 1, Jan. 1966.

3. Cottle, R. W. and G. B. Dantzig, "Complementary Pivot Theory of Math. Prog", Linear Alg. and its Appns Appl., Vol. 1, pp. 103-125, 1968.

4. Cottle, R. W. and G. B. Dantzig, "Positive (semi-) Definite Prog", Nonlinear Programming, J. Abadie, (Ed), North-Holland, Amsterdam, pp. 55-73, 1967.

5. Cottle, R. W. and G. B. Dantzig, "A Generalization of the Linear Complementarity Problem", Jl. of Combinatorial Theory", 1969.

6. Cottle, R. W., G. J. Habetler and C. E. Lemke, "Quadratic Forms Semi-definite over Convex Cones," Proc. Intl. Symposium on Math. Prog., Princeton, 1967, to appear.

7. Cottle, R. W., G. J. Habetler and C. E. Lemke, "On Classes of Co-positive Matrices", Lin. Alg. and its Appl., to appear.

8. Dantzig, G. B., "Linear Programming and Extensions", Princeton Press, 1963.

9. Eaves, B. C., "The Linear Complementarity Problem", Working paper 275, Center for Research in Management Science, Univ. of Calif., Berkeley, Calif., Aug. 1969.

10. Fiedler, M., and Ptak, V., "Some Generalizations of Positive Definite and Monotonicity", Numer. Mathematik, 9, pp. 163-172, 1966.

11. Fiedler, M., and Ptak, V., "On Matrices with non-positive Off-diagonal Elements and Pos. Principal Minors", Czech Math. Jl, 12, pp. 382-400, 1962.

12. Habetler, G. J., and A. L. Price, "Existence Theory for Generalized Nonlinear Complementarity Problems", Jl Optimization Theory and Appl., to appear.

13. Habetler, G. J., and A. L. Prince, "An Iterative Scheme for Complementarity Problems", JOTA, to appear.

14. Howson, J. T., Jr., "Equilibria of Polymatrix Games", Management Science, Science Series, to appear.

15. Karamardian, S. "The Non-linear Complementarity Prob. with Applications", JOTA, Vol. 4, No. 2, 1969.

16. Karamardian, S., "The Non-linear Complementarity Prob. with Applns: Part 2", JOTA, Vol. 4, No. 3, 1969.

17. Kuhn, H. W., "Simplicial Approx. of Fixed Points", Proc. Natl. Acad. of Sci., No. 61, 1968.

18. Lemke, C. E., "On Complementary Pivot Theory", Math. of the Decision Sciences; Amer. Math. Soc., G. B. Dantzig and A. F. Veinott, Jr., (Eds), 1968.

19. Mangasarian, O. L., "Nonlinear Programming", McGraw-Hill, 1969.

20. McCammon, S. R., "Complementary Pivoting", Ph.D. Thesis, Rensselaer Polytechnic Inst., 1970.

21. Murty, K. G., "On a Characterization of P-Matrices," Tech. Rep. No. 69-20, Dept. Ind. Eng., Univ. of Mich., May, 1969.

22. Scarf, H., "An Algorithm for a Class of Non-convex Prog. Problems", Cowles Commission Discussion Paper No. 211, Yale Univ., July, 1966.

23. Scarf, H., and T. Hansen, "On the Applications of a recent Combinatorial Algorithm," Cowles Commission Discussion Paper No. 272, April 1969.

24. Wilson, R., "Computing Equilibria of N-person Games", working Paper No. 163, Grad. Sch. of Bus., Stanford Univ., 1969.

This research was partially supported by the National Science Research Grant NSF-GP-15031.

Nonlinear Nondifferentiable Programming in Complex Space

BERTRAM MOND

ABSTRACT

The following transposition type theorem in complex space is established:

Theorem. Let $M \in R$, $A \in C^{m \times n}$, $p \in C^n$, $b \in C^m$, $B \in C^{n \times n}$ be positive semi-definite Hermitian, $S \subset C^m$ be a polyhedral convex cone. Assume that the set $K = \{x \mid Ax - b \in S\}$ is non-empty. Then the following are equivalent.

(a) $Ax - b \in S$ implies $f(x) \equiv \text{Re}[(x^*Bx)^{\frac{1}{2}} + p^*x] \geq M$

(b) $A^*y = Bz + p$, $Az - \lambda b \in S$, $y \in S^*$, $\text{Re } y^*b \geq M$, $\lambda \in R_+$, $z^*Bz \leq 1$ has a solution.

This result is then used to establish duality theorems between the following nonlinear nondifferentiable programs in complex space.

PRIMAL   Minimize $f(x) \equiv \text{Re}[(x^*Bx)^{\frac{1}{2}} + p^*x]$

Subject to
$$Ax - b \in S$$
$$x \in T$$

DUAL   Maximize $g(y) = \text{Re } b^*y$

Subject to
$$-A^*y + Bz + p \in T^*$$
$$z^*Bz \leq 1$$
$$y \in S^*$$

where $A \in C^{m \times n}$, $p \in C^n$, $b \in C^m$, $B \in C^{n \times n}$, B positive semi-definite Hermitian, $S \subset C^m$ and $T \subset C^n$ are polyhedral convex cones.

Preliminaries

Mathematical Programming in Complex Space was initiated by Levinson in [12] where Duality Theorems for Complex Linear Programs are given and the basic theorems of Linear Inequalities are extended to complex space. These results were extended to Polyhedral Cones in Complex Space by Ben-Israel [4] and Abrams and Ben-Israel [3].

Nonlinear Mathematical Programming in Complex Space is treated by Bhatia and Kaul [5], Hanson and Mond [10], Abrams [1], Abrams and Ben-Israel [2]. A partial survey of Mathematical Programming in Complex Space can be found in [2] where additional references are given. Three engineering applications of Mathematical Programming in Complex Space are listed in [1] and [2].

Introduction

In [7], Eisenberg proves the following: Let $A \varepsilon R^{m \times n}$, $b \varepsilon R^m$, $p \varepsilon R^n$, $M \varepsilon R$, $B \varepsilon R^{n \times n}$ be positive semi-definite symmetric. Assume that the set $K = \{x \mid Ax - b \geq 0\}$ is non-empty. Then the system of equations and inequalities

$$A^t y = p + Bz, \quad y^t b \geq M, \quad Az \geq \lambda b, \quad \lambda \geq 0, \quad y \geq 0, \quad z^t Bz \leq 1 \quad (1)$$

is consistent, if and only if

$$Ax \geq b \quad \text{implies} \quad p^t x + (x^t Bx)^{\frac{1}{2}} \geq M . \qquad (2)$$

Setting B = 0 yields

$$A^t y = p, \quad y^t b \geq M, \quad y \geq 0 \tag{3}$$

is consistent, if and only if

$$Ax \geq b \text{ implies } p^t x \geq M. \tag{4}$$

(The requirement that $Az \geq \lambda b$, $\lambda \geq 0$, may be dropped from (3) since it is always satisfied by $z = 0$, $\lambda = 0$. In fact, we will show that it can also be eliminated from (1), i.e., that Eisenberg's result holds with or without the restriction $Az \geq \lambda b$, $\lambda \geq 0$) The last result is the non-homogeneous form [6] of the well-known Farkas Lemma [9] that is useful in establishing the duality theorems of mathematical programming (see e.g. [16], p. 292).

Here we extend Eisenberg's result to convex polyhedral cones in complex space, thus obtaining generalizations of results of Farkas [9], Kaul [11], Eisenberg [8], Mond [14], Levinson [12], Abrams and Ben-Israel [3] in addition to [6] and [7].

These extensions are then applied to obtain duality theorems for a pair of nonlinear programming problems in complex space, generalizing results of Sinha [15], Bhatia and Kaul [5], Levinson [12] and Ben-Israel [4].

Notation and Definitions

F   a field, here either

        R - the real field or

        C - the complex field

F^n the n-dimensional vector space over F

$F^{m \times n}$ the m × n matrices over F.

$R_+^n = \{x \in R^n, x_i \geq 0 \ (i = 1, \ldots, n)\}$ the non-negative orthant of R^n. For any $x, y \in R^n$, $x \geq y$ denotes $x - y \in R_+^n$. $S \subset C^n$ is a polyhedral convex cone if there is a positive integer K and a matrix $A \in C^{n \times k}$ such that

$$S = AR_+^K = \{Ax : x \geq 0\}.$$

(An alternate, and equivalent definition, S is a polyhedral convex cone if it is the nonempty intersection of finitely many closed half spaces, each having 0 in its boundary.) The polar of $S \subset C^n$ is defined by

$$S^* = \{y \in C^n, x \in S \text{ implies } Re(x^*y) \geq 0\}.$$

For a listing of some of the properties and examples of polyhedral convex cones and their polars that will be pertinent here, see [4]. Superscript t denotes transpose, superscript *, when applied to matrices and vectors, will denote conjugate transpose.

0 will denote matrices or vectors of appropriate dimension with 0 in every position. The meaning and dimension will be clear from the context.

Extension to Complex Space

We list a number of previously established results that will be needed in proving our theorems.

Lemma 1 [11], [14]. Let $B \in C^{n \times n}$ be positive semi-definite Hermitian. Then

$$Re \ x^* Bz \leq (x^* Bx)^{\frac{1}{2}} (z^* Bz)^{\frac{1}{2}}.$$

We note that equality holds if for $\lambda \geq 0$, $Bx = \lambda Bz$.

Lemma 2 [5]. Let $B \in C^{n \times n}$ be positive semi-definite Hermitian. Then

$$[(x_0 + x)^* B(x_0 + x)]^{\frac{1}{2}} \leq (x_0^* B x_0)^{\frac{1}{2}} + (x^* B x)^{\frac{1}{2}}.$$

Theorem 1 [14]. Let $A \in C^{m \times n}$, $b \in C^n$, $B \in C^{n \times n}$ be positive semi-definite Hermitian, $S \subset C^m$ be a polyhedral convex cone. Then the following are equivalent:

(a) $A^* y = Bz + b$, $y \in S^*$, $z^* Bz \leq 1$, $Az \in S$

has a solution.

(b) $Ax \in S$ implies $\text{Re}[(x^* Bx)^{\frac{1}{2}} + b^* x] \geq 0$.

Theorem 2. Let $M \in R$, $A \in C^{m \times n}$, $p \in C^n$, $b \in C^m$, $B \in C^{n \times n}$ be positive semi-definite Hermitian, $S \subset C^m$ be a polyhedral convex cone. Assume that the set $K = \{x \mid Ax - b \in S\}$ is non-empty. Then the following are equivalent.

(a) $Ax - b \in S$ implies $f(x) = \text{Re}[(x^* Bx)^{\frac{1}{2}} + p^* x] \geq M$

(b) $A^* y = Bz + p$, $Az - \lambda b \in S$, $y \in S^*$, $\text{Re } y^* b \geq M$,

$$\lambda \in R_+, \quad z^* Bz \leq 1$$

has a solution.

Proof: (b) implies (a):

$$x^* A^* y = x^* Bz + x^* p.$$

Also

$$\text{Re}[y^* Ax - y^* b] \geq 0.$$

But,

$$\operatorname{Re} x^*A^*y = \operatorname{Re} y^*Ax \quad \text{and} \quad \operatorname{Re} x^*p = \operatorname{Re} p^*x.$$

Hence, utilizing Lemma 1, we have

$$\operatorname{Re}[(x^*Bx)^{\frac{1}{2}} + p^*x] \geq \operatorname{Re}[(x^*Bx)^{\frac{1}{2}}(z^*Bz)^{\frac{1}{2}} + p^*x]$$

$$\geq \operatorname{Re}[x^*Bz + x^*p] = \operatorname{Re} x^*A^*y = \operatorname{Re} y^*Ax \geq \operatorname{Re} y^*b \geq M.$$

(We note for later use that the condition $Az - \lambda b \,\varepsilon\, S$, $\lambda \,\varepsilon\, R_+$ in (b) was not used in the above part of the proof.)

(a) implies (b):

Assume (a) holds. We assert that the following is true.

$$Ax - \mu b \,\varepsilon\, S, \quad \mu \,\varepsilon\, R_+ \quad \text{implies} \quad \operatorname{Re}[(x^*Bx)^{\frac{1}{2}} + p^*x] \geq \mu M. \quad (5)$$

If $\mu > 0$, $A(x/\mu) - b \,\varepsilon\, S$ and by (a) which was assumed true, this implies $f(x/\mu) = (1/\mu)f(x) \geq M$.

If $\mu = 0$, assume (5) does not hold, i.e., for some $x \,\varepsilon\, C^n$, $Ax \,\varepsilon\, S$, $f(x) \equiv \operatorname{Re}[(x^*Bx)^{\frac{1}{2}} + p^*x] < 0$. Recalling that $K \neq \phi$, let x_0 satisfy $Ax_0 - b \,\varepsilon\, S$. Now, for any $\alpha \geq 0$, $x_0 + \alpha x$ satisfies

$$A(x_0 + \alpha x) - b \,\varepsilon\, S \quad \text{and, utilizing Lemma 2, we have}$$

$$f(x_0 + \alpha x) = \operatorname{Re}[(x_0 + \alpha x)^*B(x_0 + \alpha x)^{\frac{1}{2}} + p^*(x_0 + \alpha x)] \leq$$

$$\operatorname{Re}[(x_0^*Bx_0)^{\frac{1}{2}} + \alpha(x^*Bx)^{\frac{1}{2}} + p^*x_0 + \alpha p^*x] = f(x_0) + \alpha f(x).$$

By choosing α sufficiently large, we can assure that $f(x_0 + \alpha x) < M$ contradicting the assumption (a).
Thus (a) implies (5). Rewrite (5) as follows:

$$\left.\begin{array}{l} \begin{bmatrix} A - b \\ 0 & I \end{bmatrix} \begin{bmatrix} x \\ \mu \end{bmatrix} \varepsilon\, S \times R_+ \quad \text{implies} \\ \\ \operatorname{Re}\left\{ \left[\begin{bmatrix} x \\ \mu \end{bmatrix}^* \begin{bmatrix} B & 0 \\ 0 & 0 \end{bmatrix} \begin{bmatrix} x \\ \mu \end{bmatrix}\right]^{\frac{1}{2}} + \begin{bmatrix} p \\ -M \end{bmatrix}^* \begin{bmatrix} x \\ \mu \end{bmatrix} \right\} \geq 0 \end{array}\right\} \quad (6)$$

By Theorem 1, (6) is equivalent to the existence of a solution to the system

$$A^* y = Bz + p, \quad y \varepsilon\, S^*, \quad -b^* y + y_0 = -M,$$

$$z^* Bz \leq 1, \quad Az - \lambda b \varepsilon\, S,$$

$$\lambda \varepsilon\, R_+, \quad y_0 \varepsilon\, R_+^* .$$

$y_0 \varepsilon\, R_+^*$ is the same as $\operatorname{Re} y_0 \geq 0$. Thus

$$\operatorname{Re} b^* y = \operatorname{Re} y^* b \geq M .$$

This proves the theorem.

Theorem 3. Let M, A, p, b, B, S and K be as in Theorem 2. Assume that $K \neq \phi$. Then the following are equivalent:

(a) $Ax - b \varepsilon\, S$ implies $f(x) = \operatorname{Re}[(x^* Bx)^{\frac{1}{2}} + p^* x] \geq M$

(b) $A^*y = Bz + p$, $y \in S^*$, $\operatorname{Re} y^*b \geq M$, $z^*Bz \leq 1$,

has a solution.

Proof: (b) implies (a):

The proof follows exactly as in Theorem 2.

(a) implies (b):

(a) implies the existence of a solution in Theorem 2 (b); such a solution satisfies Theorem 3 (b).

Complex Nonlinear Programming

Consider the following two programs in complex space.

PRIMAL

$$\text{Minimize } f(x) \equiv \operatorname{Re}[(x^*Bx)^{\frac{1}{2}} + p^*x]$$

Subject to

$$Ax - b \in S \qquad (7)$$

$$x \in T \qquad (8)$$

DUAL

$$\text{Maximize } g(y) = \operatorname{Re} b^*y$$

Subject to

$$-A^*y + Bz + p \in T^* \qquad (9)$$

$$z^*Bz \leq 1 \qquad (10)$$

$$y \in S^* \qquad (11)$$

where $A \in C^{m \times n}$, $p \in C^n$, $b \in C^m$, $B \in C^{n \times n}$, B positive semi-definite Hermitian, $S \subset C^m$ and $T \subset C^n$ are polyhedral convex cones.

Theorem 4. **The infimum of the primal is greater than or equal to the supremum of the dual.**

Proof: Assume x satisfies (7) and (8); (y, z) satisfies (9), (10) and (11).

From (11) and (7),

$$\text{Re}[y^*Ax - y^*b] = \text{Re}[x^*A^*y - b^*y] \geq 0$$

or

$$\text{Re } x^*A^*y \geq \text{Re } b^*y . \tag{12}$$

From (8) and (9),

$$\text{Re}[-x^*A^*y + x^*Bz + p^*x] \geq 0$$

or

$$\text{Re}[x^*Bz + p^*x] \geq \text{Re } x^*A^*y . \tag{13}$$

From Lemma 1, (11), (12) and (13), we have

$$f(x) = \text{Re}[(x^*Bx)^{\frac{1}{2}} + p^*x] \geq \text{Re}[(x^*Bx)^{\frac{1}{2}}(z^*Bz)^{\frac{1}{2}} + p^*x] \geq$$

$$\text{Re}[(x^*Bz + p^*x] \geq \text{Re } x^*A^*y \geq \text{Re } b^*y = g(y) .$$

Theorem 5. **If there exists an optimal solution x_0 of the primal, then there exists an optimal solution (y, z) of the dual and $f(x_0) = g(y)$.**

Proof: Let $f(x_0) = M$. Since x_0 is optimal, we have

$$\begin{bmatrix} Ax - b \\ Ix - 0 \end{bmatrix} \varepsilon\, S \times T \text{ implies } f(x) = \text{Re}[(x^*Bx)^{\frac{1}{2}} + p^*x] \geq M.$$

By Theorem 2, a solution, therefore, exists to the system

$$A^*y + y_0 = Bz + p, \quad Az - \lambda b \,\varepsilon\, S, \quad y \,\varepsilon\, S^*, \quad y_0 \,\varepsilon\, T^*$$

$$\lambda \,\varepsilon\, R_+, \quad z^*Bz \leq 1, \quad \text{Re } y^*b \geq M.$$

Eliminating y_0 gives

$$-A^*y + Bz + p \,\varepsilon\, T^*.$$

Thus, there exists (y, z) that is a feasible solution of the dual and $\text{Re } y^*b \geq M$. By Theorem 4, for any feasible (y, z), $\text{Re } y^*b \leq M$. Thus (y, z) is optimal with

$$g(y) = \text{Re } y^*b = M = f(x_0).$$

We assume now that the constraints (9), (10) and (11) satisfy a (Kuhn-Tucker) constraint qualification [1], [2].

Theorem 6. *If there exists an optimal solution (y_0, z_0) of the dual, then there exists an optimal solution x of the primal and a real number $\lambda \geq 0$ such that $Bx = \lambda B z_0$ and $f(x) = g(y_0)$.*

Proof: Since (y_0, z_0) is optimal, by the corresponding Kuhn-Tucker conditions in complex space [1], [2], there exists (x, λ) satisfying

$$Ax - b \in S$$

$$x \in T$$

$$\lambda(z_0^* B z_0 - 1) = 0 \qquad (14)$$

$$\lambda \geq 0$$

$$Bx = \lambda B z_0 \qquad (15)$$

$$\operatorname{Re}[p^* x + x^* B z_0 + \lambda(z_0^* B z_0 - 1)] = \operatorname{Re} b^* y_0 . \qquad (16)$$

x is obviously a feasible solution of the primal. We now show that it is also optimal. (14) and (16) give

$$\operatorname{Re}[p^* x + x^* B z_0] = \operatorname{Re} b^* y_0 . \qquad (17)$$

Since (15) is the condition for equality in Lemma 1

$$\operatorname{Re} x^* B z_0 = (x^* B x)^{\frac{1}{2}} (z_0^* B z_0)^{\frac{1}{2}}$$

From (14), either $\lambda = 0$ or $z_0^* B z_0 = 1$. In either case, we have

$$\operatorname{Re} x^* B z_0 = (x^* B x)^{\frac{1}{2}} .$$

Thus,

$$f(x) = \operatorname{Re}[p^* x + (x^* B x)^{\frac{1}{2}}] = \operatorname{Re}[p^* x + x^* B z_0] =$$

$$= \operatorname{Re} b^* y_0 = g(y_0)$$

and, by virtue of Theorem 4, x is optimal.

Special Cases

Our Theorems yield, as special cases, a number of well-known results.

If $B = 0$, Theorem 3 gives the extension to complex space of the non-homogeneous form of Farkas Lemma given by Abrams and Ben-Israel [3].

If we choose the cone S as

$$S = \{x \in C^n : |\arg x| \le \alpha\} \text{ for the given } \alpha \in R^n_+, \alpha \le \pi/2, \tag{18}$$

the polar S^* is easily [4] seen to be

$$S^* = \{z \in C^n : |\arg z| \le \pi/2 - \alpha\}.$$

(Here, and subsequently, $\pi/2$ means the vector of appropriate dimension with $\pi/2$ in each position.) Theorem 2 then yields, with the set

$$K = \{x : |\arg(Ax - b)| \le \alpha\} \tag{19}$$

assumed non-empty, the equivalence of the following:

(a) $A^*y = Bz + p$, $|\arg y| \le \pi/2 - \alpha$, $z^*Bz \le 1$,

$\qquad\qquad |\arg Az - \lambda b| \le \alpha$ $\qquad\qquad$ (20)

$\qquad\qquad |\arg \lambda| \le 0$, $\operatorname{Re} b^*y \ge M$

has a solution.

(b) $|\arg(Ax - b)| \le \alpha$ implies $\operatorname{Re}[(x^*Bx)^{\frac{1}{2}} + p^*x] \ge M$. (21)

This is the non-homogeneous version of the theorem of Kaul [11]. Theorem 3 gives this result with the condition

$$|\arg(Az - \lambda b)| \leq \alpha, \quad |\arg \lambda| \leq 0$$

eliminated from (20).

If also $B = 0$, Theorem 3 gives, with the assumption that K defined by (19) is non-empty, the equivalence of the following:

(a) $\qquad A^* y = p, \quad |\arg y| \leq \pi/2 - \alpha, \quad \text{Re } b^* y \geq M$

has a solution

(b) $\qquad |\arg(Ax - b)| \leq \alpha \text{ implies Re } b^* y \geq M.$

This result is the extension to complex space of the non-homogeneous form of Farkas Lemma given by Levinson [12].

Let all vectors and matrices in Theorems 2 and 3 be real. Take $S = S^* = R_+^m$. Theorem 2 then gives Eisenberg's result [7] ((1) and (2)). If also $B = 0$, Theorem 3 yields the non-homogeneous form of the Farkas Lemma [6] ((3) and (4)).

Finally, we note that if $b = 0$ in Theorems 2 and 3, (and, of course, in all of the special cases) we obtain the corresponding homogeneous result. The assumption that the set K be nonempty is now always satisfied by $x = 0$.

Thus, with $b = 0$, Theorems 2 and 3 give the results of Mond [14]. If also $B = 0$, Theorem 3 gives the extension of Farkas Lemma [9] to complex space of Ben-Israel [4]. With $b = 0$, (20) and (21) yield Kaul's [11] result. If also $B = 0$, we obtain the complex extension of Farkas Lemma of Levinson [12]. If $b = 0$, all vectors and matrices are real, and $S = S^* = R_+^m$, Theorem 2 yields the homogeneous result in real space of Eisenberg [8]. If also $B = 0$, Theorem 3 yields the Farkas Lemma [9].

Let the cone S be chosen as in (18) and the cone T as

$$T = \{z \varepsilon C^n : |\arg z| \leq \beta\} \text{ for the given } \beta \varepsilon R_+^n, \beta \varepsilon \pi/2 \tag{22}$$

the polar T^* is easily [4] seen to be

$$T^* = \{w \varepsilon C^n : |\arg w| \leq \pi/2 - \beta\}.$$

Our primal and dual problems then yield the dual non-linear programs in complex space of Bhatia and Kaul [5]. If all matrices and vectors are real and $S = S^* = R_+^m$, $T = T^* = R_+^n$, we obtain the nonlinear dual programs in real space of Sinha [15].

If, in our primal and dual problems, $B = 0$, we obtain the dual linear programs in complex space of Ben-Israel [4] and Abrams and Ben-Israel [2]. If, also, S and T are chosen as in (18) and (22), we obtain the dual linear programs in complex space of Levinson [12]. If $B = 0$, all matrices and vectors are real, $S = S^* = R_+^m$, $T = T^* = R_+^n$, we obtain a pair of dual linear programs in real space.

Remarks

Note that the duality theorems for non-linear complex programming of [1], [2] and [10] are not applicable to our primal problem since the requisite differentiability conditions are not satisfied. However, the results of [1] and [2] are applicable to the dual problem since the factor $(z^*Bz)^{\frac{1}{2}}$ does not appear.

Our converse duality theorem (Theorem 6) is thus obtained in simple fashion without the corresponding restrictions of [5] and [15] and includes the relationship between the optimal solutions of the primal and the dual (15). Observe that if B has an inverse, i.e., B is positive definite, (15), and the equation preceeding it, yield

$$x = \lambda z_0, \lambda \geq 0.$$

REFERENCES

1. Abrams, R. A., Nonlinear Programming in Complex Space, Ph.D. dissertation, Northwestern University, Evanston, Illinois, August, 1969.

2. Abrams, R. A. and Ben-Israel, A., Complex Mathematical Programming, Report No. 69-11, Northwestern University, Evanston, Illinois, 60201, U.S.A., November, 1969.

3. Abrams, R. A. and Ben-Israel, A., On the Key Theorems of Tucker and Levinson for Complex Linear Inequalities, J. Math. Anal. Appl., Forthcoming.

4. Ben-Israel, A., Linear Equations and Inequalities on Finite Dimensional Real or Complex Vector Spaces: A unified theory, J. Math. Appl. 27, 367-368.

5. Bhatia, D. and Kaul, R. N., Nonlinear Programming in Complex Space, J. Math. Anal. Appl. 28, 144-152 (1969).

6. Duffin, R. J., Infinite Programs, in Linear Inequalities and Related Systems, Annals of Math. Studies No. 38, Princeton University Press, Princeton, N. J. (1956).

7. Eisenberg, E., A Gradient Inequality for a class of Nondifferentiable Functions, Oper. Res. 14, 157-163 (1966).

8. Eisenberg, E., Supports of a Convex Function, Bull. Amer. Math. Soc. 68, 192-195 (1962).

9. Farkas, J., Über die Theorie der Einfachen Ungleichungen, J. Reine Angew. Math. 124, 1-17 (1902).

10. Hanson, M. A. and Mond, B., Duality for Nonlinear Programming in Complex Space, J. Math. Anal. Appl. 28, 52-58 (1969).

11. Kaul, R. N., On an Extension of Farkas' Theorem, to appear in American Mathematical Monthly.

12. Levinson, N., Linear Programming in Complex Space, J. Math. Anal. Appl. 14, 44-62 (1966).

13. Mehndiratta, S. L., A Generalization of a Theorem of Sinha on Supports of a Convex Function, Aust. J. of Stat. 11, 1-6 (1969).

14. Mond, B., An Extension of the Transposition Theorems of Farkas and Eisenberg, J. Math. Anal. Appl. To appear.

15. Sinha, S. M., A Duality Theorem for Nonlinear Programming, Man. Sci. 12, 385-390 (1966).

16. Thrall, R. M. and Tornheim, L., Vector Spaces and Matrices, John Wiley Inc., N.Y. (1957).

Duality Inequalities of Mathematics and Science

R. J. DUFFIN

ABSTRACT

The problem of minimizing a scalar functional $u(x)$ under a set of constraints S in the vector variable x is termed a <u>program</u>. It often results that there is an associated program of maximizing a scalar functional $v(y)$ under a set of constraints T on the vector y. These programs are termed <u>dual</u> if it can be shown that the functional $u(x)$ exceeds the functional $v(y)$. Then there exists a constant M such that

$$u(x) \geq M \geq v(y)$$
$$x \varepsilon S \qquad y \varepsilon T.$$

The virtue of this duality inequality is that it permits estimating M with a known bound on the error. Inequalities of this form appear in various areas of mathematics, science, engineering, and economics. This paper points out several such duality inequalities and their interrelationships.

1. Introduction

In many areas of mathematics and science problems of the following nature occur. There is given a real valued functional u(x) of a vector variable x. The problem is to minimize u(x) subject to a certain set of constraints S on the vector x. Such a problem is here termed a *program*. For example, u(x) could be the cost of an engineering design as a function of the design parameter $x_1, \ldots, x_n$.

It often happens that with a given program there is an associated program of maximizing a function v(y) under a set of constraints T on the vector y. Such programs are here termed *dual* if it can be shown that u(x) ≥ v(y) for any x ϵ S and any y ϵ T. Since the variables x and y are unrelated there exists a constant M termed a *program value* such that

$$u(x) \geq M \geq v(y)$$

x ϵ S   y ϵ T .

This is termed a *duality inequality*.

If the program value M is not uniquely determined by the duality inequality then we say there is a *duality gap*. If there is no duality gap, then the duality inequality permits estimating M with a known bound on the error. In some problems the determination of M is not the main goal but rather the goal is finding optimal x and optimal y. Nevertheless the error in M gives a measure of the lack of optimality of x and y.

This paper describes certain of these duality inequalities which have been studied in the past by me and my colleagues. It is of interest to compare these inequalities and to seek common features because they arise in quite varied contexts.

Further details will be found in the papers cited at the end. Moreover, these papers make reference and interrelation to work of other researchers in duality theory. Such reference is omitted in this paper.

In many of the problems discussed here the program value M has the interpretation of the joint conductance of an electrical network. The present treatment complements the survey paper "Network Models", [25], which also is concerned with joint conductance and network duality.

2. Finite linear programming

Consider the linear functionals

$$u(x) = \sum_{j=1}^{n} b_j x_j \text{ and } v(y) = \sum_{i=1}^{m} c_i y_i .$$

Term $u(x)$ the primal functional and $v(y)$ the dual functional. Then the linear programming inequality is of the form

$$\sum_{1}^{n} b_j x_j \geq M \geq \sum_{1}^{m} c_i y_i .$$

$$y \in S \qquad y \in T$$

Let the primal constraints S be of the form:

$$S. \quad \sum_{1}^{n} a_{ij} x_j \geq c_i \quad i = 1, \ldots, m .$$

Then the dual constraints T are:

T. $\sum_{1}^{m} y_i a_{ij} = b_j \quad j = 1, \ldots, n$.

and

$$y_i \geq 0 \quad i = 1, \ldots, n.$$

It follows from S and T that

$$\sum_{1}^{n} b_j x_j = \sum_{1}^{n} \sum_{1}^{m} y_i a_{ij} x_j = \sum_{1}^{m} y_i \sum_{1}^{n} a_{ij} x_j \geq \sum_{1}^{m} y_i c_i .$$

The interchange of the order of summation is justified because m and n are supposed to be finite positive integers. This proves the duality inequality. It is well known that a duality gap can never occur in finite linear programming.

A linear program is specified by a finite sequence of numbers and this permits treatment by a predesigned computer code. A geometric program is also specified by a finite sequence of numbers and likewise may be treated by a predesigned computer code. In fact a linear program is a special case of a geometric program [26].

The duality inequality of geometric programming is termed the "main lemma". The proof is a direct consequence of the classical inequality stating that

arithmetic mean $\geq$ geometric mean .

The details are given in references [15, 18, 26] and will not be repeated here.

3. A duality gap

To obtain an example of a linear program with a duality gap we impose an infinite number of inequality constraints on the primal program. Thus let the constraints be:

S. $\quad x_1/2 + 0 \geq -1,$

$\quad\quad x_1/2^2 + x_2/4^2 \geq 0,$

$\quad\quad x_1/2^3 + x_2/4^3 \geq 0,$ etc.

T. $\quad y_1/2 + y_2/2^2 + y_3/2^3 + \ldots = 1,$

$\quad\quad 0 + y_2/4^2 + y_3/4^3 + \ldots = 0,$

$\quad\quad y_i \geq 0 \quad\quad i = 1, 2, \ldots.$

Then the duality inequality is simply

$$x_1 \geq M \geq -y_1.$$

It is clear that the primal constraints require that $x_1 \geq -x_2/2^k$. For large k this means that $x_1 \geq 0$. On the other hand, the dual constraints require that $y_2 = 0$, $y_3 = 0$, etc. Then the first constraint gives $y_1 = 2$. Thus the duality inequality has a gap of 2 units. This example was given in a paper with Karlovitz [14]. Other examples were given by Kretschmer [7, 10] and Duffin [17].

4. Infinite programming

The constraints of linear programming are based on a matrix a_{ij} and its adjoint. Such constraints can be put in an abstract setting by replacing the matrix a_{ij} by a transformation A between suitable vector spaces. Then the duality inequality can be expressed as

$$(b, x) \geq M \geq (c, y)$$
$$x \in S \qquad y \in T$$

where (b, x) and (c, y) denote real linear functionals on reflexive vector spaces. The constraints S and T are taken to be:

S. $Ax \geq c$.

T. $A^*y = b$ and $y \geq 0$.

Here A^* is the adjoint transformation of A. The relation $Ax \geq c$ is interpreted to mean that $Ax - c$ is in a certain arbitrary cone K. Likewise $y \geq 0$ is interpreted to mean that y is in the apolar cone $\overline{K^*}$. This means that

$$(k, k^*) \geq 0 \text{ if } k \in K \text{ and } k^* \in K^*.$$

Thus S and T give

$$(b, x) = (A^*y, x) = (Ax, y) \geq (c, y).$$

This proves the duality inequality.

To avoid the duality gap we ascribe locally convex topologies to the various spaces. This insures that the "separation theorem" holds. In other words a closed convex set and a point not in it can be strictly separated by a closed hyperplane. Then the following duality theorem can be proved [6].

<u>A program is consistent and has a finite value if and only if the dual program is subconsistent.</u>

"Subconsistency" means that the constraints can be approximately satisfied with an arbitrary small error in the appropriate topology. For example T is subconsistent if there is a sequence $\{y_n\}$ such that

$$y_n \geq 0, \quad \lim Ay_n = b.$$

Then the duality inequality becomes

$$(b, x) \geq M \geq \lim(c, y_n).$$

It results that the program value M is uniquely defined so there is no duality gap. However this theorem does not have the symmetry belonging to the duality theorem of finite programming.

If the constraint inequalities can all be satisfied strictly then the program is said to be <u>superconsistent.</u> This requires that the cone K have interior points. Of course this can be true only for certain special cones and topologies. If one of the programs is superconsistent it was shown that the duality theorem holds without recourse to the concept of subconsistency [6].

Various ramifications of the above duality theory were developed by Kretschmer [7, 10], Karlovitz [13], and by Karlovitz and Duffin [14]. Peterson and Duffin [15] made extensive use of the concepts of subconsistency and superconsistency in the theory of geometric programming.

The above mentioned theory of infinite programming includes finite linear programming as a special case. However to obtain the full strength of the finite duality theorem it is necessary to adjoin a compactness argument. This flaw was removed in a new analysis developed for infinite programming [17]. The complete duality theorem of finite programming is a special case of this new theory.

5. Application of infinite programming to analysis

Kretschmer and the author [7] developed several applications of infinite programming to classical analysis. This investigation was continued by Karlovitz [13, 20]. The idea of this application can be explained as follows. Many problems of classical analysis are actually stated as infinite programs but in a somewhat disguised form. Having unmasked such a problem we try to formulate a dual program. Then the duality theorem gives a property of this dual problem. Such a procedure can lead to new theorems because the dual problem may not be as well known as the primal problem.

As an example of this procedure consider the following well known inequality of A. Markoff for the derivative of a polynomial [1]. Let a polynomial p(t) of degree n satisfy the inequality $|p(t)| \leq 1$ for $0 \leq t \leq 2$. Then the derivative of p satisfies $|p'(0)| \leq n^2$ and this inequality is best possible. Markoff's theorem may be formulated as a linear program by writing

$$p(t) = \sum_0^n x_j t^j .$$

Then the primal program is: <u>Seek the supremum of x_1 subject to the constraints</u>

S. $\quad \sum_0^n x_j t^j \leq 1 \quad 0 \leq t \leq 2$

$\quad -\sum_0^n x_j t^j \leq 1 \quad 0 \leq t \leq 2 .$

The dual program is the following moment problem. <u>Seek the infimum of</u> $\int_0^2 |y(t)| dt$ <u>where</u> y(t) <u>is a continuous function subject to the constraints</u>

T. $\quad \int_0^2 ty(t)dt = 1$

$\quad \int_0^2 t^j y(t)dt = 0, \quad j = 0, 2, 3, \ldots, n .$

Theorem. <u>The duality inequality is</u>

$$x_1 \leq n^2 \leq \int_0^2 |y(t)| dt .$$

Proof:

$$x_1 = \sum_{j=0}^{n} x_1 \int_0^2 t^j y(t)dt = \int_0^2 p(t)y(t)dt \leq \int_0^2 |y(t)|dt.$$

This is seen to give the proof. The primal program is superconsistent, so there is no duality gap. Hence

$$\inf \int_0^2 |y(t)|dt = n^2$$

is the solution of the moment problem [7].

Another application of infinite programming is to prove the following orthogonality theorem.

Let $f_1, f_2, \ldots, f_n$ <u>be a set of continuous functions in a closed region</u> R. <u>Then there exists a continuous positive function</u> y <u>which is orthogonal to each</u> f_i <u>over the region</u> R <u>if and only if every linear combination of the</u> f_i <u>either changes sign in</u> R <u>or else vanishes identically.</u>
This is a generalization of a theorem of Dines [17].

6. Infinite programming and convex programming

Consider the convex constraint

$$x_1^2 + x_2^2 \leq 9.$$

Clearly this constraint is equivalent to the infinite set of linear constraints

$$x_1 \cos\theta + x_2 \sin\theta \leq 3, \quad 0 \leq \theta < 2\pi.$$

Such use of support lines is readily generalized. It results that any nonlinear convex program can be reformulated as an infinite linear program by use of support planes.

In transforming a problem to infinite programming form it is necessary to choose a topology and to determine the nature of the associated functionals. This requires some study and analysis.

My motivation to study infinite programming arose in this way. I hoped to treat certain convex problems connected with plasticity. Thus, Charnes, Dorn and Greenberg noted that linear programming had application to plasticity. Moreover, they were able to give a complete theory for a finite structure composed of beams subject to plastic failure. A much more difficult problem is the determination of the equilibrium state of a continuum subject to plastic failure. I wished to formulate such a problem as an infinite linear program and to find the dual. I believe this is still an open question.

7. The extremal length and width of a network

The problem of concern now is the joint conductance of an electrical network between two given terminals. The standard way to solve this problem is to write down Kirchhoff's voltage law and Kirchhoff's current law and solve the resulting system of equations. We shall not proceed that way but instead we shall express the joint conductance as the value of a program. The idea of doing this was suggested by the work of Ahlfors on the geometric approach to complex function theory.

Consider a network having n branches such as shown in Figure 1. Let the jth branch have conductance $g_j > 0$. Let M be the joint conductance between terminals A and B. Then the following duality inequality holds [12]

$$\frac{\sum_1^n g_j x_j^2}{x_0^2} \geq M \geq \frac{y_0^2}{\sum_1^n y_j^2/g_j}$$

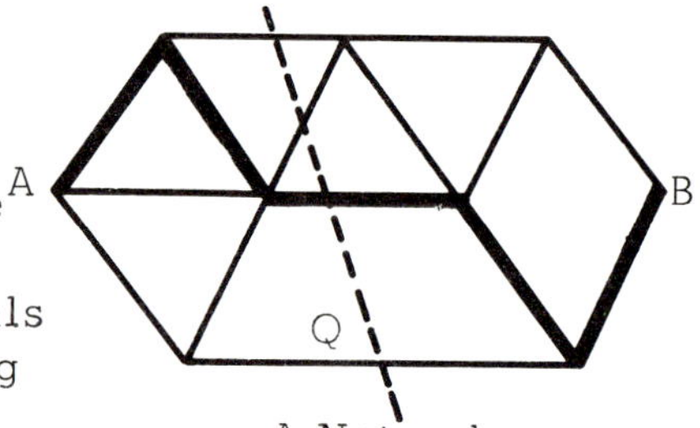

A Network

Figure 1

DUALITY INEQUALITIES

Here x and y satisfy the constraints:

S. $\quad x_j \geq 0,$

$$x_0 \leq \sum_P p^x_y$$

for any path P connecting the terminals A and B.

T. $\quad y_j \geq 0,$

$$y_0 \leq \sum_Q Q y_j$$

for any cut Q separating the terminals A and B. A path P is shown in Figure 1 by the heavy line. A cut Q is indicated in Figure 1 by the branches crossed by the dotted line.

There is no duality gap so

$$M = \max_y \frac{y_0^2}{\sum_1^n y_j^2/g_j}.$$

The expression on the right is termed the <u>extremal width</u>. Likewise

$$M^{-1} = \max_x \frac{x_0^2}{\sum_1^n g_j x_j^2}.$$

The expression on the right is termed the <u>extremal length</u>.
In the duality inequality let $y_j = g_j x_j$. Then it follows that

$$\sum_1^n x_j y_j \geq x_0 y_0 \quad x \in S, \ y \in T.$$

411

This is called the <u>width-length inequality</u>. The g_j are arbitrary so the x_j and y_j are arbitrary. It follows that the width-length inequality is simply a topological property of the network graph.

These inequalities for networks have analogies for continuous bodies. In the two-dimensional case they have been extensively studied by Ahlfors, Beurling and others. These authors are aided by complex function theory. The three-dimensional case is more difficult because complex function theory is not available. However, progress has recently been made by William R. Derrick in his paper, "A weighted volume-diameter inequality for N-cubes". He has thereby answered my conjecture in the affirmative.

8. Optimum heat transfer

Again we are concerned with the joint conductance of a network but now an economic constraint is introduced in addition to the physical constraints of Kirchhoff. The problem is expressed in the form of a program. It results that the solution is obtained by minimizing a norm over a vector space. This norm is of mixed type. That is, part of the norm comes from the economics and part comes from the physics.

Consider a network such as shown in Figure 1 but suppose that the branches are divided into two sets, α and β. In set α the conductances $g_j \geq 0$ are fixed. In set β the conductances are allowed to vary but are subject to the constraint

$$\sum_\beta g_j \leq K, \quad g_j \geq 0 .$$

This constraint corresponds to limiting the total weight of the β branches. The problem is to distribute the material so as to maximize the joint conductance relative to terminals A and B.

Let M denote the maximum joint conductance. Then M satisfies the following duality inequality [21]

$$\|x\|_P^2 \geq M \geq \|y\|_D^{-2}$$
$$x \in S \qquad\qquad y \in T$$

Here $\|x\|_P$ is a norm on the vector $(x_1, x_2, \ldots, x_n)$ defined as

$$\|x\|_P^2 = \sum_\alpha g_j x_j^2 + K \max_\beta x_j^2 \; .$$

On the right $\|y\|_D$ is a norm on the vector $(y_1, y_2, \ldots, y_n)$ defined as

$$\|y\|_D^2 = \sum_\alpha g_j^{-1} y_j^2 + K^{-1}(\sum_\beta |y_j|)^2 \; .$$

The constraints on the vectors x and y are: S. $x_1, \ldots, x_n$ are a possible set of potential differences on the branches corresponding to unit potential difference across the input terminals. T. $y_1, \ldots, y_n$ are a possible set of branch currents corresponding to a unit flow of current at the input terminals. In other words the x_i satisfy the Kirchhoff voltage law and the y_j satisfy the Kirchhoff current law.

It is seen from the above definition that the norm $\|x\|_P$ is a mixture of a Euclidean norm and a Tchebychef norm. Likewise the norm $\|y\|_D$ is a mixture of a Euclidean norm and an L_1 norm. These "mixed norms" would seem somewhat artificial if it were not for the fact that they express the essence of a natural problem of economics.

These inequalities for networks have analogies for continuous bodies. Thus Duffin and McLain [8, 22] have thereby determined the most efficient design for a cooling fin.

9. Upper and lower networks

Next consider the standard problem of computing the conductance of a flat plate. In particular let M be the conductance between the top edge A and the bottom edge B of a comensurable L-shaped plate such as is shown in Figure 2. It shall be assumed that the other edges are insulated and that unit area has unit conductance.

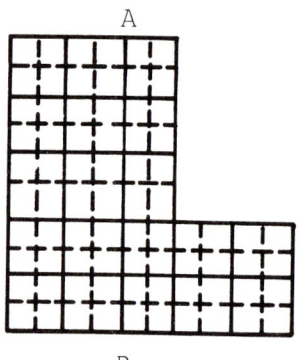

B
An L-shaped plate
Figure 2

This is a Dirichlet problem and the standard method of solving such a problem with the digital computer is to replace Laplace's equation with a difference equation. When this difference equation is examined it is not difficult to see that it is equivalent to Kirchhoff's equations for a wire screen network. Let the L-shaped area be decomposed into small squares as shown by the full lines in Figure 2. These lines compose the screen. Let this be termed the <u>upper network</u> and let M^u be the joint conductance between edges A and B when the internal edges of squares are ascribed unit conductance and when boundary edges are ascribed 1/2 unit conductance.

The positioning of the screen is arbitrary, so it could also be moved to the half-way position indicated by the dotted lines in Figure 2. Let this be termed the <u>lower network</u> and let M^L be the joint conductance between edges A and B when the half edges of the screen are ascribed 2 units of conductance. Then the following duality inequality holds [9]

$$M^u \geq M \geq M^L .$$

Thus the upper network furnishes an upper bound to conductance and the lower network furnishes a lower bound to conductance.

Various reamifications of this idea using other forms of networks have been developed [16, 23]. Indirect arguments indicate that there is no duality gap, but the literature does not contain a direct proof of this.

10. Problems which are not self-adjoint

Many questions in applied mathematics are solved by resorting to variational methods. These questions include qualitative behavior as well as computational procedures. To apply the variational method to a problem it must be of self-adjoint form.

Unfortunately on the surface of the Earth there are two ubiquitous forces which destroy self-adjointness. These are the Coriolis force and the magnetic force. These forces are usually small but rigorously they cannot be neglected. We consider here a common problem in which the Earth's field causes the trouble.

Consider the problem of estimating the electrical conductance M between two surface areas A and B of a conducting body R having constant conductivity σ. Also assume that there is a uniform magnetic field $\vec{H}$ present. Then if f is the electric potential the current density $\vec{j}$ is given by

$$\vec{j} = -\sigma \nabla f + \rho \vec{H} \times \nabla f .$$

Here, the first term on the right is Ohm's law and the second term is Hall's law. But div $\vec{j} = 0$ in R so it follows that $\nabla^2 f = 0$. In other words f is a harmonic function.

A body R is shown in Figure 3. Then it is required to satisfy the boundary conditions: $f = 1$ on A, $f = 0$ on B, and $\vec{j} \cdot \vec{n} = 0$ on C, the remainder of the boundary of R. Here $\vec{n}$ is the interior normal. Then it is easy to show that

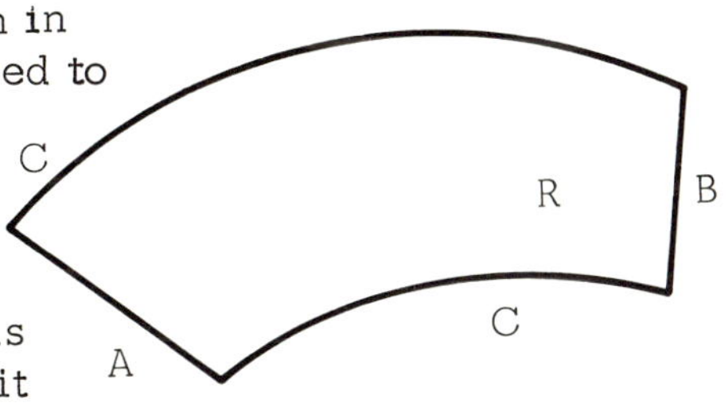

A Conducting Body

Figure 3

$$M = \sigma \iiint_R (\nabla f)^2 dv$$

if f is the equilibrium potential.

415

If there is no magnetic field then Dirichlet's principle gives the following duality inequality [9]

$$\sigma \iiint (\nabla X)^2 dv \geq M \geq \frac{\sigma}{\iiint \vec{Y}^2 dv}.$$

Constraints on X and $\vec{Y}$ are

S. X is a smooth function such that X = 1 on A and X = 0 on B

T. $\vec{Y}$ is a smooth vector field such that

$$\text{div } \vec{Y} = 0, \quad \iint_A \vec{n} \cdot \vec{Y} \, ds = 1, \text{ and } \vec{n} \cdot \vec{Y} = 0 \text{ on } C.$$

Unfortunately Dirichlet's principle is false if there is a magnetic field present. The reason for this is that the boundary conditions are not of the self-adjoint form. However there is a second duality inequality which remains valid [19]. It is

$$M \geq \iint_A (\vec{n} \cdot \vec{J}) ds + \iint_C (\vec{n} \cdot \vec{J})^- ds + \sigma \iiint (-\nabla^2 X)^- dv$$

$$M \leq \iint_A (\vec{n} \cdot \vec{J}) ds + \iint_C (\vec{n} \cdot \vec{J})^+ dx + \sigma \iiint (-\nabla^2 X)^+ dv.$$

Here $f^+ = f$ if $f \geq 0$ and $f^+ = 0$ if $f \leq 0$. Here X is any smooth function which takes on the value 1 on A and the value 0 on B and

$$\vec{J} = -\sigma \nabla X - \rho \vec{H} \times \nabla X.$$

This second duality inequality is not based on Dirichlet's principle. Instead it stems from the maximum principle for elliptic equations. Thus whether or not a magnetic field is present the function f is harmonic so its maximum and minimum values are on the boundary of R.

DUALITY INEQUALITIES

11. Legendre transforms and nonlinear networks

The networks and conducting bodies considered above are assumed to obey Ohm's law. In other words there is a linear relation between current flow and potential difference. Now it is desired to treat nonlinear networks, and these will be discussed in the light of the Legendre transform.

Let $u(x)$ be a smooth convex function defined on an open region R of n-dimensional space E. Then each point x of R is mapped into a point y by the relation

$$y_j = \frac{\partial u}{\partial x_j} \qquad j = 1, 2, \ldots, n.$$

Let R^* be this map of R. Then the Legendre transform is defined as

$$v = u(x) - \sum_1^n x_j y_j .$$

It is easily shown that v is a single valued function of y in R^*.

The following duality inequality holds between a function u and its Legendre transform

$$u(x) \geq M \geq v(y) .$$
$$x \in S \qquad y \in T$$

Here the constraints are:

S. $\qquad x \in R \cap K ,$

T. $\qquad y \in R^* \cap K^* .$

Here K is a subspace of E and K^* is the orthogonal complementary subspace.

This simple duality inequality has many applications. In particular it was used to treat certain geometric programs [11]. Another important application is to nonlinear networks [3, 5]. Thus let a network have n branches. Then y_j represents current in branch j and x_j represents voltage drop across branch j. Then for each branch there is a functional relation of the form

$$y_j = g_j(x_j).$$

In a common situation the branch conductances are such that $g_j(\ )$ is a continuous monotone increasing function. I have termed such networks "quasi-linear" while Minty has termed them "monotone".

Let the function $u(x)$ be defined as

$$u(x) = \sum_1^n G_j(x_j)$$

where

$$G_j = \int_0^x g_j(x)dx.$$

Let r_j be the inverse function to g_j, so $x_j = r_j(y_j)$. Then it is easy to check that

$$v(y) = \sum_1^n [G_j(r_j(y_j)) - y_j r_j(y_j)]$$

is the Legendre transform.

The subspace K is the Kirchhoff voltage space. The subspace K^* is the Kirchhoff current space. The orthogonality of current and voltage is an important concept due to Weyl. Weyl's concept was further developed by Bott and Duffin [4, 5] and by Tellegen.

There is no duality gap, but the value M does not have an especially important meaning for nonlinear networks.

12. Chemical equilibrium

Consider the chemical reaction

$$2H_2 + O_2 = 2H_2O \text{ (steam)}$$

which is supposed to occur in a reaction chamber which has a constant volume V and which is maintained at constant temperature. Let x_1 be the number of molecules of H_2, let x_2 be the number of molecules of O_2, and let x_3 be the number of molecules of H_2O. It is assumed that under the given condition there are only a negligible number of other types of molecules present. Then the following duality inequality maintains

$$F(x_1, x_2, x_3) \geq M \geq F^*(y_1, y_2)$$

where F is the free energy function of Helmholtz. The vector (x_1, x_2, x_3) must satisfy the mass balance constraints:

S. $2x_1 + 2x_3 = e_1$

$2x_2 + x_3 = e_2$.

Here e_1 is the total number of hydrogen atoms and e_2 is the total number of oxygen atoms. Zener and Duffin [24] have termed F^* the anti-Helmholtz function and defined it as

$$F^*(y_1, y_2) = y_1 e_1 + y_2 e_2 - P(y_1, y_2)V$$

where P is the pressure as a function of y_1 and y_2. The variable y_1 is the chemical potential of atomic hydrogen, that is $y_1 = \partial F/\partial e_1$. Likewise y_2 is the chemical potential of atomic oxygen. There is no constraint on y.

It was proved in [24] that the above dual programs are perfectly general. These programs become especially tractable for perfect gases because then the explicit form of the functions F and F^* are known.

An important extension of this problem concerns the case of equilibrium between phases maintained at the same pressure. For example suppose that in the above reaction the H_2O could be present in two phases, as a liquid and as a vapor. It was found by Avriel, Passy, and Wilde that geometric programming was especially suited to treat such phase equilibrium problems [18, p. 255]. This is a rather strange coincidence of science, because geometric programming was developed from the point of view of engineering economics.

Chemical equilibrium is, of course, a topic in thermodynamics. But thermodynamics may be deduced from statistical mechanics. Moreover, Darwin and Fowler applied statistical method to treat chemical equilibrium. The Darwin-Fowler method relies heavily on contour integeration in the complex plane. Following their procedures, Zener and I found a relation between the duality inequality and the method of steepest descent in the complex plane [27].

13. Schrödinger's equation

Consider Schrödinger's equation

$$\Delta \Psi + (\lambda - v)\Psi = 0$$

where v is a given function. It is required to find a solution Ψ in a region R and such that Ψ vanishes on the boundary. Let M be the least eigenvalue λ. Then the following duality inequality holds [2]

$$\frac{\iiint_R [(\nabla X)^2 + vX^2] d\tau}{\iiint_R X^2 d\tau} \geq M \geq \min_R \left(\frac{-\Delta Y}{Y} + v\right).$$

The constraints are:

S. X is a smooth function vanishing on the boundary of R

T. Y is a smooth function such that $Y \geq 0$ in R.

REFERENCES

1. R. J. Duffin and A. C. Schaeffer, "A refinement of an inequality of the Brothers Markoff", Trans. Amer. Math. Soc. 50 (1941), 517-528.

2. R. J. Duffin, "Lower bounds for eigenvalues", Phys. Rev. 71 (1947), 827-828.

3. R. J. Duffin, "Nonlinear networks, IIa", Bull. Amer. Math. Soc. 53 (1947), 963-971.

4. R. Bott and R. J. Duffin, "On the algebra of networks", Trans. Amer. Math. Soc. 74 (1953), 99-109.

5. R. J. Duffin, "Impossible behavior of nonlinear networks", Jour. of Appl. Phys. 26 (1955), 603-605.

6. R. J. Duffin, "Infinite programs", Linear Inequalities and Related Systems, edited by H. W. Kuhn and A. W. Tucker, Princeton, (1956), 157-170.

7. K. S. Kretschmer, "Linear programming in locally convex spaces and its use in analysis", Thesis, Carnegie-Mellon University, 1958.

8. R. J. Duffin, "A variational problem relating to cooling fins", Jour. of Math. and Mech. 8 (1959), 47-56.

9. R. J. Duffin, "Distributed and lumped networks", Jour. of Math. and Mech. 8 (1959), 793-826.

10. K. S. Kretschmer, "Programmes in paired spaces", Can. Jour. of Math. 13 (1961), 221-238.

11. R. J. Duffin, "Dual programs and minimum cost", SIAM Jour. 10 (1962), 119-123.

12. R. J. Duffin, "The extremal length of a network", Jour. Math. Anal. and Appl. 5 (1962), 200-215.

13. L. A. Karlovitz, "Theory and application of duality in infinite programming", Thesis, Carnegie-Mellon University, 1964.

14. R. J. Duffin and L. A. Karlovitz, "An infinite linear program with a duality gap", Management Science 12 (1965), 122-134.

15. R. J. Duffin and E. L. Peterson, "Duality theory for geometric programming", SIAM Jour. 18 (1966), 1307-1349.

16. R. J. Duffin and T. A. Porsching, "Bounds for the conductance of a leaky plate via network models", Proc. Symposium on Generalized Networks, Polytechnic Institute of Brooklyn, 1966.

17. R. J. Duffin, "An orthogonality theorem of Dines related to moment problems and linear programming", Jour. of Combinatorial Theory, 2 (1967), 1-26.

18. R. J. Duffin, E. L. Peterson, and C. Zener, Geometric Programming, John Wiley and Sons, 1967.

19. R. J. Duffin, "Estimating Dirichlet's integral and electrical resistance for systems which are not self-adjoint", Arch. Rat. Mech. and Anal. 30 (1968), 90-101.

20. R. J. Duffin and L. A. Karlovitz, "Formulation of linear programs in analysis, I: Approximation theory", SIAM Jour. 16 (1968), 662-675.

21. R. J. Duffin, "Optimum heat transfer and network programming", Jour. Math. and Mech. 17 (1968), 759-768.

22. R. J. Duffin and D. K. McLain, "Optimum shape of a cooling fin on a convex cylinder", Jour. Math. and Mech. 17 (1968) 769-784.

23. R. J. Duffin, "Potential theory on a rhombic lattice", Jour. of Comb. Theory, 5 (1968), 258-272.

24. R. J. Duffin and C. Zener, "Geometric programming, chemical equilibrium, and the antientropy function", Proc. Nat. Acad. Sciences 63 (1969), 629-636.

25. R. J. Duffin, "Network models", Proceedings of the Symposium on Mathematical Aspects of Electrical Network Theory, American Mathematical Society, 1969.

26. R. J. Duffin, "Linearizing geometric programs", SIAM Review, 1970.

27. R. J. Duffin and C. Zener, "Geometric programming and the Darwin-Fowler method in statistical mechanics", Jour. of Phys. Chem., 1970.

Prepared under Research Grant DA-AROD-31-124-G680, Army Research Office, Durham, North Carolina.

Programming Methods in Statistics and Probability Theory

OLAF KRAFFT

ABSTRACT

One of the main problems in mathematical statistics is to find procedures which satisfy certain optimality conditions. A great variety of these conditions turns out to be equivalent to programming problems. Hence methods and results of the theory of linear and nonlinear programming can be applied to a set of statistical problems. The paper is a survey of such applications. Since there is always a time lag between the development of new methods in one field (programming) and their applications in another field (statistics), we are mainly concerned with applications of (infinite) linear programming. As another field of applications we discuss a problem in probability theory which is connected with the moment problem.

1. Introduction

Soon after the first publications on linear programming had appeared, its methods were recognized as basic for many problems in other fields and adopted there. This regards computational methods as well as the structure underlying programming problems. This paper gives a survey of such applications of programming methods in statistics and probability theory with stress on the structural similarities, in particular on the duality theory which seems to become a powerful tool - especially in statistics - for describing the structure of optimal solutions of certain problems and for a better understanding of the connections between seemingly unconnected concepts.

Since there is always a time lag between the development of new methods in one field and their applications in another field, most of the examples concern linear programming, but some examples of nonlinear programming methods are discussed, too.

The survey is in no sense exhaustive. As regards the selection of the material, we excluded all dynamic and stochastic programming methods and all examples which are not in the strict sense from statistics or probability theory, in particular all examples from game theory and control theory. Since most of the author's research dealt with problems indicated by the title of the paper, his own work covers more space than it deserves.

When writing the paper, the author had two aims in mind. On the one hand, in trying to apply the known results from the theory of programming to optimization problems in statistics or probability theory he found that most of the results are not directly applicable, since they are not tailored to the problems of those fields. One example shall illustrate that fact. One class of objects one deals with in those fields is the class of probability densities with respect to a measure μ, i.e. a subset of L_1^μ. But in almost all duality theorems of infinite programming, which could be useful for applications, cf. Duffin [8], Kretschmer [21], Rockafellar [26], Dieter [6], it is assumed that a certain subset of L_1^μ has an interior point. In many of the applications this condition is not satisfied, so that one has to reprove the theorems by imposing other conditions. Perhaps this paper will stimulate some further research in that direction. A first step towards this end has been done by van Slyke and Wets [28]. On the other hand, programming methods do not seem to be known enough to all decision scientists. Perhaps some results obtained by those methods and described here will provoke the one or the other to become interested in the theory of programming. This aim is also emphasized in the papers by Wegner [31] and Dieter [7] and in the appendix of the book of Morgenstern [24].

2. The structure of most powerful tests

One of the main problems in testing hypotheses is the following: Let $P = \{P_\vartheta : \vartheta \in \theta\}$ be a class of probability measures on a measurable space $(\mathfrak{X}, \mathfrak{B})$. P is assumed to be dominated by a σ-finite measure μ. The class of μ-densities f_ϑ, corresponding to P is denoted by $\mathfrak{F}$. Let g be a μ-density not belonging to $\mathfrak{F}$. The problem is to determine a $\mathfrak{B}$-measurable function φ, $0 \leq \varphi(x) \leq 1$ for all $x \in \mathfrak{X}$, such that

$$\int_\mathfrak{X} g(x)\varphi(x)d\mu = \sup_{\varphi \in \Phi_\alpha} , \qquad (1)$$

where for a prescribed number α, $0 < \alpha < 1$, Φ_α is defined to be the set

$$\Phi_\alpha = \{\varphi : \int_X f_\vartheta(x)\varphi(x)d\mu \le \alpha \text{ for all } \vartheta \in \Theta\}.$$

A solution of (1) is called a most powerful test for testing H: $f \in \mathfrak{F}$ against K: g at level α. It is immediately seen that problem (1) can be regarded as a problem of infinite linear programming to which corresponds the following dual program (for a more rigorous treatment of how to embed the problem in a general theory of infinite linear programming cf. Krafft and Witting [20] and Witting [34]):

$$\alpha\lambda(\Theta) + \int_X v(x)d\mu = \inf$$

subject to

$$\int_\Theta f_\vartheta(x)d\lambda + v(x) \ge g(x) \qquad \mu\text{- a.e.} \qquad (2)$$

$$\lambda(C) \ge 0 \qquad \text{for all } C \in \mathfrak{J}$$

$$v(x) \ge 0 \qquad \mu\text{- a.e.}$$

Here it is assumed that there is a σ-ring $\mathfrak{J}$ on Θ such that $f_\vartheta(x)$ is $\mathfrak{J} \times \mathfrak{B}$-measurable, λ denotes a σ-additive set function on $(\Theta, \mathfrak{J})$ and $v(x)$ a $\mathfrak{B}$-measurable function. Using all constraints of the primal and dual program, we get the following chain of inequalities, which corresponds to the weak duality theorem:

$$\int_X g(x)\varphi(x)d\mu \le \int_X \varphi(x)[\int_\Theta f_\vartheta(x)d\lambda + v(x)]d\mu$$
$$\le \int_\Theta \int_X \varphi(x)f_\vartheta(x)d\mu d\lambda + \int_X v(x)d\mu \le \alpha\lambda(\Theta) + \int_X v(x)d\mu. \qquad (3)$$

By means of a technique due to Fan [10] it can be shown that a duality theorem holds for problems (1) and (2), i.e. that the sup in (1) is equal to the inf in (2). Thus, if φ^* is an optimal solution of (1) and (λ^*, v^*) is an optimal solution of (2), then for $\varphi^*, \lambda^*, v^*$ equality holds in all inequalities (3). That implies the following result: If there exists a solution λ^* (v^* then has the form $v^*(x) = \max(0, g(x) - \int_\theta f_\vartheta(x) d\lambda^*)$ μ - a.e.) of problem (2), then a necessary and sufficient condition for φ^* to be most powerful is that

$$\varphi^*(x) = \begin{cases} 1, & g(x) > \int_\theta f_\vartheta(x) d\lambda^* \\ 0, & g(x) < \int_\theta f_\vartheta(x) d\lambda^* \end{cases}, \mu - a.e.$$

and

$$\int_\theta f_\vartheta(x) \varphi^*(x) d\mu = \alpha \quad \lambda^* - a.e.$$

Before introducing the programming technique those conditions were only known as sufficient.

Historical remarks and extensions: The use of programming techniques for the theory of testing hypotheses was first observed by Barankin [1]. Weiss [32] showed how to use the simplex method for calculating minimax decision functions. Duality theory as described above was first used by Witting [33] and for the more general class of maximin tests by Krafft and Witting [20]. Schaafsma [27] then extended the results to the class of most stringent tests and Baumann [2] introduced a setting of the problem where the theory of infinite linear programming becomes directly applicable. Finally Krafft and Schmitz [19] showed that this method can be employed for multi-decision problems, too.

If the set θ consists of finitely many points, problem (1) is strongly connected with a problem the solution of which is known as Neyman-Pearson lemma. From the vast literature on that lemma we will mention only the papers by Dantzig and Wald [5], Chernoff and Scheffé [4], Vyrsan [29] and the monograph by Beckenbach and Bellmann [3], where a method is used which is very similar to that in the example above.

3. A new optimality concept

If λ is an optimal solution of problem (2), then $\lambda^* = \lambda/\lambda(\theta)$ is called a least favorable distribution at level α. This notion does not permit an easy statistical interpretation, if one considers only problem (2). The interpretation is the following: Let $\tilde{\lambda}$ be a measure on $(\theta, \mathfrak{J})$ with norm one, $f_{\tilde{\lambda}}$ be the "mixture" of $\{f_\vartheta, \vartheta \in \theta\}$ w.r.t. $\tilde{\lambda}$, i.e. $f_{\tilde{\lambda}}(x) = \int_\theta f_\vartheta(x) d\tilde{\lambda}$, and $\varphi^*_{\tilde{\lambda}}$ be a most powerful test for testing $H: f_{\tilde{\lambda}}$ against $K: g$ at level α. If $\tilde{\lambda}^*$ is such that $\varphi^*_{\tilde{\lambda}^*}$ has minimal power, i.e. $\tilde{\lambda}^*$ is the measure which is least favorable for testing H against K, then one can show that $\tilde{\lambda}^*$ coincides with λ^*.

The concept of a least favorable distribution seems to become more transparent, if one introduces a notion of distance between $f_{\tilde{\lambda}}$ and g. To that end let for some $h > 0$ $\tilde{\lambda}^*$ be a solution of

$$\|g - h f_{\tilde{\lambda}}\| = \inf,$$

where $\|\ \|$ denotes the L_1^μ-norm and the inf is with respect to all measures $\tilde{\lambda}$ on $(\theta, \mathfrak{J})$ with norm one. Then $\tilde{\lambda}^*$ can be obtained by normalizing the solution of the following program:

$$\int_{\mathfrak{X}} v(x) d\mu = \inf$$

subject to

$$v(x) + \int_\theta f_\vartheta(x) d\lambda \geq g(x) \quad \mu\text{-a.e.}$$

$$\lambda(\theta) \leq h$$

$$\lambda(C) \geq 0 \quad \text{for all } C \in \mathfrak{J}$$

$$v(x) \geq 0 \quad \mu\text{-a.e.}$$

(4)

It can be shown that solving the dual program to (4) is equivalent to determining a test φ^* such that

$$\int_X g(x)\varphi(x)d\mu - h \sup_{\vartheta \in \Theta} \int_X f_\vartheta(x)\varphi(x)d\mu = \sup. \tag{5}$$

The sup on the right-hand side of (5) is with respect to all tests φ. Hence duality theory has led to a new concept of optimal tests, namely to tests for which a weighted sum of the maximal error probabilities is a minimum. For proofs and for a discussion of that kind of tests cf. Krafft [16].

4. An application of geometric programming

In this section we will show that the duality theory of geometric programming reveals an interesting connection between statistics and information theory. For that purpose let the parameter space Θ in section 3 be a one-point set, $\Theta = \{\vartheta_0\}$, and φ_n^* be a solution of (5), where the subscript n indicates that we consider the n observation version of the test problem, i.e. $g(x)$ is replaced by $g_n(\underset{\sim}{x}) = \prod_{i=1}^{n} g(x_i)$, $f_{\vartheta_0}(x)$ by $f_{\vartheta_0,n}(\underset{\sim}{x}) = \prod_{i=1}^{n} f_{\vartheta_0}(x_i)$, μ by the product measure $\mu^{(n)}$ and $h = h_n$ may depend on n. For the asymptotic behavior of the performance of φ_n^* one can prove the following theorem, cf. Krafft and Plachky [18]: Let the sequence $\{h_n\}$ be such that $\lim h_n^{1/n} = h_0 > 0$ and assume that $h_0^t \rho(t) = h_0^t \int_X f_{\vartheta_0}^t(x) g^{1-t}(x) d\mu$ attains its minimum at an interior point of the interval $\{t : \rho(t) < \infty\}$. If the set $\{x : f_{\vartheta_0}(x) g(x) > 0\}$ has positive μ-measure and if $g(x)/f_{\vartheta_0}(x)$ is not equal to h_0 μ - a.e., then it holds that

$$\lim_{n \to \infty} [1 + \int (h_n f_{\vartheta_0,n}(\underset{\sim}{x}) - g_n(\underset{\sim}{x}))\varphi_n^*(\underset{\sim}{x})d\mu^{(n)}]^{1/n} =$$
$$= \inf_{0 \leq t \leq 1} h_0^t \rho(t). \tag{6}$$

431

Since $h_0^t\rho(t)$ can be written as

$$h_0^t\rho(t) = \int_X g(x) u^{\log(h_0 f_{\vartheta_0}(x)/g(x))} d\mu, \quad u = e^t,$$

the problem of determining the inf in (6) can be regarded as an infinite version of a geometric program, cf. Duffin, Peterson and Zener [9], p. 78, namely:
Find the minimum value of

$$\int_X g(x) u^{\log(h_0 f_{\vartheta_0}(x)/g(x))} d\mu$$

subject to the constraints $u > 0$ and $e^{-1} u \leq 1$, $u^{-1} \leq 1$. The infinite analogue to dual program B in [9] is for this example:
Find the maximum value of

$$\exp\{\int_X \delta(x) \log(g(x)/\delta(x)) d\mu - \delta_1\}$$

subject to the constraints $\delta(x) \geq 0$, $\delta_1 \geq 0$, $\delta_2 \geq 0$ and

$$\int_X \delta(x) d\mu = 1, \quad \int_X \delta(x) \log(h_0 f_{\vartheta_0}(x)/g(x)) d\mu + \delta_1 - \delta_2 = 0.$$

It is near at hand to assume that a duality theorem holds for this problem. Since until now no general theory of infinite geometric programming exists, we will give a direct proof of the duality theorem for this special example.

For any t, $0 \leq t \leq 1$, and any $(\delta(x), \delta_1, \delta_2)$ satisfying the constraints of the dual program we have by Jensen's inequality

$$\int_X \delta(x)\log(g(x)/\delta(x))d\mu - \delta_1$$

$$= t[\int_X \delta(x)\log(h_0 f_{\vartheta_0}(x)/g(x))d\mu + \delta_1 - \delta_2]$$

$$+ \int_X \delta(x)\log(g(x)/\delta(x))d\mu - \delta_1$$

$$= \int_X \delta(x)\log[\exp\{t \log(h_0 f_{\vartheta_0}(x)/g(x))\}g(x)/\delta(x)]d\mu \qquad (7)$$

$$- [t\delta_2 + (1-t)\delta_1]$$

$$\leq \log \int_X g(x)\exp\{t \log(h_0 f_{\vartheta_0}(x)/g(x))\}d\mu - [t\delta_2 + (1-t)\delta_1]$$

$$\leq \log \int_X g(x)\exp\{t \log(h_0 f_{\vartheta_0}(x)/g(x))\}d\mu = \log h_0^t \rho(t),$$

with equality in both inequalities if

$$\exp\{t \log(h_0 f_{\vartheta_0}(x)/g(x))\}g(x)/\delta(x) = \text{const} \quad \mu - \text{a.e.}$$

and

$$t\delta_2 + (1-t)\delta_1 = 0.$$

Let t^* be a solution of the primal program and let $H(t)$ be the first derivative of $h_0^t \rho(t)$. If $t^* = 0$, then $\lim_{t \to 0+} H(t) = \log h_0 + \int_X g(x)(f_{\vartheta_0}(x)/g(x))d\mu \geq 0$; if $t^* = 1$, then

$\lim_{t \to 1-} H(t) = h_0[\log h_0 + \int f_{\vartheta_0}(x)\log(f_{\vartheta_0}(x)/g(x))d\mu] \leq 0$ and if $0 < t^* < 1$, then $H(t^*) = 0$. Hence $\delta^*(x) = f_{\vartheta_0}^{t^*}(x)g^{1-t^*}(x)/\rho(t^*)$ together with $\delta_1^* = 0$, $\delta_2^* = \int_X g(x)\log(h_0 f_{\vartheta_0}(x)/g(x))d\mu$, for

433

$t^* = 0$, $\delta_1^* = \int_X f_{\vartheta_0}(x)\log(g(x)/h_0 f_{\vartheta_0}(x))d\mu$, $\delta_2^* = 0$, for $t^* = 1$,
$\delta_1^* = \delta_2^* = 0$, for $0 < t^* < 1$, are feasible solutions of the dual program such that, according to (7), the primal program and the dual program have the same value of the objective function.

Now for any two μ-densities f_1 and f_2 the integral $\int_X f_1(x)\log(f_1(x)/f_2(x))d\mu = I(f_1 : f_2)$ is known in information theory as the information measure between f_1 and f_2, cf. e.g. Kullback [23], p. 5. Hence, if $0 < t^* < 1$, determining the limes in (6) is equivalent to determining a probability density $\delta^*(x)$ which maximizes $\exp\{-I(\delta : g)\}$ under all probability densities $\delta(x)$ for which $\log h_0 = I(\delta : f_{\vartheta_0}) - I(\delta : g)$.

Remarks. (i) A possible course of how to extend geometric programming to the infinite version is indicated in [17]. (ii) A similar connection between the measure of information and probabilities of large deviations has been derived by Hoeffding [12], lemma 1. (iii) Relation (6) represents another application of the duality theory of linear programming. For - as it is shown in [18], lemma 1, and indicated in section 3 - by dualizing the problem of determining φ_n^*, one obtains

$$1 + \int (h_n f_{\vartheta_0, n}(\underset{\sim}{x}) - g_n(\underset{\sim}{x}))\varphi_n^*(\underset{\sim}{x})d\mu^{(n)}$$

(8)

$$= \sup_{0 \leq u_n \leq 1} \{u_n + h_n - \int \max(g_n(\underset{\sim}{x}), h_n f_{\vartheta_0, n}(\underset{\sim}{x}))d\mu^{(n)}\}.$$

An adequate treatment of the integral on the right-hand side of (8) leads to (6).

5. Tchebycheff bounds

Let $(\Omega, \mathcal{A})$ be a measurable space, $f(\omega) = (f_1(\omega), \ldots, f_m(\omega))$ be a Borel measurable mapping of Ω into

R^m and $\mathcal{P}$ be a class of probability measures on $(\Omega, \mathcal{Q})$ with respect to each of which $f(\omega)$ is integrable. Further, let $c = (c_1, \ldots, c_m) \in R^m$ and $A \in \mathcal{Q}$ be given. A number α (resp. β) for which $P(A) \leq \alpha$ (resp. $\geq \beta$) holds for all

$$P \in \mathcal{P}_c = \{P \in \mathcal{P}: E_P f = \int_\Omega f(\omega) dP(\omega) = c\}$$

is called an upper (lower) Tchebycheff bound for the problem (f, A, c). The problem of determining a sharp upper (or lower) Tchebycheff bound is easily seen to be a problem of infinite linear programming, if the class $\mathcal{P}$ can be described by means of linear relations. For simplicity we assume that $\mathcal{P}$ is the class of all probability densities p with respect to a σ-finite measure μ. Using the set characteristic function $\chi_A(\omega)$, we can write $P(A) = \int_\Omega \chi_A(\omega) p(\omega) d\mu = E_P \chi_A(\omega)$. There is no difficulty in generalizing the objective function $E_P \chi_A(\omega)$ by letting E_P operate on an arbitrary μ-integrable function $g(\omega)$. Thus, the determination of a sharp upper bound for the problem (f, g, c) is equivalent to:

Find the maximum value of

$$\int_\Omega g(\omega) p(\omega) d\mu$$

subject to the constraints (9)

$$\int_\Omega f_i(\omega) p(\omega) d\mu = c_i, \quad i = 0, 1, \ldots, m$$

$$p(\omega) \geq 0$$

where $c_0 = 1$, $f_0(\omega) \equiv 1$ and $p \in L_1^\mu$.
The dual program to (9) is:

Find the minimum value of

$$\sum_{i=0}^{m} c_i u_i$$

subject to the constraints (10)

$$\sum_{i=0}^{m} f_i(\omega) u_i \geq g(\omega), \quad \omega \in \Omega .$$

If one can show that a duality theorem holds, then the problem of determining an extremal function is reduced by dualization to the problem of finding an extremal element of R^{m+1}, more precisely, the expected value of $g(\omega)$ under the extremal distribution can be calculated by means of the dual program. But also a statement on the extremal distribution P^* can be made, if one knows an optimal solution of the dual program: Since there is equality between the objective functions of (9) and (10) only if $p^*(\omega) = 0$ μ - a.e. on $\{\omega : \sum_{i=0}^{m} f_i(\omega) u_i^* > g(\omega)\}$, the spectrum of $P^*(\omega)$ is confined to the set

$$\{\omega : \sum_{i=0}^{m} f_i(\omega) u_i^* = g(\omega)\} .$$

This method of finding sharp Tchebycheff bounds has first been applied by Richter [25]. A more general setting of the problem and many interesting examples for different choices of $g(\omega)$ and P can be found in the book of Karlin and Studden [15]. The first, who has remarked the connections with the theory of mathematical programming was Isii [13], [14]. In all of these references one can find conditions under which the duality theorem holds and under which the sup and inf of both problems are attained.

Now it is natural to ask whether the duality theorem holds also for the case of more than finitely many constraints in (9). I cannot give a general answer to that question, but I will give two very special examples - one for the case of countably many constraints and one for the case of uncountably many constraints - where the duality theorem holds and where duality theory leads to some interesting relations for the normal distribution.

To that end let $\mathfrak{B}$ be the Borel σ-ring on R_1^+, $(\Omega, \mathcal{Q}) = (R_1^+, \mathfrak{B})$, μ be the Lebesgue measure, $f_i(x) = x^i$, $i = 0, 1, \ldots$, $c_{2i} = (2i-1)! [(i-1)! \, 2^i]^{-1}$, $i = 1, 2, \ldots$, $c_{2i+1} = (2\pi)^{-1/2} 2^i i!$, $i = 0, 1, \ldots$, $c_0 = 1/2$ and $g(x) = e^{\lambda^2/2} \chi_{[\lambda, \infty)}(x)$, $\lambda > 0$. Here the factor $e^{\lambda^2/2}$ is introduced in order to elude the function $\chi_{[\lambda, \infty)}(x)$ which would make the dual program unsolvable, cf. transformation (11). It is well known that there is precisely one $p(x)$ satisfying the constraints of (9), namely $p(x) = (2\pi)^{-1/2} e^{-x^2/2}$. Hence the optimum (sup or inf) of (9) is given by $e^{\lambda^2/2}(1-\Phi(\lambda))$, where $\Phi(\lambda)$ is the normal distribution

$$\Phi(\lambda) = (2\pi)^{-1/2} \int_{-\infty}^{\lambda} e^{-x^2/2} dx .$$

Since

$$e^{\lambda^2/2}(1 - \Phi(\lambda)) = (2\pi)^{-1/2} \int_0^{\infty} e^{-x\lambda} e^{-x^2/2} dx , \quad (11)$$

program (10) now becomes:

Find the inf (sup) value of

$$w/2 + (2\pi)^{-1/2} \sum_{i=0}^{\infty} 2^i i! \, u_i + \sum_{i=1}^{\infty} [(2i-1)!/(i-1)! \, 2^i] v_i$$

subject to the constraints (12)

$$w + \sum_{i=0}^{\infty} x^{2i+1} u_i + \sum_{i=1}^{\infty} x^{2i} v_i \geq (\leq) e^{-x\lambda}, \quad x \geq 0 .$$

437

The inequalities

$$\sum_{i=0}^{2k+1} ((-1)^i (x\lambda)^i / i!) < e^{-x\lambda}, \quad k = 0, 1, \ldots \tag{13}$$

$$\sum_{i=0}^{2k} ((-1)^i (x\lambda)^i / i!) > e^{-x\lambda}, \quad k = 0, 1, \ldots \tag{14}$$

imply that the following values of w, u_i, v_i are feasible for (12):

$$w = 1, \quad v_i = \begin{cases} \lambda^{2i}/(2i)!, & i = 1, 2, \ldots, k \\ 0, & i > k \end{cases}$$

$$u_i = \begin{cases} -\lambda^{2i+1}/(2i+1)!, & i = 0, \ldots, k \text{ (for the sup prob.)}, \\ & i = 0, \ldots, k-1 \text{ (for the inf prob.)} \\ 0, & i > k \text{ (for the sup prob.)}, \, i > k-1 \text{ (for the inf prob.)} \end{cases}$$

Substituting these solutions in the objective function of (12) and applying the weak duality theorem – which obviously holds for this example – we get the following bounds for the normal distribution which seem to be new:

$$1 - \Phi(\lambda) \le e^{-\lambda^2/2} \left[\sum_{i=0}^{k} (\lambda^{2i}/2^{i+1} i!) - (2\pi)^{-1/2} \sum_{i=0}^{k-1} (2^i \lambda^{2i+1} i!/(2i+1)!) \right] \tag{15}$$

$$1 - \Phi(\lambda) \geq e^{-\lambda^2/2} [\sum_{i=0}^{k} (\lambda^{2i}/2^{i+1} i!) - $$
$$- (2\pi)^{-1/2} \sum_{i=0}^{k} (2^i \lambda^{2i+1} i!/(2i+1)!)], \quad k = 0, 1, \ldots . \quad (16)$$

Here the last sum in (15) is defined to be zero for $k = 0$. Note that these bounds for $1 - \Phi(\lambda)$ differ only by the k-th term of the second sum, so that the bounds have some numerical appeal. Letting k tend to infinity we obtain from (15) and (16) the expansion

$$1 - \Phi(\lambda) = 1/2 - (2\pi)^{-1/2} e^{-\lambda^2/2} \sum_{i=0}^{\infty} (2^i \lambda^{2i+1} i!/(2i+1)!),$$

i.e. the duality theorem holds.

Remark. If instead of (11) we would use

$$e^{\lambda^2/2} (1 - \Phi(\lambda)) = (2\pi)^{-1/2} \int_0^{\infty} x(x^2 + \lambda^2)^{-1/2} e^{-x^2/2} dx \quad (17)$$

and instead of (13) and (14)

$$\sum_{i=0}^{k} \binom{-1/2}{i} (\frac{x}{\lambda})^{2i+1} \geq x(x^2 + \lambda^2)^{-1/2}, \text{ if } k \text{ is even,}$$

$$\sum_{i=0}^{k} \binom{-1/2}{i} (\frac{x}{\lambda})^{2i+1} \leq x(x^2 + \lambda^2)^{-1/2}, \text{ if } k \text{ is odd,}$$

we would obtain Feller's [11], p. 193, bounds for the normal distribution. Since $\sum_{i=0}^{\infty} \binom{-1/2}{i} (\frac{x}{\lambda})^{2i+1}$ converges only for $x < \lambda$, transformation (17) provides only an asymptotic expansion for $1 - \Phi(\lambda)$.

In order to discuss the case of uncountably many constraints we apply instead of the Mellin transforms $\int_0^\infty x^{i-1} f(x) dx$ from the foregoing example the Fourier cosine transform, i.e. we consider the primal problem:

Find the optimal value of

$$\int_0^\infty e^{\lambda^2/2} \chi_{[\lambda,\infty)}(x) p(x) dx$$

subject to the constraints

$$\int_0^\infty \cos(xy) p(x) dx = e^{-y^2/2}/2, \quad y \geq 0,$$

$$p(x) \geq 0 .$$

Here again we have the normal density as solution. Using (11) the dual program becomes:

Find the inf (sup) value of

$$\frac{1}{2} \int_0^\infty e^{-y^2/2} h(y) dy$$

subject to the constraints

$$\int_0^\infty \cos(xy) h(y) dy \geq (\leq) e^{-x\lambda}, \quad x \geq 0 .$$

The optimal solution of the dual program is $h^*(y) = 2\lambda/\pi(y^2 + \lambda^2)$. Since

$$e^{\lambda^2/2}(1 - \Phi(\lambda)) = \frac{\lambda}{\pi} \int_0^\infty (e^{-y^2/2}/(y^2 + \lambda^2)) dy,$$

the duality theorem holds and the Cauchy density $h^*(y)/2$ can be regarded as dual to the normal density in the sense described here.

6. Unbiased estimators with smallest variance

Let $(\mathcal{X}, \mathcal{B}, \mu)$ be a measure space, $\{f_\vartheta(x) : \vartheta \in \Theta\}$ be a class of probability densities with respect to μ and $g(\vartheta)$ be a given function of ϑ. Some effort has been made in statistics to find under all measurable functions $T(x)$ with $E_\vartheta T = g(\vartheta)$ for all $\vartheta \in \Theta$ one for which $\text{Var}_{\vartheta_0} T = E_{\vartheta_0} T^2 - E_{\vartheta_0}^2 T$ or, equivalently, $E_{\vartheta_0} T^2$ is minimal. Here $\vartheta_0 \in \Theta$ is a prescribed parameter value. This is a problem of infinite quadratic programming. We will give here a (finite) example where the solution can be obtained in an elementary manner.

Let X be a random variable such that $P(X=0) = 1 - p$, $P(X=1) = p$, $0 < p < 1$. On the basis of n independent observations of X one shall find an unbiased estimator of p with smallest variance under p_0. The sample space consists of 2^n n-tuples $(x_1, \ldots, x_n)$, where $x_i = 0$ or 1, $i = 1, \ldots, n$. There are $\binom{n}{i}$ n-tuples consisting of i ones and $n-i$ zeros. To every of those $\binom{n}{i}$ n-tuples we assign in an arbitrary manner one of the numbers $1, 2, \ldots, \binom{n}{i}$. If $(x_1, \ldots, x_n)$ is the j-th of the $\binom{n}{i}$ tuples consisting of i ones and $(n-i)$ zeros and if $T(x_1, \ldots, x_n)$ is an estimator of p, we write
$$T(x_1, \ldots, x_n) = y_{ij}, \quad i = 0, 1, \ldots, n, \quad j = 1, 2, \ldots, \binom{n}{i}.$$

For $\sum_{j=1}^{\binom{n}{i}} y_{ij}^2$ and $\sum_{j=1}^{\binom{n}{i}} y_{ij}$ we shortly write $(y_{i.})^2$ and $y_{i.}$, respectively. With those notations the problem becomes:

Find the minimum value of
$$\sum_{i=0}^{n} p_0^i (1 - p_0)^{n-i} (y_{i.})^2$$

subject to the constraints

$$\sum_{i=0}^{n} p^i (1-p)^{n-i} y_{i.} = p \text{ for all } p, \ 0 < p < 1.$$

From the expansion of $(1 + \frac{p}{1-p})^n$ and from $(1-p)^n (1 + \frac{p}{1-p})^n = 1$ it follows that the constraints are satisfied if and only if $y_{i.} = \frac{(n-1)!}{(i-1)!(n-i)!}$, $i = 1, \ldots, n$, $y_{0.} = 0$. By the Cauchy-Schwarz inequality we have for all feasible $y_{i.}$

$$\binom{n}{i} (y_{i.})^2 \geq y_{i.}^2$$

or by substitution of $y_{i.}$

$$(y_{i.})^2 \geq (\frac{i}{n})^2 \binom{n}{i}, \ i = 0, 1, \ldots, n,$$

with equality only if $y_{ij} = \frac{i}{n}$, $j = 1, 2, \ldots, \binom{n}{i}$. Hence $y_{ij}^* = \frac{i}{n}$, $i = 0, 1, \ldots, n$, $j = 1, 2, \ldots, \binom{n}{i}$ or, equivalently, $T^*(x_1, \ldots, x_n) = \frac{1}{n} \sum_{i=1}^{n} x_i$ is the best unbiased estimator of p and for any unbiased estimator $T(x_1, \ldots, x_n)$ it holds that

$$\text{Var}_{p_0} T \geq p_0 (1 - p_0)/n.$$

This last inequality is the well known Cramér-Rao inequality. Since T^* is independent of p_0, T^* is even an unbiased estimator with smallest variance uniformly for all p.

A rigorous discussion of the general problem of finding a best unbiased estimator by means of programming methods is given by Isii [14]. Applications of linear and quadratic programming in regression analysis are made by Wagner [30] and Zellner [35].

REFERENCES

1. Barankin, E. W. (1951). On systems of linear equations, with applications to linear programming and the theory of tests of statistical hypotheses. Univ. of California Publ. Stat. 1, 161-214.

2. Baumann, V. (1968). Eine parameterfreie Theorie der ungünstigsten Verteilungen für das Testen von Hypothesen. Z. Wahrscheinlichkeitstheorie und Verw. Gebiete 11, 41-60.

3. Beckenbach, E. F. and Bellmann, R. (1961). Inequalities. Springer, Berlin.

4. Chernoff, H. and Scheffé, H. (1952). A generalization of the Neyman-Pearson fundamental lemma. Ann. Math. Stat. 23, 213-225.

5. Dantzig, G. B. and Wald, A. (1951). On the fundamental lemma of Neyman and Pearson. Ann. Math. Stat. 22, 87-93.

6. Dieter, U. (1966). Optimierungsaufgaben in topologischen Vektorräumen I: Dualitätstheorie. Z. Wahrscheinlichkeitstheorie und Verw. Gebiete 5, 89-117.

7. Dieter, U. (1968). Dual extremal problems in linear spaces with examples and applications in game theory and statistics. In: Theory and Applications of Monotone Operators. Proceedings of a NATO Advanced Study Institute, Venice, 1-9.

8. Duffin, R. J. (1956). Infinite Programs. In [22].

9. Duffin, R. J., Peterson, E. L. and Zener, C. M. (1967). Geometric Programming. Wiley, New York.

10. Fan, K. (1956). On systems of linear inequalities. In [22].

11. Feller, W. (1968). An Introduction to Probability Theory and Its Applications, vol. I, 3rd ed. Wiley, New York.

12. Hoeffding, W. (1967). On probabilities of large deviations. Proc. Fifth Berkeley Symp. Math. Stat. Prob. I, 203-219.

13. Isii, K. (1963). On sharpness of Tchebycheff-type inequalities. Ann. Inst. Stat. Math. 14, 185-197.

14. Isii, K. (1964). Inequalities of the types of Chebychev and Cramér-Rao and mathematical programming. Ann. Inst. Stat. Math. 16, 277-293.

15. Karlin, S. J. and Studden, W. J. (1966). Tchebycheff Systems: With Applications in Analysis and Statistics. Interscience Publishers, New York.

16. Krafft, O. (1967). Eine symmetrische Behandlung des Testproblems. Arch. Math. (Basel) 18, 545-560.

17. Krafft, O. Geometric programming as a special case of Dieter's optimality theory. Operations Research Verfahren. In print.

18. Krafft, O. and Plachky, D. Bounds for the power of likelihood ratio tests and their asymptotic behavior. To appear in Ann. Math. Stat..

19. Krafft, O. and Schmitz, N. (1970). A symmetrical multiple decision problem and linear programming. Operations Research Verfahren 7, 126-149.

20. Krafft, O. and Witting, H. (1967). Optimale Tests und ungünstigste Verteilungen. Z. Wahrscheinlichkeitstheorie und Verw. Gebiete 7, 289-302.

21. Kretschmer, K. S. (1961). Programmes in paired spaces. Canad. J. Math. 13, 221-238.

22. Kuhn, H. W. and Tucker, A. W., ed. (1956). Linear Inequalities and Related Systems. Annals of Mathematics Studies Nr. 38, Princeton.

23. Kullback, S. (1959). Information Theory and Statistics. Wiley, New York.

24. Morgenstern, D. (1968). Einführung in die Wahrscheinlichkeitsrechnung und mathematische Statistik, 2nd ed., Springer, Berlin.

25. Richter, H. (1957). Parameterfreie Abschätzung und Realisierung von Erwartungswerten. Blätter d. Deutschen Ges. f. Versicherungsmathematik 3, 147-162.

26. Rockafellar, R. T. (1966). Extension of Fenchel's duality theorem for convex functions. Duke Math. J. 33, 81-90.

27. Schaafsma, W. (1970). Most stringent and maximin tests as solutions of linear programming problems. Z. Wahrscheinlichkeitstheorie und Verw. Gebiete 14, 290-307.

28. Van Slyke, R. M. and Wets, R. J. (1968). A duality theory for abstract mathematical programs with applications to optimal control theory. J. Math. Anal. Appl. 22, 679-706.

29. Vyrsan, K. (1967). The Neyman-Pearson lemma and linear programming. (Russian). Rev. Roumaine Math. Pures Appl. 12, 279-293.

30. Wagner, H. M. (1959). Linear programming techniques for regression analysis. J. Amer. Statist. Assoc. 54, 206-212.

31. Wegner, P. (1963). Relations between multivariate statistics and mathematical programming. Appl. Statist. 12, 146-150.

32. Weiss, L. (1961). Statistical Decision Theory. McGraw-Hill, New York.

33. Witting, H. (1966). Mathematische Statistik, Teubner, Stuttgart.

34. Witting, H. (1966). Unendliche Programme und ihre Anwendung in der Statistik. Z. Angew. Math. Mech. 46, 109-110.

35. Zellner, A. Linear regression with inequality constraints on the coefficients: An application of quadratic programming and linear decision rules. Report 6109 of the Econometric Inst. of the Netherlands School of Economics.

Applications of Mathematical Programming to ℓ_p Approximation

I. BARRODALE AND F. D. K. ROBERTS

ABSTRACT

The problem of determining a best ℓ_p approximation to discrete data is recast as a nonlinear program. For a linear approximating function the resulting program involves linear constraints and a nonlinear objective function. This objective function is concave for $0 < p < 1$ and convex for $1 < p < \infty$. For $p = 1$ or $p = \infty$ the determination of best approximations can be accomplished by linear programming.

Computational aspects of these formulations are discussed, and some numerical and theoretical results are presented.

I. Introduction

One of the most common problems in scientific computing is that of fitting a curve to a set of empirical data. In addition to the traditional method of least-squares, it is now possible to determine both minimax and least-first-power approximations by the simplex method of linear programming.

In this paper we present a unified approach to linear ℓ_p approximation, for all values of p in the range $0 < p < \infty$. This is accomplished by viewing the problems from the standpoints of linear, convex, and concave programming. By so doing, we are lead to intuitively simple proofs of some of the properties of best ℓ_p approximations. Moreover, both this theory and the accompanying algorithms can be extended with ease to cover the more general problems of multi-dimensional approximation, constrained approximation, simultaneous approximation, and combinations of these generalizations.

The practical motivation for considering the continuum of best ℓ_p approximations is, simply, that it is often useful to compare various approximations to a given set of experimental data. For example, a perusal of the parameters corresponding to best $\ell_{0.5}$, ℓ_1, $\ell_{1.5}$, and ℓ_2 approximations, by a prescribed function, may be sufficient to conclude that a meaningful interpretation of these parameters is (or is not) feasible.

APPLICATION TO ℓ_p APPROXIMATION

Following statements of the ℓ_p approximation problems in II, and of the linear, convex, and concave programming problems in III, we reformulate the best approximation problem as a mathematical programming problem in IV. Some theoretical aspects of these formulations are also presented here, and a discussion of algorithms follows in V. Section VI contains some examples, and some final remarks are made in VII.

II. Best ℓ_p approximation

The best approximation problem for a discrete point set is the following. Suppose that a given real-valued function $f(x)$ is defined on a set $X = \{x_1, x_2, \ldots, x_N\}$. (Although we shall refer to the elements x_i of X as real numbers, it is only necessary that they be points from some Euclidean space E_k. Thus the case of multi-dimensional approximation is included in all that follows). An approximating function $F(A, x)$ is chosen which is defined at least on X, but usually on some interval $I \supset X$. Here, $A = \{a_1, a_2, \ldots, a_n\} \varepsilon E_n$ is a set of parameters which we try to adjust until, for some set A^*, the function $F(A^*, x)$ provides a best approximation to $f(x)$ on X, according to some prescribed criterion.

The linear ℓ_p best approximation problem arises when $F(A, x)$ depends linearly upon all of its parameters a_j, and when the criterion for best approximation is that the ℓ_p norm $\|f(x) - F(A, x)\|_p$ be minimized. More specifically, for a given set of linearly independent functions $\{\phi_1(x), \phi_2(x), \ldots, \phi_n(x)\}$ defined on X, with $n \leq N$, we wish to find $F(A^*, x) = \sum_{j=1}^{n} a_j^* \phi_j(x)$ which minimizes

$$\{\sum_{i=1}^{N} |f(x_i) - F(A, x_i)|^p\}^{1/p} \tag{1}$$

for all choices of $A \varepsilon E_n$. This notion of best approximation may be broadened by the introduction of a specified non-negative weight function $w(x)$ into (1) in the obvious manner.

This causes no complications whatsoever in our considerations, and we may assume that $w(x) \equiv 1$ without restricting ourselves in any way.

Now the expression (1) defines a norm on $f(x) - F(A, x)$ for $1 \leq p < \infty$, while for the range $0 < p < 1$ this is only a quasi-norm (the triangle inequality is actually reversed). However the minimum of (1) obviously occurs at the minimum of

$$\sum_{i=1}^{N} |f(x_i) - F(A, x_i)|^p \qquad (2)$$

A metric can be induced by (2), and so the minimization of (2) becomes our well-posed objective.

The linear ℓ_∞ best approximation problem is to minimize

$$\max_{1 \leq i \leq N} |f(x_i) - F(A, x_i)| \qquad (3)$$

The relationship between (1) and (3) is such that the sequence of best ℓ_p approximations for $p \geq 1$ converges to a best ℓ_∞ approximation.

III. Linear, convex, and concave programming

We are concerned only with linear approximating functions in this paper. Consequently when we refer to a linear, convex, or concave programming problem herein, we shall mean a problem where the objective function to be minimized, is a linear, convex, or concave function which is subject to linear constraints.

It would be inappropriate to review the basic properties of these three mathematical programming problems here. However, it is certainly worth remarking upon the difficulty of solving the third problem. Although the minimum of a strictly concave function must occur at an extreme point of the convex set of linear constraints, such a function

APPLICATION TO ℓ_p APPROXIMATION

often possesses several local minima at other extreme points. Thus any simplex-type algorithm which descends to a minimum via a sequence of extreme points may terminate prematurely.

IV. Mathematical programming and ℓ_p approximation

In this section the minimization of (2) is expressed as a mathematical programming problem, and some properties of best ℓ_p approximations are deduced directly from this reformulation.

We introduce the nonnegative variables u_i, v_i, b_j, and c_j, and put

$$f(x_i) - \sum_{j=1}^{n} a_j \phi_j(x_i) = u_i - v_i \text{ for } i = 1, 2, \ldots, N$$

and

$$a_j = b_j - c_j \text{ for } j = 1, 2, \ldots, n.$$

Then, writing $f_i \equiv f(x_i)$ and $\phi_{j,i} \equiv \phi_j(x_i)$, we have

$$f_i - \sum_{j=1}^{n} (b_j - c_j)\psi_{j,i} = u_i - v_i \text{ (for } i = 1, 2, \ldots, N) \quad (4)$$

These N linear constraints (4) in nonnegative variables remain fixed for all ℓ_p approximations, where $0 < p < \infty$.

Now we wish to solve:

$$\text{minimize } \sum_{i=1}^{N} |u_i - v_i|^p \text{ subject to (4)} \quad (5)$$

This is equivalent to:

$$\text{minimize} \sum_{i=1}^{N} (u_i^p + v_i^p)$$

$$\text{subject to (4) and } b_j c_j = 0, \quad u_i v_i = 0, \tag{6}$$

for every such product.

Consider the simpler mathematical programming problem:

$$\text{minimize} \sum_{i=1}^{N} (u_i^p + v_i^p) \text{ subject to (4)} \tag{7}$$

Clearly, at any optimal solution to (7) it follows that each product $u_i v_i$ must be zero, and that every product $b_j c_j$ can be made equal to zero. Thus the optimal solutions to (5) and (7) are the same, and any property of an optimal solution to (7) is a property of a best ℓ_p approximation.

Problem (7) is that of minimizing a strictly concave, linear, or strictly convex objective function subject to linear constraints, depending on whether $0 < p < 1$, $p = 1$, or $1 < p < \infty$, respectively.

Firstly, we demonstrate the existence of a best linear ℓ_p approximation for any positive value of p. The objective function in (7) is continuous, but the constraints (4) form a closed convex set in E_{2n+2N} whose elements have components which are bounded from below but not from above. However, the linear independence of the n column vectors $\Phi_j = (\phi_{j,1}, \phi_{j,2}, \ldots, \phi_{j,N})^T$ in (4) implies that we can bound all the variables b_j and c_j whenever the u_i's and v_i's have upper bounds. Now at an optimal solution to (7) no u_i or v_i can exceed $\sum_{i=1}^{N} |f_i|^p$, hence all the optimal solutions are contained in a compact subset of the convex set of constraints (4). Finally, the feasible solution given by $b_j = c_j = 0$ and $u_i - v_i = f_i$ proves that this compact subset is nonvoid.

APPLICATION TO ℓ_p APPROXIMATION

The situation with respect to uniqueness of best approximations is also clear. For any p in the range $0 < p < 1$, there is a finite number of best ℓ_p approximations, since they only occur at extreme points of the convex set of constraints (4). For $p = 1$ the set of all best ℓ_p approximations is convex, and so there is either a unique best approximation or an infinity of them. The extreme points of this convex set of optimal solutions can all be determined by the simplex method. Finally, for each finite value of $p > 1$ there is exactly one best ℓ_p approximation.

The following theorem gives a characteristic property of all best ℓ_p approximations where $0 < p < 1$.

<u>Theorem 1</u>. If $\{\phi_1(x), \phi_2(x), \ldots, \phi_n(x)\}$ is a set of functions which are linearly independent on X, and $0 < p < 1$, then every best ℓ_p approximation $F(A^*, x) = \sum_{j=1}^{n} a_j^* \phi_j(x)$ to $f(x)$ satisfies $f(x_i) = F(A^*, x_i)$ for at least n points $x_i \in X$.

<u>Proof</u>. For $0 < p < 1$ any optimal solution to (7) occurs at an extreme point of the set of constraints (4). At any extreme point at most N of the variables b_j, c_j, u_i and v_i can be positive; these positive variables must also correspond to linearly independent column vectors of coefficients from (4).

Now suppose $a_j^* \neq 0$ for each $j = 1, 2, \ldots, n$. Then at most N-n of the quantities $u_i - v_i$ are nonzero and the theorem is proved.

But some of the parameters a_j^* may be zero. Suppose $k \leq n$ of them are zero, say $a_1^* = a_2^* = \ldots = a_k^* = 0$. Since $\sum_{j=k+1}^{n} a_j^* \phi_j(x)$ is a best approximation to $f(x)$, then $G(x) = \sum_{j=1}^{k} \phi_j(x) + \sum_{j=k+1}^{n} a_j^* \phi_j(x)$ is a best approximation to $g(x) = f(x) + \sum_{j=1}^{k} \phi_j(x)$. But, by the argument above, the difference $g(x) - G(x)$ has at least n zeros on X. This completes the proof.

The analogous result for $p = 1$ is the following.

<u>Theorem 2.</u> If $\{\phi_1(x), \phi_2(x), \ldots, \phi_n(x)\}$ is a set of functions which are linearly independent on X, then some best ℓ_1 approximation $F(A^*, x)$ interpolates $f(x)$ in at least n points $x_i \in X$.

<u>Proof.</u> When $p = 1$ there exists an optimal solution to (7) which occurs at an extreme point of the set of constraints (4).

For this optimal solution, suppose $a_j^* \neq 0$ for each $j = 1, 2, \ldots, n$. Then at most $N - n$ of the quantities $u_i - v_i$ are nonzero and the theorem is proved.

But some of the parameters a_j^* may be zero. Suppose $k \leq n$ of them are zero, say $a_1^* = a_2^* = \ldots = a_k^* = 0$. For each $j = 1, 2, \ldots, k$, if neither b_j nor c_j are basic variables one of them can replace a u_i or v_i of the basis without altering the optimal value of the objective function. The linear independence of the n column vectors Φ_j ensures that a suitable pivot can be chosen from only the rows corresponding to basic u_i's and v_i's, even though this may cause infeasibility in the basic b_j's and c_j's; for in this event the solution can be made feasible by interchanging b_j and c_j in the basis. Thus for every $j = 1, 2, \ldots, n$, either b_j or c_j can become a basic variable and so at most $N - n$ of the quantities $u_i - v_i$ are nonzero.

For any value of p in the range $1 < p < \infty$, the Kuhn-Tucker conditions characterize an optimal solution to the convex programming problem (7). However, these conditions are equivalent to the familiar normal equations for a best ℓ_p approximation.

Finally, for the sake of completeness we restate the best ℓ_∞ approximation problem in linear programming form.

For any choice of the parameter set A we can define $w = \max_{1 \leq i \leq N} |f(x_i) - F(A, x_i)|$. Then problem (3) is equivalent to:

$$\text{minimize } w$$

$$\text{subject to} \begin{cases} \sum_{j=1}^{n} a_j \phi_{j,i} - w \leq f_i \\ -\sum_{j=1}^{n} a_j \phi_{j,i} - w \leq -f_i \end{cases} \text{(for } i = 1, 2, \ldots, N\text{)} \quad (8)$$

The unsymmetric dual of (8) yields the following more efficient linear programming problem in nonnegative variables s_i and t_i:

$$\text{maximize } \sum_{i=1}^{N} f_i(s_i - t_i)$$

$$\text{where } \begin{cases} \sum_{i=1}^{N} (s_i + t_i) \leq 1 \\ \sum_{i=1}^{N} \phi_{j,i}(s_i - t_i) = 0 \quad \text{(for } j = 1, 2, \ldots, n\text{)} \end{cases} \quad (9)$$

We shall not carry our investigation further in this norm, for Osborne and Watson (1967) have considered the properties of optimal solutions to (9) in detail.

V. Algorithms

Best ℓ_p approximations for $0 < p < 1$ are the most difficult to determine in general. Here, problem (7) has a concave objective function that may assume local minima, which are not optimal solutions. A gradient vector cannot be determined, since the interpolatory nature of these best approximations causes the objective function to become nondifferentiable at some points. Thus although the optimal solutions are located at extreme points of the convex set of constraints (4), there appears to be no efficient way of descending to an optimal point from an extreme point at which a local minimum occurs.

The simplex method is easily modified so that at each iteration a new extreme point is generated which decreases the objective function by (say) the greatest amount possible. Another modified simplex approach, with which we have also had some experience, is to solve the ℓ_1 problem first and then let p decrease by small increments. Typically, the ℓ_1 solution allows the objective function in (7) to remain optimal for each of the first several choices of p, and then at the next value the objective function can be decreased, and one or more iterations are performed to achieve optimality again. This process is continued until the desired value of p is reached. The weakness of this method is that these "optimal" solutions may be just local minima; this is illustrated by the following remarks.

Suppose that we wish to compute a best ℓ_p approximation, for some $0 < p < 1$, by a straight line $F(A, x) = a_1 + a_2 x$ to $f(x)$ on $X = \{x_1, x_2, x_3, x_4\}$. For each distinct pair of values x_i and x_j let $F(A_{ij}, x)$ interpolate $f(x)$ on x_i and x_j. Finally, define $S_{ij}(p) = \{ \sum'_{x \varepsilon X} |f(x) - F(A_{ij}, x)|^p \}^{1/p}$ where Σ' denotes the sum of the nonzero quantities only. Jensen's inequality shows that $S_{ij}(p)$ is a strictly monotonic decreasing function for increasing p, and Figure 1 illustrates how these curves $S_{ij}(p)$ might appear in practice.

Clearly, for any value of p the best ℓ_p approximation yields a point on the lower envelope of these curves S_{ij}, i.e. a point on $\min_{i,j} S_{ij}(p)$. Thus our algorithm which allows p to decrease from $p = 1$ is an attempt to ascend this envelope from its lowest point.

An iteration of the simplex method produces a neighbouring extreme point which has one new nonbasic component. This usually produces an approximating function $F(A, x)$ which interpolates $f(x)$ in one different point from the previous approximation. This explains why our algorithm can always successfully negotiate any intersection of S_{ij} curves which differ in just one subscript (as at A in Figure 1), but will probably fail otherwise (such as at B) unless the objective function is allowed to increase temporarily.

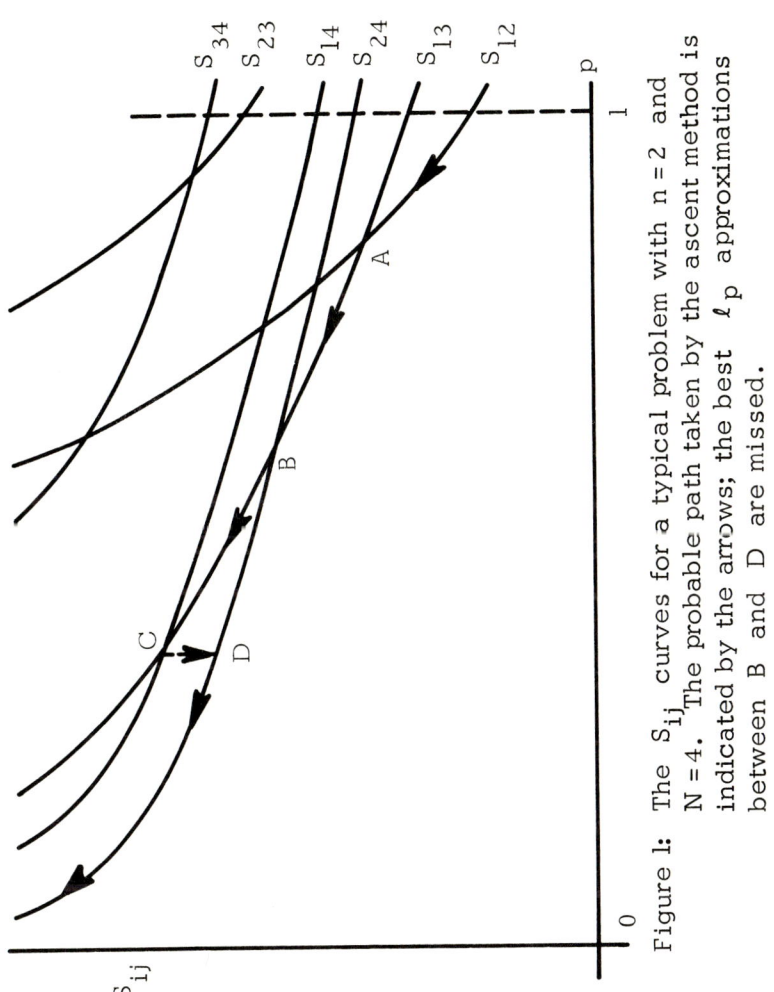

Figure 1: The S_{ij} curves for a typical problem with $n = 2$ and $N = 4$. The probable path taken by the ascent method is indicated by the arrows; the best ℓ_p approximations between B and D are missed.

Having introduced these error curves S_{ij} we now pause to state and prove a result which is suggested by Figure 1.

Theorem 3. If $\{\phi_1(x), \phi_2(x), \ldots, \phi_n(x)\}$ is a set of functions which are linearly independent on X, then for some $0 < p_1 < 1$ there exists a best ℓ_1 approximation which is also a best ℓ_p approximation for every value of p where $p_1 < p \leq 1$.

Proof. For each distinct subset $\{x_{i_1}, x_{i_2}, \ldots, x_{i_n}\}$ of n points from X, let $F(A_{i_1 \ldots i_n}, x)$ interpolate $f(x)$ on this subset. The linear independence of the set $\{\phi_1(x), \phi_2(x), \ldots, \phi_n(x)\}$ guarantees that at least one such interpolating function exists when all choices of n points from X are considered. Corresponding to each $F(A_{i_1 \ldots i_n}, x)$ which exists we define a function $S_{i_1 \ldots i_n}(p) =$
$\{\sum'_{x \in X} |f(x) - F(A_{i_1 \ldots i_n}, x)|^p\}^{1/p}$, where again Σ' denotes the sum of the nonzero quantities only. The set $\mathcal{S}$ of all such functions contains at most $\binom{N}{n}$ members, and, for any fixed value of p, Theorem 1 states that $\min_{S \in \mathcal{S}} S(p)$ corresponds to some best ℓ_p approximation. Each function $S \in \mathcal{S}$ has a Taylor series expansion about $p = 1$, and so it follows that there exists an interval $(p_1, 1]$ on which some member of $\mathcal{S}$, say S^*, satisfies $S^*(p) = \min_{S \in \mathcal{S}} S(p)$ for all $p_1 < p \leq 1$. Furthermore $S^*(1)$ must correspond to a best ℓ_1 approximation, otherwise the continuity of the norm (1) implies that there exists a number $\varepsilon > 0$ for which $S^*(p)$ cannot yield a best ℓ_p approximation for $1 - \varepsilon < p < 1$. Thus the result is established.

If the best ℓ_1 approximation in a given problem is unique, then Theorem 3 shows that the ascent algorithm starts off on the correct path. Our empirical experience to date suggests that the ascent algorithm is more successful at locating true optimal solutions to (7) than the direct modification of the simplex method to which we first referred. We have encountered examples where the ascent method is successful and the direct method is not, where the reverse situation occurs, and where both methods fail. However, for over 90% of the test examples that we have run the ascent method converges to the true optimal solution. It must be emphasized though that for all these test problems the number $\binom{N}{n}$ is relatively small: otherwise Theorem 1 cannot be employed to check the validity of the results.

For any problem where $\binom{N}{n}$ is not prohibitively large a best ℓ_p approximation can always be found by this naive approach whereby all the interpolatory schemes are examined. This corresponds to an enumeration of a subset of the set of all extreme points of the convex set of constraints (4).

For $p = 1$ problem (7) is a linear programming problem, which can be represented in a condensed simplex tableau with N rows and n columns as in Table 1.

Costs → ↓	Basic Variables		0 b_1	0 b_2	...	0 b_n
1	u_1	f_1	$\phi_{1,1}$	$\phi_{2,1}$	...	$\phi_{n,1}$
1	u_2	f_2	$\phi_{1,2}$	$\phi_{2,2}$	...	$\phi_{n,2}$
⋮	⋮	⋮	⋮	⋮		⋮
1	u_N	f_N	$\phi_{1,N}$	$\phi_{2,N}$	...	$\phi_{n,N}$

Table 1: Initial simplex tableau for the ℓ_1 algorithm where each $f_i \geq 0$

Notice that an initial basis is provided by some N of the coefficient vectors of the u_i's and v_i's, and that the current values of the remaining N variables u_i and v_i and the n variables c_j are all readily available at every stage. Thus in considering any column vector of the tableau for entry to the basis, the corresponding negative vector must also be considered.

Since this formulation appeared in Barrodale and Young (1966) a very large number of ℓ_1 problems have been solved. Based on this computational experience some modifications to the standard simplex method are recommended.

For the first n iterations the choice of pivotal column should be restricted to those columns associated with the coefficients b_j and c_j. Here, the pivotal column can be chosen as that which has the maximum positive marginal (or reduced) cost at each iteration. Thereafter almost every iteration involves the replacement of a u_i or v_i column in the basis by another u_i or v_i column. The following pivotal column selection rule is now very effective in reducing the total number of the remaining iterations required to solve a given problem.

For any stage after the first n iterations, suppose that the first n rows of the tableau correspond to basic variables b_j and c_j, and so the last N-n rows correspond to u_i's and v_i's. Then the pivotal column is chosen to be that column which maximizes the ratio of its positive marginal cost divided by the sum of its positive entries which appear in its last N-n rows. This rule is based on one that is recommended in Wolfe and Cutler (1963); it appears to minimize the effect that scaling of the original tableau can have on the total number of iterations of the algorithm. Scaling the original tableau is of course a sound practice.

For large values of N the dual of problem (7) should be solved when $p = 1$. The dual is a bounded variable linear program (see Rabinowitz (1968)) and this problem in turn may be expressed as an interval linear program (see Robers and Ben-Israel (1969)).

For $p > 1$ equation (7) is a convex programming problem, for which a variety of algorithms are available. However, solving the normal equations of (1) by Newton's method appears to be the most efficient way of determining these best ℓ_p approximations.

VI. Test examples

In order to compare some of the algorithms available for $0 < p < 1$ and $1 < p < \infty$ a set of fifteen data sets were compiled, where in each case $N \leq 100$. The approximating functions chosen were polynomials of degree five and less, although none of the algorithms used require any of the special properties of polynomial approximating functions. Twenty values of p were specified between $p = 0.1$ and $p = 20$, and we then attempted to determine best ℓ_p approximations in each case.

For $0 < p < 1$ the best ℓ_p approximations were first computed using Theorem 1. This method is bound to produce the correct answer but its limitations are obvious. Both the direct simplex method and the ascent method described in V were then used in an attempt to duplicate these results. As reported previously the failures here are relatively few.

For $1 < p < \infty$ three algorithms were employed. These are the convex simplex method of Zangwill (see Zangwill (1969)), the Frank-Wolfe algorithm (see Canon (1969)), and Newton's method applied to the normal equations of (1) (see Hebden (1969)). For our problem, and based only on this numerical study, we find that Newton's method is normally about ten times faster than the convex simplex method, which in turn is more efficient than the Frank-Wolfe algorithm.

However, for the range $1 < p < 2$, Newton's method may give rise to numerical difficulties since quantities of the form $|f(x_i) - F(A, x_i)|^q$ appear with $-1 < q < 0$. The convex simplex method can be used in this eventuality.

Table 2 contains the results obtained for one of the fifteen data sets. For $0.3 \leq p \leq 0.95$ the ascent method of V is successful except for $p = 0.4$ and $p = 0.5$; however the alternative direct method of V is successful in these cases. Generally we experienced numerical difficulties with these two methods for small values of p, and in this example the results for $p = 0.1$ and $p = 0.2$ were obtained using Theorem 1. For $1.25 \leq p \leq 20$, Newton's method and the convex simplex method produced approximations which agreed to almost five decimal places before the iterations were terminated. The Frank-Wolfe algorithm had exhibited poor convergence for $p = 10$ and no convergence for $p = 1.25$ and $p = 20$ when its alloted calculation time expired.

VII. Conclusions

The problems of best ℓ_p approximation can be conveniently expressed in mathematical programming form. This reformulation provides for some different insights into approximation theory. For example, the interpolatory nature of best ℓ_p approximations for $0 < p < 1$ is immediately apparent from the simplex method. Also, the ℓ_1 norm becomes a more natural choice than the popular ℓ_2 norm in so far as computational expediency is concerned. The problems associated with constrained best approximation are easier to deal with too.

The algorithms of mathematical programming are certainly appropriate for $p = 1$, and $p = \infty$, while for $1 < p < \infty$ they appear to be far less attractive. Finally for $0 < p < 1$ some combination of the methods discussed in V may be as successful as any other technique presently available.

Acknowledgements

We wish to acknowledge the programming assistance of Mr. B. K. Wilson, and the financial support provided by NRC Grants No. A5251 and A7143.

APPLICATION TO ℓ_p APPROXIMATION

p	a_1^*	a_2^*	a_3^*	a_4^*	a_5^*	$\sum_{i=1}^{21} \|f_i - F_i\|^p$
0.1	1.05034	0.27442	2.98295	-2.49516	0.55725	$(0.11293)10^2$
0.2	1.05034	0.27442	2.98295	-2.49516	0.55725	$(0.82058)10^1$
0.3	1.07057	-0.00180	3.78752	-3.16637	0.72728	$(0.60859)10^1$
0.4	1.07057	-0.00180	3.78752	-3.16637	0.72728	$(0.45475)10^1$
0.5	1.07057	-0.00180	3.78752	-3.16637	0.72728	$(0.34467)10^1$
0.6	1.06362	0.05408	3.94450	-3.38535	0.79144	$(0.26360)10^1$
0.7	1.06490	0.03548	4.00935	-3.44711	0.80859	$(0.20125)10^1$
0.8	1.06490	0.03548	4.00935	-3.44711	0.80859	$(0.15484)10^1$
0.9	1.06455	0.04074	3.98929	-3.42348	0.80097	$(0.12001)10^1$
0.95	1.06455	0.04074	3.98929	-3.42348	0.80097	$(0.10590)10^1$
1.0	1.09359	-0.31307	4.66321	-3.82879	0.87473	$(0.93239)10^0$
1.25	1.05961	-0.18817	4.57220	-3.83757	0.88950	$(0.49317)10^0$
1.5	1.05518	-0.24634	4.84096	-4.09045	0.95672	$(0.26230)10^0$
1.75	1.05515	-0.29885	5.02385	-4.25141	0.99829	$(0.14166)10^0$
2.0	1.05769	-0.37242	5.23514	-4.42344	1.04067	$(0.77519)10^{-1}$
3.0	1.06769	-0.60832	5.87950	-4.94066	1.16754	$(0.75509)10^{-2}$
4.0	1.07562	-0.76540	6.29590	-5.27223	1.24866	$(0.76933)10^{-3}$
6.0	1.08506	-0.92072	6.68038	-5.56663	1.31852	$(0.81045)10^{-5}$
10.0	1.09281	-1.02000	6.89448	-5.70911	1.34667	$(0.94346)10^{-9}$
20.0	1.09929	-1.09405	7.03996	-5.79785	1.36198	$(0.15645)10^{-18}$
						$\max_i \|f_i - F_i\|$
∞	1.10615	-1.16557	7.17663	-5.88170	1.37662	$(0.10615)10^0$

Table 2: Best ℓ_p approximations on $x = 0(0.1)2$ by $F(A, x) = \sum_{j=1}^{5} a_j x^{j-1}$

to $f(x) = \begin{cases} e^x & \text{on } [0, 1] \\ e^{-x} + e - e^{-1} & \text{on } [1, 2] \end{cases}$

REFERENCES

1. Barrodale, I. and A. Young (1966). "Algorithms for best L_1 and L_∞ linear approximations on a discrete set." Numer. Math., 8, 295-306.

2. Canon, M. (1969). "Nonlinear programming: a survey." Invited paper presented at 1969 ACM/SIAM/IEEE Joint Conference. Anaheim, Calif.

3. Hebden, M. D. (1969). Topics in linear approximation theory with special reference to the Chebyshev norm. Doctoral thesis, University of Leeds.

4. Osborne, M. R., and G. A. Watson (1967). "On the best linear Chebyshev approximation." Comp. J., 10, 172-177.

5. Rabinowitz, P. (1968). "Applications of linear programming to numerical analysis." SIAM Review, 10, 121-159.

6. Robers, P. D., and A. Ben-Israel (1969). "An interval programming algorithm for discrete L_1 approximation problems." J. Approx. Theory, 2, 323-336.

7. Wolfe, P., and L. Cutler (1963). "Experiments in linear programming." from Recent advances in mathematical programming, edited by R. L. Graves and P. Wolfe, McGraw-Hill, New York, 177-200.

8. Zangwill, W. I. (1969). Nonlinear programming - a unified approach. Prentice-Hall, Englewood Cliffs, N. J.

Theoretical and Computational Aspects of Nonlinear Regression

R. R. MEYER

ABSTRACT

Nonlinear regression problems can often be stated in the form: minimize $\|y - f(\theta)\|^2$, where y is a fixed n-vector of observations and f is a nonlinear mapping from E^r to E^n. Such problems constitute an important class of nonlinear unconstrained minimization problems. Theoretical and computational aspects of some numerical method for nonlinear regression are described with emphasis on the relationship to mathematical programming.

1. Introduction

The nonlinear regression problem to be dealt with is conveniently stated in the form,

$$\operatorname*{minimize}_{\theta} \|y - f(\theta)\|^2, \qquad (1.1)$$

where y is a fixed n-vector of "observations", f is a continuously differentiable nonlinear mapping from E^p to E^n, θ is the vector variable (the "parameters") in E^p, and the norm is the Euclidean norm in E^n. (In engineering and statistical work, it is customary to write $g(X_k, \theta)$ in place of $f_k(\theta)$ $(k = 1, \ldots, n)$, where g is a mapping (the "model") from E^{m+p} to E^1 and X_k is the setting of the independent variables at the kth "experiment". In this analysis of nonlinear regression, however, references to the function $g(X, \theta)$ and the values X_k would tend to complicate and unnecessarily restrict the scope of the presentation.) Theoretical and computational aspects of iterative techniques for solving nonlinear regression problems will be discussed in this paper. Section 2 contains a description of some previously developed techniques, starting with a procedure proposed by Gauss [1], and introduces a new method, the Modified Damped Least Squares (MDLS) algorithm, devised by the author. Some numerical aspects and convergence properties of MDLS are proved in Section 3, and compared with those of other methods. Section 4 deals with computational devices that have been found in practice to improve the efficiency and the reliability of MDLS. Finally, an evaluation of numerical results for several difficult problems is given in Section 5.

Some of the notation and notational conventions to be used throughout the paper will be now described. With the exception of gradients, which are always taken to be row vectors, all vectors without the superscript ^{T}(denoting transpose) will be column vectors. Hence we may write the nonlinear regression problem (1.1) as

$$\underset{\theta}{\text{minimize}} \ (f(\theta) - y)^T (f(\theta) - y) . \qquad (1.2)$$

For notational purposes it is convenient to define the vector error function and the sum of squared errors function:

$$e(\theta) \equiv f(\theta) - y \qquad (1.3)$$

$$s(\theta) \equiv e^T(\theta) e(\theta) = \|y - f(\theta)\|^2 . \qquad (1.4)$$

It is easily verified that the gradient of $s(\theta)$, to be denoted by $s'(\theta)$ (a row vector) is given by

$$s'(\theta) = 2e^T(\theta) f'(\theta) , \qquad (1.5)$$

where $f'(\theta)$ is the $n \times p$ Jacobian matrix $(\partial f_i(\theta)/\partial \theta_j)$. By differentiating (1.5) we obtain the $p \times p$ Hessian matrix of second partials of $s(\theta)$,

$$s''(\theta) \equiv 2f'(\theta)^T f'(\theta) + 2e^T(\theta) f''(\theta) , \qquad (1.6)$$

where

$e^T(\theta)f''(\theta)$ denotes the summation

$$\sum_{\ell=1}^{n} e_\ell(\theta) f_\ell''(\theta) . \qquad (1.7)$$

(Note that the expression $e^T(\theta)f''(\theta)$ is merely a symbolic representation for the matrix given by (1.7). Dimensionally the expression makes no sense when n and p are greater than 1.)

2. Iterative Methods for Nonlinear Regression

Since nonlinear regression problems form a class of unconstrained minimization problems, it is natural to seek iterative methods for their solution that take advantage of their particular structure. The first such method appears to have been described by Gauss [1] in 1809. (All of the methods to be presented in this section are iterative algorithms that assume that an initial guess θ^0 for the parameters is given. Hence, for each method we will simply describe the basic iterative procedure that is used to derive the $(i+1)$st iterate, θ^{i+1}, from the ith iterate, θ^i. Given an initial guess θ^0, a sequence $\theta^1, \theta^2, \ldots$ of iterates can thus, in theory, be derived. The convergence properties of these sequences are derived in the next section of the paper.) An iteration of the Gauss-Newton method, as it is usually referred to, essentially consists of linearizing <u>within</u> the norm signs of (1.1), and solving the resulting linear least squares problem.

Gauss-Newton Method

1) Determine the solution of the linear least squares problem:

$$\underset{\Delta}{\text{minimize}} \; \| f(\theta^i) + f'(\theta^i)\Delta - y \|^2 \qquad (2.1)$$

2) Let $\theta^{i+1} = \theta^i + \Delta^*$, where Δ^* is the solution of (2.1).
(Assuming that the matrix

$$F \equiv f'(\theta^i)^T f'(\theta^i) \qquad (2.2)$$

is nonsingular, the solution of (2.1) is given by

$$\Delta^* = -F^{-1}(f'(\theta^i)^T e(\theta^i)) \qquad (2.3)$$

This is the usual assumption in the Gauss-Newton method; the procedure may be directly extended to the case in which F is singular by using the generalized inverse of $f'(\theta^i)$ as suggested by Ben-Israel [2].) Note that this reduces to Newton's method when $p = n$, since in such a case nonsingularity of F implies $F^{-1} = (f'(\theta^i))^{-1}(f'(\theta^i)^T)^{-1}$ and thus $\Delta^* = -(f'(\theta^i))^{-1} e(\theta^i)$, the correction of Newton's method.

As might be expected, the solution Δ^* of (2.1) will in general not satisfy the monotonicity or "stability" property

$$s(\theta^i + \Delta^*) < s(\theta^i) . \qquad (2.4)$$

Consequently, convergence of the Gauss-Newton iterates to a stationary point will not occur unless the initial guess θ^0 is, in some sense, "good". Pereyra [3] and Ben-Israel [2] have stated sufficient conditions on θ^0 that guarantee convergence to a stationary point. These conditions will be discussed in the next section.

The Gauss-Newton method persisted as one of the standard methods for the solution of nonlinear regression problems until 1944 when Levenberg [4] developed the method of Damped Least Squares (DLS) in order to obtain better convergence properties. An iteration of DLS is based on the solution of problems of the form

$$\underset{\Delta}{\text{minimize}} \ \|f(\theta^i) + f'(\theta^i)\Delta - y\|^2 + \lambda \Delta^T W \Delta , \quad (2.5)$$

where λ is a non-negative scalar (the <u>damping factor</u>) and W is a positive definite diagonal matrix. The term $\lambda \Delta^T W \Delta$ has the effect of "damping" the size of the solution of the unconstrained problem. For $\lambda > 0$ the problem (2.5) (without any assumptions on F) will have the unique solution $\Delta^*(\lambda, W)$ given by the formula

$$\Delta^*(\lambda, W) = -(F + \lambda W)^{-1}(f'(\theta^i)^T e(\theta^i)) . \quad (2.6)$$

Damped Least Squares (DLS) Method

1) Solve problems of the form (2.5) for increasing values of the damping factor λ until a value λ_i^* is determined for which $s(\theta^i + \Delta^*(\lambda_i^*, W))$ is an "approximate minimum" of $s(\theta^i + \Delta^*(\lambda, W))$ considered as a function of λ.

2) Let $\theta^{i+1} = \theta^i + \Delta^*(\lambda_i^*, W)$.

The procedures by which an "approximate minimum" is determined in DLS and in the two remaining methods of this section will be described in detail in sections 3 and 4. Here we shall point out only that Levenberg showed that for any positive definite diagonal matrix W, the inequality

$$s(\theta^i + \Delta^*(\lambda, W)) < s(\theta^i) \quad (2.7)$$

is satisfied for all sufficiently large λ, so that one of the criteria for λ_i^* is that (2.7) be satisfied when $\lambda = \lambda_i^*$. Note, however, that if the initial guess for λ_i^* does not satisfy (2.7), then the modified least squares problem (2.5) must be solved with successively larger values of λ until monotonicity is achieved.

In 1961, Hartley [5] described an approach to stability based upon a one-dimensional search in the direction of Δ^*, the solution of (2.1).

Hartley's Method

1) Determine the solution Δ^* of the linear least squares problem (2.1).
2) Determine a <u>step-length factor</u> γ_i^* such that $s(\theta^i + \gamma_i^* \Delta^*)$ is an "approximate minimum" of $s(\theta^i + \gamma \Delta^*)$ for $\gamma \geq 0$.
3) Let $\theta^{i+1} = \theta^i + \gamma_i^* \Delta^*$

It is easily shown that if $s'(\theta^i) \neq 0$, then $s'(\theta^i)\Delta^* < 0$ and consequently for all sufficiently small step-length factors we have

$$s(\theta^i + \gamma \Delta^*) < s(\theta^i) . \qquad (2.8)$$

Hence, Hartley's method also (in theory) leads to a set of monotone decreasing function values, but the solution of only one least squares problem ((2.1)) is required at each iteration.

The modified Damped Least Squares (MDLS) method to be described below extends the DLS method in a manner analogous to Hartley's extension of the Gauss-Newton method. In MDLS the matrix of weights W for the ith iteration is taken to be the diagonal matrix D whose diagonal elements are those of the matrix F. The diagonal entries of D are all positive unless one or more columns of $f'(\theta^i)$ are identically 0, hence the underlying assumption of MDLS is that no such column vanishes. (This is one of two suggestions for W advanced by Levenberg. The other called for setting $W = I$, the identity matrix. The DLS method in the case $W = D$ is often referred to in the literature as <u>Marquardt's method</u>. Levenberg's original paper apparently received very little attention, and it was not until after DLS was rediscovered by Marquart [6] that the use of the procedure became widespread.) Note that adding the matrix λD to F is equivalent to multiplying the diagonal elements of F by the factor $(1 + \lambda)$.

Modified Damped Least Squares (MDLS) Method

1) Choose a damping factor λ_i
2) Determine the solution $\Delta^*(\lambda_i, D)$ of the corresponding damped least squares problem.
3) Determine a step-length factor $\bar{\gamma}_i$ such that $s(\theta^i + \bar{\gamma}_i \cdot \Delta^*(\lambda_i, D))$ is an "approximate minimum" of $s(\theta^i + \gamma \Delta^*(\lambda_i, D))$ for $\gamma \geq 0$.
4) Let $\theta^{i+1} = \theta^i + \bar{\gamma}_i \Delta^*(\lambda_i, D)$.

Of course, if λ_i is taken to be 0 for all i the iterative procedure reduces to Hartley's method. It will be shown in the next section that $s'(\theta^i) \neq 0$ implies that $s'(\theta^i) \Delta^*(\lambda, W) < 0$ for any positive definite matrix W and any $\lambda \geq 0$. Hence the iterates of MDLS will have the property that

$$s(\theta^{i+1}) < s(\theta^i) \qquad (2.9)$$

unless θ^i is a stationary point of $s(\theta)$. The algorithm by which λ_i is chosen at the ith iteration will be described in section 4.

3. Theoretical Comparison

In this section we shall derive some theoretical properties of the methods to use as the basis for further comparison. The following three properties of the problem (2.5) are obtained under the assumptions that D is positive definite and that $W = D$.

a) If $s'(\theta^i) \neq 0$, then for $\lambda > 0$

$$s'(\theta^i) \Delta^*(\lambda, D) < 0, \qquad (3.1)$$

where $\Delta^*(\lambda, D) = -(F + \lambda D)^{-1} (f'(\theta^i)^T e(\theta^i))$. $\qquad (3.2)$

b) For all $\lambda > 0$, $\Delta^*(\lambda, D)$ is the unique solution of (2.5).

c) The condition number (here taken to be the ratio of the maximum over the minimum eigenvalue) of $F + \lambda D$ is a nonincreasing function of λ.

The proof of a) follows easily from the facts that $(F + \lambda D)^{-1}$ is positive definite for $\lambda > 0$ and that by using the equation (1.5) for the gradient of $s(\theta)$ we have

$$\Delta^*(\lambda, D) = -(1/2)(F + \lambda D)^{-1} s'(\theta^i)^T \quad (3.3)$$

To prove property b) note that $F = f'(\theta^i)^T f'(\theta^i)$ is either positive semi-definite or positive definite, so that $F + \lambda D$ is positive definite and hence nonsingular for $\lambda > 0$. By setting $W = D$ and taking the gradient with respect to Δ in (2.5), it is easily verified that the unique solution is given by (3.2). If F^{-1} exists, then (3.2) is also valid for $\lambda = 0$ and reduces to the formula for Δ^* in this case. Note that it is not necessary to assume nonsingularity of F when $\lambda > 0$, so that MDLS can be applied in some cases in which Hartley's method cannot. Even when F^{-1} does exist, the use of a positive damping factor improves the <u>conditioning</u> of the system of equations that must be solved, as asserted in property c). In order to demonstrate this fact denote the maximum eigenvalues of D and $F + \lambda D$ by d_1 and $\alpha_1(\lambda)$ respectively, and the minimum eigenvalues of D and $F + \lambda D$ by d_p and $\alpha_p(\lambda)$ respectively. If $\lambda'' > \lambda' \geq 0$, then the following inequalities result from applying the theorem that the field of values of a normal matrix is the convex hull of the spectrum:

$$\frac{\alpha_1(\lambda'')}{\alpha_p(\lambda'')} \leq \frac{\alpha_1(\lambda') + (\lambda'' - \lambda')d_1}{\alpha_p(\lambda') + (\lambda'' - \lambda')d_p}$$
$$\leq \frac{\alpha_1(\lambda') + (\lambda'' - \lambda')(1 + \lambda')^{-1}\alpha_1(\lambda')}{\alpha_p(\lambda') + (\lambda'' - \lambda')(1 + \lambda')^{-1}\alpha_p(\lambda')}$$
$$= \frac{\alpha_1(\lambda')}{\alpha_p(\lambda')}$$

Let us now compare the methods of the previous section in the light of properties a) - c). It has already been observed that the original Gauss-Newton method can be guaranteed to converge only when the starting point θ^0 is sufficiently close to a stationary point. Because of the monotonicity property of the other three methods (in the case of MDLS, (3.1) guarantees monotonicity), they are, under rather weak assumptions, globally convergent. That is, regardless of the starting point, they will converge to a stationary point of the original problem if a stationary point exists. This idea will be made precise later in this section. DLS has the disadvantage that an evaluation of the effect of different damping factors at each iteration is inefficient (relative to Hartley's method or MDLS) because a number of systems of equations (instead of just one) must be solved. Lastly, according to properties b) and c), the case in which the matrix F is singular or ill-conditioned is readily handled in MDLS, but not in Hartley's method.

In order to establish the convergence properties of MDLS, it is necessary to rigorously define what is meant by an "approximate minimum" of the one-dimensional search performed at each iteration. One approach to the definition is to require a certain fixed percentage $\beta \in (0, 1)$ of "optimality" at each iteration. That is, $\bar{\gamma}_i$ is taken to be the first integer power of $1/2$ for which the following inequality is satisfied:

$$s(\theta^i + \bar{\gamma}_i \Delta^*(\lambda_i, D)) \leq s(\theta^i) + \beta \cdot \bar{\gamma}_i \cdot s'(\theta^i)\Delta^*(\lambda_i, D) \quad (3.4)$$

We shall now prove that a $\bar{\gamma}_i$ satisfying (3.4) does indeed exist. If $s'(\theta^i) = 0$, then $\Delta^*(\lambda_i, D) = 0$ and the inequality is trivially satisfied, so assume $s'(\theta^i) \neq 0$. Differentiability of $s(\theta)$ at θ^i implies

$$s(\theta^i + \gamma\Delta) \leq s(\theta^i) + \gamma s'(\theta^i)\Delta + \epsilon(\|\gamma\Delta\|)\|\gamma\Delta\|, \quad (3.5)$$

for all γ and Δ, where $\epsilon(x) \to 0$ as $x \to 0$. In particular, when Δ is taken to be $\Delta^*(\lambda_i, D)$ and γ is sufficiently small,

$$\epsilon(\|\gamma\Delta^*(\lambda_i, D)\|) \leq -(1-\beta)s'(\theta^i)\Delta^*(\lambda_i, D)\|\Delta^*(\lambda_i, D)\|^{-1}.$$

Substituting the latter inequality in (3.5) and replacing γ by $\bar{\gamma}_i$ and Δ by $\Delta^*(\lambda_i, D)$, we obtain (3.4). (Note that when β is chosen very close to 0 the inequality (3.4) is almost equivalent to

$$s(\theta^i + \bar{\gamma}_i\Delta^*(\lambda_i, D)) < s(\theta^i). \qquad (3.6)$$

In fact, from a computational standpoint, since only a finite number of iterations can be performed, the satisfaction of (3.6) for the iterates actually obtained implies the existence of a range of values of β for which the iterates obtained would also satisfy (3.4). However, the inequality (3.6) by itself is not strong enough to guarantee the desired theoretical convergence properties.)

Theorem 1: Let $\{\theta^i\}$ be a sequence of iterates generated by MDIS using the rule given above for the selection of the step-length factor at the ith iteration. If there exists a subsequence $\{\theta^{n_i}\}$ converging to some set of parameters θ^* and if the corresponding subsequence $\{f'(\theta^{n_i})^T f'(\theta^{n_i}) + \lambda_{n_i} D_{n_i}\}$ converges to some positive definite matrix P (D_{n_i} denotes the diagonal matrix whose diagonal coincides with that of $f'(\theta^{n_i})^T f'(\theta^{n_i})$), then $s'(\theta^*) = 0$.

Proof: We will assume that $s'(\theta^*) \neq 0$ and show that this leads to a contradiction. Let $\Delta^{n_i} = -(f'(\theta^{n_i})^T f'(\theta^{n_i}) + \lambda_{n_i} D_{n_i})^{-1}(f'(\theta^{n_i})^T e(\theta^{n_i}))$ and $\Delta^{**} = \lim \Delta^{n_i} = -P^{-1}(f'(\theta^*)^T e(\theta^*))$. Since $s'(\theta^*)$ is assumed not to vanish and P is positive definite, $s'(\theta^*)\Delta^{**} < 0$. Let γ^{**} be the smallest power of $1/2$ satisfying

$$s(\theta^* + \gamma^{**}\Delta^{**}) < s(\theta^*) + \beta \cdot \gamma^{**} \cdot s'(\theta^*)\Delta^{**}.$$

By continuity it follows that for n_i sufficiently large we also have

$$s(\theta^{n_i} + \gamma^{**}\Delta^{n_i}) < s(\theta^{n_i}) + \beta \cdot \gamma^{**} \cdot s'(\theta^{n_i})\Delta^{n_i},$$

from which we conclude that

$$s(\theta^{n_i+1}) = s(\theta^{n_i} + \bar{\gamma}_{n_i}\Delta^{n_i}) < s(\theta^{n_i}) + \beta \cdot \gamma^{**} \cdot s'(\theta^{n_i})\Delta^{n_i}.$$

By the monotonicity property (2.9) it follows that $\lim s(\theta^{n_i+1}) = \lim s(\theta^{n_i}) = s(\theta^*)$. Hence, taking limits on both sides of the last inequality yields $s(\theta^*) \leq s(\theta^*) + \beta \cdot \gamma^{**} \cdot s'(\theta^*)\Delta^{**}$, which is impossible since $\beta \cdot \gamma^{**} \cdot s'(\theta^*)\Delta^{**} < 0$.

There are a number of ways of guaranteeing that there will be a convergent subsequence of iterates. One of the simplest is to assume that for any $\bar{\theta} \in E^p$ the set $L(\bar{\theta}) \equiv \{\theta \mid s(\theta) < s(\bar{\theta})\}$ is compact. This hypothesis implies that for an arbitrary starting point θ^0 the sequence generated by MDLS will lie in a compact set. The next theorem states that the addition of two further hypotheses leads to convergence of the whole sequence to a stationary point.

Theorem 2: If (a) $L(\bar{\theta})$ is compact for each $\bar{\theta} \in E^p$, (b) $s(\theta)$ has at most a finite number of stationary points having any given function value, (c) $f'(\theta)^T f'(\theta)$ is positive definite for all θ, and (d) there is an upper bound on the damping factor to be used at any iteration, then, given an aribtrary starting point θ^0, the iterates $\{\theta^i\}$ generated by MDLS will converge to a stationary point of $s(\theta)$.

Proof: It follows from hypothesis (a) and the monotonicity of the function values of the iterates that $\{\theta^i\}$

lies in a compact set. Hypotheses (c) and (d) imply by theorem 1 that every accumulation point of $\{\theta^i\}$ will be a stationary point of $s(\theta)$, and (b) implies that there can only be a finite number of these accumulation points, since they must all have the same function value. Using the formula (3.3) and hypotheses (c) and (d), it is easily shown that $\Delta^*(\lambda_i, D) \to 0$. Suppose that the sequence $\{\theta^i\}$ has more than one accumulation point, and let ϵ^* be the minimum distance between any two accumulation points. Since $\{\theta^i\}$ lies in a compact set, there exists an integer N such that for all $i \geq N$ θ^i lies within a ball of radius $\epsilon^*/4$ about some accumulation point. On the other hand, there exists an $N' \geq N$ such that $\|\Delta^*(\lambda_i, D)\| < \epsilon^*/4$ for $i \geq N'$. Hence, for $i \geq N'$ all θ^i must lie within a ball of radius $\epsilon^*/4$ about one particular accumulation point, contradicting the assumed existence of more than one accumulation point.

Note that theorem 2 is a **global** convergence result that applies to Hartley's method (by setting the upper bound on the damping factor to 0) as well as MDLS. A similar convergence result could also be obtained for DLS if the rule for the selection of the quasi-optimal damping factor at each iteration was analogous to the rule given for $\overline{\gamma}_i$. Recall that the Gauss-Newton method does not have this type of global convergence property.

It is clear from the expression (1.6) for the second derivative $s''(\theta)$ that the sum of squared errors function $s(\theta)$ will in general be non-convex. Hence there is no guarantee that a stationary point located by MDLS (or any other algorithm considered in this paper) will be the global solution of the problem of minimizing $s(\theta)$. The following theorem deals with the rate of convergence to stationary points that will at least be "local minima." Since the theorem below assumes that $\lambda_i \to 0$ (as is usually the case in MDLS), the same convergence rate estimate applies to <u>all</u> of the methods described in section 2 even though it is stated for MDLS.

Theorem 3: Let $\{\theta^i\}$ be a sequence generated by MDLS which converges to a point θ^*. Let m be the minimum eigenvalue of $f'(\theta^*)^T f'(\theta^*)$ and M be the maximum of the absolute values of the eigenvalues of $e^T(\theta^*) f''(\theta^*)$. If $r \equiv M/m < 1$, $\beta < (1-r)/2$, and $\lambda_i \to 0$, then $\bar{\gamma}_i = 1$ for all sufficiently large i and $\lim \sup \|\theta^{i+1} - \theta^*\| / \|\theta^i - \theta^*\| \le r$. Moreover, $s(\theta^*)$ will be an isolated local minimum of the function $s(\theta)$.

Proof: (For notational convenience in this proof Δ will be used to denote $\Delta^*(\lambda_i, D)$.) A second-order approximation of the change of function value is given by the equation

$$s(\theta^i + \Delta) - s(\theta^i) = s'(\theta^i)\Delta + 1/2\, \Delta^T s''(\bar{\theta})\Delta, \qquad (3.7)$$

where $\bar{\theta} = \theta^i + \delta\Delta$, $\delta \in (0, 1)$. In order to have $\bar{\gamma}_i = 1$, the following inequality is required:

$$\beta \cdot s'(\theta^i)\Delta - (s(\theta^i + \Delta) - s(\theta^i)) \ge 0 \qquad (3.8)$$

Using the relations $s'(\theta^i) = -\Delta^T(2F + 2\lambda_i D)$ and (3.7), the expression on the LHS of the inequality takes the form

$$(1 - \beta)\Delta^T(2F + 2\lambda_i D)\Delta - 1/2\Delta^T s''(\bar{\theta})\Delta =$$

$$\Delta^T [2 \cdot (1 - \beta)F - 1/2 s''(\theta^i) + 2 \cdot (1 - \beta)\lambda_i D -$$

$$1/2(s''(\bar{\theta}) - s''(\theta^i))]\Delta =$$

$$\Delta^T [(1 - 2\beta)F - e^T(\theta^i) f''(\theta^i) + G_i]\Delta,$$

where $G_i = 2 \cdot (1 - \beta)\lambda_i D - 1/2(s''(\bar{\theta}) - s''(\theta^i))$.

Since $G_i \to 0$, in order to verify (3.8) for large i, it is necessary only to prove that $(1 - 2\beta)F - e^T(\theta^i)f''(\theta^i)$ converges to a positive definite matrix. But the minimum eigenvalue of

$$(1 - 2\beta)f'(\theta^*)^T f'(\theta^*) - e^T(\theta^*)f''(\theta^*)$$

is bounded from below by the value $(1 - 2\beta)m - M = m(1 - 2\beta - r)$, which is positive under the assumption $\beta < (1 - r)/2$.

To prove the estimate on the rate of convergence note that for large i

$$\theta^{i+1} - \theta^* = \theta^i - \theta^* - 1/2(F + \lambda_i D)^{-1} s'(\theta^i)^T$$

$$= \theta^i - \theta^* - 1/2(F + \lambda_i D)^{-1}[s''(\theta^i)(\theta^i - \theta^*)$$

$$+ s'(\theta^i)^T + s''(\theta^i)(\theta^* - \theta^i)]$$

$$= -1/2(F + \lambda_i D)^{-1}[2e^T(\theta^i)f''(\theta^i)(\theta^i - \theta^*)$$

$$- 2\lambda_i D(\theta^i - \theta^*) + s'(\theta^i)^T + s''(\theta^i)(\theta^* - \theta^i)] .$$

Taking norms on both sides yields

$$\|\theta^{i+1} - \theta^*\| \leq 1/2 \|F^{-1}\| [2 \|e^T(\theta^i)f''(\theta^i)\| \: \|\theta^i - \theta^*\|$$

$$+ 2\lambda_i \|D\| \: \|\theta^i - \theta^*\| + \|s'(\theta^i)^T + s''(\theta^i)(\theta^* - \theta^i)\|]$$

Since

$$\|s'(\theta^i)^T + s''(\theta^i)(\theta^* - \theta^i)\| = \|s'(\theta^i)^T +$$

$$+ s''(\theta^i)(\theta^* - \theta^i) - s'(\theta^*)^T\| \leq \epsilon_i \|\theta^* - \theta^i\| ,$$

where $\epsilon_i \to 0$, dividing through by $\|\theta^i - \theta^*\|$ yields

$$\|\theta^{i+1} - \theta^*\|/\|\theta^i - \theta^*\| \leq \|F^{-1}\|[\,\|e^T(\theta^i)f''(\theta^i)\|$$

$$+ \lambda_i \|D\| + 1/2\epsilon_i]$$

From which it follows immediately that $\limsup \|\theta^{i+1} - \theta^*\|/\|\theta^i - \theta^*\| \leq m^{-1}M = r$.

Since $s'(\theta^*) = 0$, the final conclusion of the theorem is equivalent to the demonstration that $s''(\theta^*)$ is positive definite. The matrix $s''(\theta^*)$ is positive definite since it is the sum

$$2f'(\theta^*)^T f'(\theta^*) + 2e^T(\theta^*)f''(\theta^*),$$

whose minimum eigenvalue is bounded from below by $m - M > 0$.

As noted previously, both Pereyra and Ben-Israel have also derived convergence results for the Gauss-Newton method. These, however, are sufficient conditions for convergence rather than global convergence results such as theorem 2. In addition, the estimated rates of convergence given by Pereyra and Ben-Israel involve upper bounds on first and second derivatives in a sphere about the initial guess, θ^0. The convergence rate estimate of theorem 3 is more closely related to a result of Daniel [7] for Newton-like methods.

From a qualitative standpoint, theorem 3 indicates that the efficiency of the algorithms of section 2 is related to the magnitude of the error, $e(\theta^*)$, at the stationary point to which the iterates converge. In particular, if $e(\theta^*) = 0$, then $M = 0$ and therefore $r = 0$, so that the rate of convergence is superlinear. This is not surprising, since $e(\theta^*) = 0$ implies (in the context of theorem 3) that $f'(\theta^i)^T f'(\theta^i) \to s''(\theta^*)$, so that the search directions $\Delta^*(\lambda_i, D)$ are a good approximation to the corrections $-s''(\theta^i)^{-1}s'(\theta^i)^T$

of Newton's method for determining a zero of $s'(\theta)$. In this instance, the limiting behavior is similar in principle to the well-known Davidon-Fletcher-Powell method [8] and other quasi-Newton methods that generate matrices that converge to the Hessian matrix (or its inverse) in the limit. In the quasi-Newton methods the approximation to the Hessian is constructed using information from previous iterations, whereas algorithms derived from the Gauss-Newton method take advantage of the special form of the problem to construct an approximation to the Hessian based only on first derivatives at the most recent set of parameter values. Since the reliance of the quasi-Newton methods upon information from previous iterations has been observed to lead to numerical difficulties in some cases [9], the Gauss-Newton approach appears to be preferable in nonlinear regression problems in which a good "fit" of the data is expected. Results of numerical experiments by Pitha and Jones [10] and Bard [11] agree with this conclusion.

4. Computational Devices

Computational experience with certain difficult problems revealed numerical difficulties in applying the MDLS method as outlined in section 2. In order to resolve these difficulties, various computational safeguards were added to the computer program implementing MDLS. This section describes the most important of these safeguards, as well as certain devices introduced to increase efficiency.

a) The program user has the option of introducing <u>explicit</u> bounds on the components of the correction $\theta^{i+1} - \theta^i$ in addition to the <u>implicit</u> bounds resulting from the damping term. (The size of the correction cannot be accurately controlled by damping alone.)

b) One or more univariate searches are performed in the event that numerical difficulties prevent the satisfaction of (3.6). A number of other "alternate" search direction schemes for this type of emergency have also been tested and found to be successful. A further extension of this concept would be to establish at each iteration a set of directions along which searches would be performed until a set of criteria specified by the user were satisfied. In a time-sharing system, the user would even be allowed to establish the order in which directions were tried at each iteration and to exercise his own judgement on the acceptability of the result of a search.

c) The initial damping factor, λ_0, is set to some nominal value such as 1/100 or 1/1000. At the ith iteration, λ_i is set to $4\lambda_i$ if a univariate search was required to reduce the sum of squares function at the $(i-1)^{st}$ iteration; otherwise λ_i is set to $\lambda_{i-1}/4$. This is similar to the rule proposed by Marquardt. Note that if the basic MDLS method as described in section 2 is successful in reducing the sum of squares at each iteration, then $\lambda_i \to 0$.

d) In order to improve efficiency, the method of determining $\bar{\gamma}_i$ presently being used in the program is a modified quadratic interpolation scheme. A number of similar schemes are described by Bard [11].

5. Computational Experience

The computer program implementing MDLS has been successfully tested on a large number of problems and is presently being used on a "production" basis. In this section some representative computational results will be presented.

Rosenbrock's Problem

This is a two variable test problem originally discussed by Rosenbrock [12] and now frequently cited in the literature. In the notation of section 1, the problem is given by $y_1 = y_2 = 0$, $f_1(\theta) = (\theta_2 - \theta_1^2)$, and $f_2 = 0.1(1-\theta_1)$, so that $s(\theta) = (\theta_2 - \theta_1^2)^2 + 0.01(1 - \theta_1)^2$. The starting point is given by $\theta^0 = (-1.2, 1.0)$, and the iterates are forced to "track" along a steep-sided parabolic valley to the solution $\theta = (1, 1)$. The solution was obtained in 17 iterations (with a total of 31 function evaluations) by MDLS, in 17 iterations (with a total of 32 function evaluations) by Hartley's method, and in 38 iterations (with a total of 61 function evaluations) by DLS. The performance of the former two methods compares favorably with that of the Fletcher-Powell method as cited in [8]. The Fletcher-Powell method required 18 iterations to obtain a solution, but the more accurate unidirectional searches required in the Fletcher-Powell method imply that the total number of function evaluations in 18 iterations would be much greater than 31.

Thermistor Problem

The data for this problem were furnished to us by J. H. Badley of Shell Development Company. They represent the resistance of a thermistor as a function of temperature.

y	T
34,780	50
28,610	55
23,650	60
19,630	65
16,370	70
13,720	75
11,540	80
9,744	85
8,261	90
7,030	95
6,005	100
5,147	105
4,427	110
3,820	115
3,307	120
2,872	125

The model is given by $f_j(\theta) = \theta_1 \cdot \exp \theta_2/(T_j + \theta_3)$, and the initial values used were $\theta_1^o = 0.02$, $\theta_2^o = 4,000$, and $\theta_3^o = 250$.

This problem is of interest because it was the only one tested in which MDLS failed to obtain in 51 iterations a solution when started with a non-zero damping factor. When the damping factor was set initially to 0 (yielding

Hartley's method), however, convergence in 7 iterations to the optimal solution, $\theta_1^* = 0.005609$, $\theta_2^* = 6,181$, $\theta_3^* = 345.2$ was obtained. DLS also failed on this problem.

Spectroscopic Data Problem Set

This is a set of 11 representative problems assembled by D. D. Tunnicliff of Shell Development Company. Each problem in the set involves the fitting of spectrographic data by a sum of nonlinear functions. The simplest function used in the fit has the form

$$\sigma \cdot \exp - [(x - \mu)/\omega]^2,$$

where σ, μ, and ω are parameters and x is the independent variable. The performances of Hartley's method and MDLS on this problem set were essentially the same except for the most difficult problem in the set, which involved 35 parameters. On this problem Hartley's method failed to converge to the optimal set of parameters, whereas the iterates of MDLS did converge to the correct values. When applied to this problem set, the DLS method failed on the most difficult problem, as well as two others in the set. Quasi-Newton methods were not tested on this problem set, but the results of Pitha and Jones [10] on similar spectroscopic problems indicated that the performance of DLS was superior to that of the Fletcher-Powell method for such problems. (Hartley's method was not tested by Pitha and Jones.)

In conclusion, for the types of problems dealt with, Hartley's method and MDLS proved to be superior to the other techniques tested. Except for the two problems cited above, Hartley's method and MDLS seemed to be comparable with respect to efficiency in obtaining solutions. MDLS has a theoretical advantage over Hartley's method in that it can be applied to problems in which matrix singularity rules out Hartley's method, but no such problems were actually tested in this study.

Acknowledgement

I am indebted to Susan E. Post for her efforts in writing the computer program implementing MDLS and in running a large number of test problems.

REFERENCES

1. K. F. Gauss, Theoria Motus Corporum Coelestiam, (1809).

2. Victor Pereyra, "Iterative Methods for Solving Nonlinear Least Squares Problems," SIAM J. Numer. Anal., 4 (1967), pp. 27-36.

3. Adi Ben-Israel, "On Iterative Methods for Solving Nonlinear Least Squares Problems over Convex Sets," Israeli J. of Math. (1967), pp. 211-224.

4. Kenneth Levenberg, "A Method for the Solution of Certain Nonlinear Problems in Least Squares," Quart. App. Math., 2 (1944), pp. 164-168.

5. H. O. Hartley, "The Modified Gauss-Newton Method for the Fitting of Nonlinear Regression Functions by Least Squares," Technometrics, 3 (1961), pp. 269-280.

6. D. W. Marquardt, "An Algorithm for the Least Squares Estimation of Nonlinear Models," J. of SIAM, 11 (1963), pp. 431-441.

7. J. W. Daniel, "On Newton-like Methods," The University of Wisconsin Computer Sciences Department Technical Report No. 21, Madison, Wisconsin, 1968.

8. R. Fletcher and M. J. D. Powell, "A Rapidly Convergent Descent Method for Minimization," *Computer J.*, 6 (1963), pp. 163-168.

9. Garth P. McCormick and John D. Pearson, "Variable Metric Methods and Unconstrainted Optimization," paper presented at the Joint Conference on Optimization, University of Keele, March 25-28, 1968.

10. J. Pitha and R. Norman Jones, "A Comparison of Optimization Methods for Fitting Curves to Infrared Band Envelopes," *Canadian Journal of Chemistry*, 44 (1966), pp. 3031-3050.

11. Yonathan Bard, "Comparison of Gradient Methods for the Solution of Nonlinear Parameter Estimation Problems," IBM New York Scientific Center Technical Report No. 320-2955, September, 1968.

12. H. H. Rosenbrock, "An Automatic Method for Finding the Greatest or Least Value of a Function," *The Computer Journal*, 7 (1960), p. 175.

Index

A

algorithm, 455
  Beale's, 159
  conceptual, 276
  Cottle-Dantzig, 157
  exchange, 128
  Frank-Wolfe, 26, 282, 461
  implementable, 276
  Levenberg-Marquardt, 38, 162, 471
  numerical comparison, 60
approximation
  best, 449, 453
  constrained, 448
  least-first-power, 448
  linear, 111
  linear ℓ_p, 448, 452
  multi-dimensional, 448
  second derivative, 32
  simultaneous, 448

B

Banach space, 340
basis, 88

C

complementarity, 155, 349
complementary slackness, 235
complex space, 388
condition
  Holder, 179
  Kuhn-Tucker, 70, 211, 335
  Lipschitz, 240
  necessary, 328, 330, 331, 333, 334, 335
  second order Kuhn-Tucker, 208, 211, 212
  Slater, 296
  sufficient, 328, 329, 330, 331, 333, 334, 335, 341
conditioning, 473
cone, 393
  dual, 336
  convex, 388
  polar, 376
  polyhedral, 386
conjugacy relation, 105
conjugate function, 294
consistency, 406, 407

constraint
    active, 88
    qualification, 394
control, optimal, 341
convergence, 50, 62
    rate of, 179
    superlinear, 59, 178, 240
convex
    faithfully, 294
    function, 48
    quasi, 263
    Z, 325
cut, 411

D

data fitting, 448
decomposition
    Cholesky, 150, 151
    LU, 126, 128, 137
    principle, 296
    QR, 142, 155
    singular value, 160
direction
    feasible, 247
    generators, 94
directional
    derivatives, 299
    differential, 326, 332
Dirichlet
    problem, 414
    principle, 416
distribution, 430, 437
dual, 460
    unsymmetric, 455

duality, 386
    gap, 402, 404, 407
    inequality, 402
    network, 403
    theorem, Fenchel, 314

E

eigensystems, 167
eigenvalues, 170, 420
equilibrium, 419, 420
errors, round-off, 131

F

Farkas lemma, 387
F-distance, 4
function
    Lagrangian, 68
    penalty, 68
    upperbounding, 12

H

Householder transforms, 142, 153, 161, 163
hypotheses testing, 427

I

interpolation, 462
inverse, product-form, 136
iterative refinement, 141, 153

J

Jacobian, positively bounded, 374

INDEX

L

least squares, 151, 162
    damped, 469, 472
Legendre transform, 295, 417
linear program
    bounded variable, 460
    interval, 460

M

Markoff inequality, 408
matrix
    Hessian, 177, 178
    projection, 69
    sparse, 139
maximum principle, 416
method
    convex simplex, 208
    Davidon-Fletcher-Powell, 481
    feasible direction, 21, 208
    Gauss-Newton, 468
    Goldfarb's variable metric gradient projection 99, 114 269
    gradient, 420
    Hartley's, 471
    Newton's, 30, 460, 469
    of centers, 4, 8, 24
    of modified centers, 280
    reduced gradient, 208
    Rosen's gradient projection, 27, 99, 101, 125, 208, 247, 248, 268, 271
    simplex, 140, 456, 461
    steepest descent, 177, 178, 188
    SUMT, 74
    variable metric, 33, 78, 106, 178, 269
    variable reduction, 208
    variational, 415
minimax, 448
minimum, local, 451
monotone, 375, 377
multipliers, Lagrange, 68, 88

N

network, 410, 412, 414, 417
norm
    Euclidean, 413
    ℓ_p, 450, 453, 462
    minimum, 412
    mixed, 413
    Tchebycheff, 413

O

orthogonalization, 146, 149

P

path, 411
perturbations, 294
phase coordinates, 341
pivot
    parametric, 359
    principle, 157, 362
polar, 388

problem
  dual, 294, 336, 337, 338, 339
  well-posed, 327, 330
  test, 73
procedures
  adaptive, 283
  open loop, 286
programming
  complex nonlinear, 392
  concave, 448
  convex, 295, 327, 448
  geometric, 295, 317, 404, 407, 418, 420, 431
  infinite, 405, 407, 428
  linear, 448
  linearly constrained nonlinear, 109
  quadratic, 88, 142, 151, 441
  quasiseparable, 307
  unconstrained, 104
projection, 146

R

rank-one
  formula, 78, 84
  modification, 167
rank-two formula, 84
regression, 466
relaxation, 255

S

Schrödinger equation, 420
search routine, 267
stepsize, 250, 253, 254, 258, 263, 267

T

Tchebycheff bound, 435

U

unbiased estimator, 441
unimodal, 263

Z

zigzagging, 88